COMPREHENSIVE SERIES IN PHOTOCHEMISTRY
& PHOTOBIOLOGY

Series Editors

Donat P. Häder
Professor of Botany

and

Giulio Jori
Professor of Chemistry

European Society for Photobiology

COMPREHENSIVE SERIES IN PHOTOCHEMISTRY
& PHOTOBIOLOGY

Series Editors: Donat P. Häder and Giulio Jori

Titles in this Series

COMPREHENSIVE SERIES IN PHOTOCHEMISTRY
& PHOTOBIOLOGY – VOLUME 2

Photodynamic Therapy

Editor

Thierry Patrice
Department Laser-Neurochirurgie
CHU Régional et Universitaire de Nantes
Hôpital Guillaume et René Laënnec
44093 Nantes Cedex 01
France

advancing the chemical sciences

ISBN 0-85404-306-3

A catalogue record for this book is available from the British Library

Published by The Royal Society of Chemistry,
Thomas Graham House, Science Park, Milton Road,
Cambridge CB4 0WF, UK

Registered Charity Number 207890

For further information see our web site at www.rsc.org

Typeset by Alden Bookset, Northampton, UK
Printed and bound by Sun Fung Offset Binding Co. Ltd., Hong Kong

Preface for the ESP series in photochemical and photobiological sciences

"Its not the substance, it's the dose which makes something poisonous!" When Paracelsius, a German physician of the 14th century made this statement he probably did not think about light as one of the most obvious environmental factors. But his statement applies as well to light. While we need light for example for vitamin D production too much light might cause skin cancer. The dose makes the difference. These diverse findings of light effects have attracted the attention of scientists for centuries. The photosciences represent a dynamic multidisciplinary field which includes such diverse subjects as behavioral responses of single cells, cures for certain types of cancer and the protective potential of tanning lotions. It includes photobiology and photochemistry, photomedicine as well as the technology for light production, filtering and measurement. Light is a common theme in all these areas. In recent decades a more molecular centered approach changed both the depth and the quality of the theoretical as well as the experimental foundation of photosciences.

An example of the relationship between global environment and the biosphere is the recent discovery of ozone depletion and the resulting increase in high energy ultraviolet radiation. The hazardous effects of high energy ultraviolet radiation on all living systems is now well established. This discovery of the result of ozone depletion put photosciences at the center of public interest with the result that, in an unparalleled effort, scientists and politicians worked closely together to come to international agreements to stop the pollution of the atmosphere.

The changed recreational behavior and the correlation with several diseases in which sunlight or artificial light sources play a major role in the causation of clinical conditions (e.g. porphyrias, polymorphic photodermatoses, Xeroderma pigmentosum and skin cancers) have been well documented. As a result, in some countries (e.g. Australia) public services inform people about the potential risk of extended periods of sun exposure every day. The problems are often aggravated by the phototoxic or photoallergic reactions produced by a variety of environmental pollutants, food additives or therapeutic and cosmetic drugs. On the other hand, if properly used, light-stimulated processes can induce important beneficial effects in biological systems, such as the elucidation of several aspects of cell structure and function.

Novel developments are centered around photodiagnostic and phototherapeutic modalities for the treatment of cancer, artherosclerosis, several autoimmune diseases, neonatal jaundice and others. In addition, classic research areas such as vision and photosynthesis are still very active. Some of these developments are unique to photobiology, since the peculiar physico-chemical properties of electronically excited biomolecules often lead to the promotion of reactions which are characterized by high levels of selectivity in space and time. Besides the biologically centered areas, technical developments have paved the way for the harnessing of solar energy to produce warm water and electricity or the development of environmentally friendly techniques for addressing problems of large social impact (e.g. the decontamination of polluted waters). While also in use in Western countries, these techniques are of great interest for developing countries.

The European Society for Photobiology (ESP) is an organization for developing and coordinating the very different fields of photosciences in terms of public knowledge and scientific interests. Due to the ever increasing demand for a comprehensive overview of the photosciences the ESP decided to initiate an encyclopedic series, the "Comprehensive Series in Photochemical and Photobiological Sciences". This series is intended to give an in-depth coverage over all the very different fields related to light effects. It will allow investigators, physicians, students, industry and laypersons to obtain an updated record of the state-of-the-art in specific fields, including a ready access to the recent literature. Most importantly, such reviews give a critical evaluation of the directions that the field is taking, outline hotly debated or innovative topics and even suggest a redirection if appropriate. It is our intention to produce the monographs at a sufficiently high rate to generate a timely coverage of both well established and emerging topics. As a rule, the individual volumes are commissioned; however, comments, suggestions or proposals for new subjects are welcome.

Donat P. Häder and Giulio Jori
Spring 2002

Volume preface

Light has been used to cure various diseases for centuries. However it is only recently that a new medical field has arisen under the name of photodynamic therapy (PDT), also called photochemotherapy. This modality, initially devoted to cancer care, is now also used for cancer diagnosis and the treatment of various diseases of vascular origins.

Photodynamic therapy had for a long time been confined to laboratories simply because the sensitizers used could not be protected. It then slowly emerged as a promising treatment technique in various clinical fields. The PDT story is emblematic of difficulties faced by new technologies. Several scientific organisations played a major role gathering scientists to help in PDT development. Among them is the European Society for Photobiology, ESP, and it is not surprising that it supported this PDT book. It appeared necessary to us to sum up not all the knowledge but the knowledge really required to understand what PDT is. Not all sensitizers are cited and neither are all the PDT applications, but through the examples given one can begin to understand PDT or imagine new PDT applications. This is the purpose of this comprehensive book, which also provides pertinent references to aid in understanding the subject and the safe treatment of patients. I asked to some famous authors to contribute to this book not only for their knowledge in this field but also for the ability that I personally noticed to easily explain complicated things. I hope that readers will find within these pages all they need to start or teach PDT. I wish that they will have as much pleasure as I have to work in this very fascinating field.

Thierry Patrice
May 2003

Contributors

Ludovic Bourré
Départment Laser
Hôpital Laënnec
44093 Nantes
France

Nicole Brasseur
Conseillére à la recherche,
secteur Sciences de la santé
Bureau de la recherche
Faculté de médecine
Université de Sherbrooke
12e Av. Nord
Fleurimont, Quèbec
Canada, J1H 5N4

A. Degen
Dept. OB/GYN
University Hospital of Zürich
8091 Zürich
Switzerland

E.F. Gudgin Dickson
Department of Chemistry and
Chemical Engineering
Royal Military College of Canada
PO Box 17000
Station Forces
Kingston ON
Canada, K7K, 7B

Angela Ferrario
Clayton Centre for Ocular Oncology,
Children's Hospital Los Angeles and
Keck School of Medicine
University of Southern California
Los Angeles, CA
USA

C. Fritsch
Department of Dermatology
Heinrich Heine University
Moorenstr. 5
D-40225 Düsseldorf
Germany

Prakash S. Gill
Clayton Centre for Ocular Oncology
Children's Hospital Los Angeles and
Keck School of Medicine
University of Southern California
Los Angeles, CA
USA

Charles J. Gomer
Clayton Centre for Ocular Oncology
Children's Hospital Los Angeles and
Keck School of Medicine
University of Southern California
Los Angeles, CA
USA

Tayyaba Hasan
Wellman Laboratories of
Photomedicine
Massachusetts General Hospital
Wellman 224
Boston, MA 02114
USA

J.C. Kennedy
Department of Oncology
Queen's University
Kingston ON
Canada, K7L 3N6

Herwig Kostron
Department of Neurosurgery
University of Innsbruck
A-6020 Innsbruck
Austria

K. Lang
Department of Dermatology
Heinrich Heine University
Moorenstr. 5
D-40225 Düsseldorf
Germany

P. Lehman
Department of Dermatology
Heinrich Heine University
Moorenstr. 5
D-40225 Düsseldorf
Germany

J. Moan
Institute for Cancer Research
0310 Montebello
Oslo
Norway

Anne C.E. Moor
Wellman Laboratories of
Photomedicine
Massachusetts General Hospital
Wellman 224
Boston, MA 02114
USA
Present address: medac GmbH, Wedel,
Germany

W. Neuse
Department of Dermatology
Heinrich Heine University
Moorenstr. 5
D-40225 Düsseldorf
Germany

Bernhard Ortel
Wellman Laboratories of
Photomedicine
Massachusetts General Hospital
Wellman 224
Boston, MA 02114
USA

Thierry Patrice
Department Laser-Neurochirurgie CHU
Régional et Universitaire de Nantes
Hôpital Guillaume et René Laënnec
44093 Nantes Cedex 01
France

Q. Peng
Institute for Cancer Research
0310 Montebello
Oslo
Norway

R.H. Pottier
Department of Chemistry and
Chemical Engineering
Royal Military College of Canada
PO Box 17000
Station Forces
Kingston ON
Canada, K7K, 7B4

Nathalie Rousset
Départment Laser
Hôpital Laënnec
44093 Nantes
France

Natalie Rucker
Clayton Centre for Ocular Oncology
Children's Hospital Los Angeles and
Keck School of Medicine
University of Southern California
Los Angeles, CA
USA

T. Ruzicka
Department of Dermatology
Heinrich Heine University
Moorenstr. 5
D-40225 Düsseldorf
Germany

K.W. Schulte
Department of Dermatology
Heinrich Heine University
Moorenstr. 5
D-40225 Düsseldorf
Germany

Margaret A. Shwartz
Clayton Centre for Ocular Oncology
Children's Hospital Los Angeles and
Keck School of Medicine
University of Southern California
Los Angeles, CA
USA

G. Sutedja
Department of Pulmonology
Academic Hospital Vrije Universiteit
PO Box 7057
1007 MB Amsterdam
The Netherlands

Sonia Thibaud
Départment Laser
Hôpital Laënnec
44093 Nantes
France

Karl von Tiehl
Clayton Centre for Ocular Oncology
Children's Hospital Los Angeles and
Keck School of Medicine
University of Southern California
Los Angeles, CA
USA

Kenneth K. Wang
Division of Gastroenterology and
Hepatology
Mayo Clinic and Foundation
Rochester, MN 55905
USA

Brian C. Wilson
Ontario Cancer Institute
University of Toronto
Canada

Pius Wyss
Dept. OB/GYN
University Hospital of Zürich
8091 Zürich
Switzerland

Contents

Chapter 1

An outline of the history of PDT

J. Moan and Q. Peng

Table of Contents

1.1 Introduction

Photodynamic therapy, PDT, has now reached the level of being an accepted treatment for a number of diseases, among which are several forms of cancer. Many countries have approved its use. The number of articles on PDT published in a year, both clinical and basic, seems to be steadily increasing. It has, however, been observed that many of the investigators are obviously unaware of the early work done in this field and hence, repeat many of the experiments reported earlier (before the internet and the modern database were established). Therefore, in the present historical review, the early work is weighted more heavily than the recent work that is more easily accessible to the readers.

1.2 The 'photodynamic action'

The term 'photodynamic action' (*'photodynamische Wirkung'*) was introduced in 1904 by one of the pioneers of photobiology: Professor Hermann von Tappeiner, director of the Pharmacological Institute of the Ludwig-Maximilians University in Munich [1]. It is not clear why he called the process 'dynamic'; it might have been to distinguish this biological phenomenon from the reactions taking place in the photographic process that had been discovered a few years earlier. Actually, von Tappeiner was not completely happy with his term, as he says in the foreword of his book *Die Sensibilisierende Wirkung Fluorescierender Substanzen*: 'Whether or not the name is to be used further or dropped, must be left to the discretion of my colleagues' [2]. Also, Blum in his textbook on photosensitization [3] expresses objections to the term: 'The choice of 'photodynamic action' is not altogether a happy one, but has advantage of priority and usage'. Feeling the need for a more correct and descriptive term we have tried to replace 'photodynamic therapy', PDT, by 'photochemotherapy', PCT without overwhelming success [4]. There may be several reasons for this. First, PDT has the advantage of priority. Secondly, photochemotherapy has a wider definition (see later) including processes not requiring oxygen. Thirdly, PCT is an abbreviation for *porphyria cutanea tarda*, a disease that involves photosensitivity. However, the name 'photochemotherapy' would have paved the way for clinical applications better than the name 'photodynamic therapy' since all oncologists are familiar with chemotherapy, while 'photodynamic' may give associations to 'biodynamic' which, at the best, is regarded as a quasi-scientific term. According to the original definition, as well as to Blum's later recommendation, the term 'photodynamic action' should be used only for photosensitized reactions requiring oxygen. It should be remarked that oxygen is not only involved in photosensitization of Type II, but also usually involved in photosensitization of Type I. A Type I reaction is a *radical* or *redox reaction* in which a photosensitizer, excited to the triplet state (^{3}S), interacts with a neighbouring molecule (A) by exchange of an electron or a hydrogen atom [5]:

$$^3S + A \rightarrow S^{\cdot -} + A^{\cdot +} \tag{1}$$

followed by

$$A^{\cdot +} + {}^3O_2 \rightarrow A_{ox} \tag{2}$$

or

$$S\cdot^- + O_2 \rightarrow S_0 + O_2\cdot^- \tag{3}$$

$$O_2\cdot^- + A \rightarrow A_{ox}, \tag{4}$$

in both cases giving an oxidized biomolecule A_{ox}.

An alternative type I reaction pathway might be:

$$^3S + {}^3O_2 \rightarrow S^+ + O_2\cdot^- \tag{5}$$

followed by

$$S^+ + A \rightarrow S_0 + A\cdot^+ \tag{6}$$

Reaction (4) may follow after reaction (5), and reaction (2) may follow after reaction (6). Type I reactions may also be independent of oxygen, as is the case for the reactions of psoralens with DNA.

A photosensitized process of Type II is by definition an *energy transfer process* [5]. The most common Type II process is oxidation via singlet oxygen ($^1O_2$) formation:

$$^3S + {}^3O_2 \rightarrow S_0 + {}^1O_2 \tag{7}$$

$$^1O_2 + A \rightarrow A_{ox} \tag{8}$$

A further description of $^1O_2$ reactions in biological systems can be found in ref. 5.

Many authors use the expression 'photodynamic action' and 'PDT' synonymously with '$^1O_2$ reactions' and even with: 'Type II reactions', but according to the definitions given above this is not strictly correct. However, in most practical cases it may be acceptable since most PDT sensitizers act via $^1O_2$ which is formed in an electron exchange process [6].

1.3 Phototherapy

Phototherapy can be defined as the use of light alone for therapeutic purposes. However, endogenous sensitizers are usually involved, so phototherapy often relies on photodynamic processes. Solar light has been used to treat a number of disorders such as vitiligo, psoriasis, rickets, skin cancer and even psychosis [7]. Heating, as well as psychological effects mediated by vision, may have played roles in these therapies, but the effect of visible light and ultraviolet radiation on the skin was probably more important. Phototherapy has been applied by humans for 3000 years and was known by the Egyptians, the Indians and the Chinese [7]. In Greece, Herodotes called it 'heliotherapy' and recommended it for 'restoration of health' in the 2nd century BC. In the 18th century the effect of sunlight on rickets was known. In 1815, Carvin wrote that sunlight had a curing effect on 'scrofula', rickets, rheumatism, scurvy, paralysis and muscle weakness [8]. The Polish physician Sniadecki documented in 1822 the importance of sun exposure for the prevention of rickets [9]. Later, in 1903, the Dane, Niels Finsen, was awarded the Nobel Prize for his work on the use of light from the carbon arc in the treatment of lupus vulgaris (skin tuberculosis) [10] and was acknowledged as the founder of modern phototherapy.

He also treated smallpox with red light and found that this treatment prevented suppuration of pustules.

In the 1950s Richard Cremer in Essex, England, after listening to an observant nurse, introduced phototherapy as a treatment of jaundice in newborn babies [11]. This is the most widely used form of phototherapy today. However, light as a therapeutic agent for depression and for maintenance of biological rhythms may in the future become equally important. It seems that, in the latter context, light may act not only via vision but also via absorption in skin [12].

Since several of the above-mentioned forms of phototherapy may belong to photochemotherapy or even PDT, contact between PDT scientists and scientists in the field of phototherapy might be mutually beneficial.

1.4 Photochemotherapy

In one of India's sacred books *Atharava-veda* (1400 BC) it is described how seeds of the plant *Psoralea corylifolia* can be used for the treatment of vitiligo. Psoralens are the photoactive components of these seeds, just as in the extracts of the plant, *Ammi majus*, which grows on the banks of the Nile, was used by the Egyptians to treat vitiligo. For centuries photochemotherapy made no further progress until 1974 when PUVA (i.e. treatment with psoralens and UVA radiation) was reported to be an efficient treatment of psoriasis [13]. Photochemotherapy can also be tailored to act on the immune system such as in extracorporeal photophoresis of *mucoses fungoides*, cutaneous T-cell lymphoma [14].

1.5 The early days of photodynamic therapy

The very first attempts to apply PDT to treatment of tumors and other skin diseases, such as lupus of the skin and chondylomata of the female genitalia, were performed by the group of von Tappeiner in 1903–1905 [15,16]. They tried a number of dyes: eosin, fluorescein, sodium dichloroanthracene disulfonate and 'Grubler's magdalene red'. The dyes were in most cases topically applied, but intratumoral injections were also attempted. Favorable results were reported, but there was no long-term follow-up, and PDT was soon forgotten, probably because of the advent of ionizing radiation in cancer therapy.

The story of how this German group hit upon the idea of using dyes as biological sensitizers for light is fascinating: one of von Tappeiner's students, Oscar Raab, was investigating the toxic effects of acridine on paramecia. In one experiment the paramecia survived incubation with a given acridine concentration for about 1.5 h, while in another experiment they survived for about 15 h under identical conditions, except, as recorded by the observant student, during one of the experiments there was a heavy thunderstorm. So he started to wonder whether light might have played a role—this resulted in the discovery of photodynamic action [17,18]. Characteristic of that time: the supervisor [17] published the finding before the student [18]. The group performed a large amount of work on photosensitization, discovered that oxygen was required for the photodynamic effect [19] and summarized their work in a book [20]. At about the same time as Raab made his discovery, a French neurologist administered eosin orally for epilepsy and found that the patient got dermatitis on light-exposed areas of the body [21].

1.5.1 Hematoporphyrin and hematoporphyrin derivative

No history of PDT can be told without mentioning hematoporphyrin. This compound was first produced (in an impure form) by Scherer who removed iron from dried blood in 1841 by treatment with sulfuric acid [22]. The spectrum of this red substance, as well as its fluorescence, was described by Thudichum in 1867 [23], and the name 'hematoporphyrin' was dedicated to it by Hoppe-Seyler in 1871 [24]. In the period 1908–1913, a number of photobiological experiments were carried out with hematoporphyrin, demonstrating how it sensitized paramecia, erythrocytes, mice [25], guinea pigs [26] and humans [27] to light. The German doctor Friedrich Meyer Betz became extremely photosensitive for more than two months after injecting 200 mg of hematoporphyrin into himself [27]. This dose is similar to that used of HPD and Photofrin for PDT today, and one may wonder whether the hematoporphyrin used by Meyer Betz was HPD rather than pure hematoporphyrin which turns out to be a rather poor photosensitizer due to its high water solubility. Hans Fischer, who was awarded the Nobel Prize for his work on porphyrins, reported that uroporphyrin was almost as phototoxic as hematoporphyrin [28]. This is surprising in view of the fact that the water solubility of uroporphyrin is even greater than that of hematoporphyrin.

The first observation of porphyrin fluorescence from tumors was published by Policard in 1924 [29]. The red fluorescence from endogenously produced porphyrins in experimental rat sarcomas was attributed to bacteria infecting the tumors. A few years later Kördler detected a similar fluorescence in breast carcinomas and some other superficial tumors but found no evidence for the involvement of bacteria [30]. Several investigators reported preferential accumulation of porphyrins and porphyrin precursors in neoplastic tissue [31–33]. Of particular interest is 5-aminolevulinic acid (ALA), a precursor to heme synthesis; Rubino and Rasetti [32] stated that the preferential tumor accumulation of porphyrins was not related to an elevated activity of the enzymes of heme synthesis. Later work showed that this may not be a universal observation since the activity of porphobilinogen deaminase seems to be high in some tumors while the activity of ferrochelatase as well as the concentration of iron seems to be low [34]. Auler and Banzer were probably the first to study the accumulation of injected porphyrins in tumors [35]. They reported that hematoporphyrin injected in rats accumulated in primary and metastatic tumors as well as in lymph nodes. Animal tumors were treated with light after injection of this dye, and promising results were obtained [35]. Figge et al. [36] showed that a number of porphyrins have a selective affinity, not only for neoplastic tissue, but also for embryonic and regenerating tissues in rodents. Surprisingly, it was observed that tumor-bearing mice tolerated larger doses of ^{65}Zn hematoporphyrin than mice without tumors [37]. This observation probably indicated that the tumors accumulated a significant fraction of the dye, thus protecting sensitive organs from radiation damage. This group was probably the first to demonstrate that hematoporphyrin also had a tumor-localizing ability in a variety of human malignancies [38]. A large amount of data on the tumor-localizing ability of porphyrins was published by Lipson et al. at the Mayo Clinic during 1960–1967 [39–44]. Their work was inspired by that of Dr. Samuel Schwartz [45], who was interested in porphyrins from the point of view of radiosensitization. Some of the results of Schwartz and co-workers are remarkable and deserve to be mentioned even though they concern ionizing radiation. Some 153 mice-bearing rhabdomyosarcoma tumors were injected with different doses of crude hematoporphyrin, ranging from 10 to 1250 μg [46]. Three hours later the tumors were exposed to ionizing radiation. Twenty seven of the mice that had got 50 μg were all cured, while among the remaining 126 animals no cures were observed. The conclusion was that low or intermediate concentrations of hematoporphyrin have a stronger sensitizing effect than either very low (10 μg) or very high (> 250 μg) doses. We once looked for the radiation modifying effect of hematoporphyrin, but found no such effect, possibly because we used too large concentrations [47].

Schwartz realized that commercial samples of hematoporphyrin were impure and tried to purify them [45]. Surprisingly, he found that pure hematoporphyrin was a poor tumor-localizer. Treatment with acetic–sulfuric acid mixture gave some components which had better properties with respect to tumor-localization. These components came to be known as 'hematoporphyrin derivative', HPD. They were later used by Lipson and co-workers and by a large number of clinical investigators, both for diagnostic and therapeutic purposes [48]. HPD contains several porphyrin monomers as well as dimers and oligomers. A debate about its composition lasted for several years. Among others we studied HPD by means of HPLC and fluorescence methods and found three different groups of components, monomers (hematoporphyrin stereoisomers, hematoporphyrin vinyl deuteroporphyrin isomers and protoporphyrin) with a high fluorescence quantum yield but with a poor tumor uptake, dimers with a lower fluorescence quantum yield but with a higher tumor uptake and non-fluorescent aggregates with the best tumor-localizing properties [49–53]. The cellular uptake of these components, and of other related porphyrin compounds, increased with increasing lipophilicity, and so did the quantum yield of cell inactivation [53]. The non-fluorescent fraction, the aggregates, had a low photosensitizing ability although they seemed to be slightly more photoactive than one might expect on the basis of their weak or non-existent fluorescence. The aggregates seemed to be the best tumor-localizers [49]. Several investigators proposed that the aggregates might decompose to fluorescent and photoactive components after being taken up by the tumor, but this was never proven. Kessel and others continued the investigation to find the chemical identity of HPD and concluded that the dimers and oligomers were coupled by ether as well as ester linkages [54]. Dougherty and his co-workers partly purified HPD by removing the monomers. The resulting product was called Photofrin and is still the most widely used sensitizer for clinical PDT.

The modern era of PDT was founded in the 1970s with the pioneering work of Dougherty and co-workers at the Roswell Park Memorial Cancer Institute in Buffalo who used HPD/Photofrin. This history is well known and has already been elegantly described in several articles [55,56].

1.5.2 Other photosensitizers introduced for PDT

As soon as it was realized that hematoporphyrin was difficult to purify and that HPD was a mixture of a number of porphyrins with widely different properties, the search for pure substances started. Winkelman introduced tetraphenyl porphine sulfonate (TPPS) and claimed that it had a better tumor-localizing ability than HPD [57,58]. The phthalocyanines were introduced by Ben-Hur and co-workers [59]. TPPS and aluminum phthalocyanin sulfonates (AlPcS) can be produced with different numbers of sulfonate groups, ranging from 1 to 4, attached to the aromatic ring structure. This results in different water solubilities, and these compounds have been very useful for experiments carried out to elucidate structure–function relationships with respect to a number of parameters: specificity and absolute tumor uptake, intratumoral and intracellular localization, and quantum yield of photo-inactivation of cells and tumors [60–65]. The phthalocyanines are also relevant for clinical use, mainly due to their strong absorption in the far-red spectral band ($\lambda \sim 670$ nm) where strong dye lasers are available and where tissue is more transparent than at around 630 nm. The meso-substituted chlorine, tetra(*meta*-hydroxyphenyl)chlorin (mTHPC) was introduced by Berenbaum and co-workers [66] and is being used in clinical trials. Recently, some other new photosensitizers have also been introduced for clinical trials [67], which are summarized in Table 1. The use of endogenous protoporphyrin IX (PpIX) induced by exogenous ALA is a novel concept of clinical PDT and photodetection [68]. This was experimentally introduced by Malik et al. in 1987 in the treatment of erythroleukemic cells in vitro [69]. In the meantime, based upon their own data on ALA-mediated PpIX production in the normal skin of mice [70], Kennedy and Pottier successfully treated human skin tumors with topically ALA-based PDT [71]. We were among the first workers to suggest the use of porphyrin precursors for PDT [72]. It should be pointed out that the knowledge obtained from research on porphyric diseases plays an important role in understanding the mechanisms of ALA-based PDT and in developing this new technique.

1.6 Why do some photosensitizers localize selectively in tumors?

This question has been addressed by a number of researchers and their results have been reviewed elsewhere [60,65]. The selective tumor uptake is probably not due to special properties of tumor cells [73], but rather to differences in the physiology between tumors and normal tissues: (1) tumors have a larger interstitial volume than normal tissues, (2) tumors often contain a larger fraction of macrophages than normal tissues, (3) tumors have a leaky microvasculature, (4) tumors have poor lymphatic drainage, (5) the extracellular pH is low in tumors, (6) tumors contain a relatively large amount of newly synthesized collagen and (7) tumor tissue contains many receptors for lipoproteins.

Table 1. Summary of new photosensitizers and protoporphyrin IX precursors used in PDT and PD clinical trials

Compound	Trade Mark	Producer	Wavelength (nm) used	Current status of product development
Porfimer sodium	Photofrin	QLT Photo Therapeutis Ltd.	632	**April 19, 1993 Canadian Health Protection Branch** approved marketing of Photofrin for PDT of recurrent superficial papillary bladder cancer. This was the first approval for PDT in the world and should be regarded as a milestone in PDT history **Axcan Pharma regulatory approvals from other countries:** **April 11, 1994** The Netherlands for superficial lung cancer and palliation of advanced lung and esophageal cancer **October 5, 1994** Japan for inoperab e superficial esophageal and gastric cancer, early stage lung and cervical cancer and cervical dysplasia **July 13, 1995** Canada for palliation of advanced esophageal cancer **December 27, 1995** US FDA for palliation of esophageal cancer in patients unsuitable for Nd:YAG therapy **April 9, 1996** France for recurrent lung and esophageal cancer **July 30, 1997** Germany for early stage lung cancer **January 9, 1998** US FDA for early stage lung cancer **December 22, 1998** US FDA for late stage lung cancer **December 22, 1998** UK for advanced lung and esophageal cancer **February 15, 1999** Finland for advanced lung and esophageal cancer **Pending Photofrin trials for regulatory approvals:** early stage esophageal cancer, Barrett's esophagus, head and neck cancer, superficial bladder cancer and adjuvant therapy procedures for brain tumors, intrathoracic tumors (pleural mesothelioma) and intraperitoneal tumors.

Table 1. (*cont.*)

Compound	Trade Mark	Producer	Wavelength (nm) used	Current status of product development
Benzoporphyrin derivative monoacid ring A (BPD-MA)	Verteporfin Visudyne	QLT Inc. & Novartis Opthalmic	690	**Systemic use:** Approved worldwide Phase III: multiple basal cell carcinoma
Tetra(meta-hydroxyphenyl) chlorin (mTHPC)	Foscan	Biolitee	652	**Systemic use:** Phase III: head and neck cancer Phase II: chest, gastrointestinal, pancreas Phase I/II: adjuvant therapy for late stage cancer
Tin ethyletiopurpurin (SnET2)	Purlytin	Miravant Medical Technologies	660	**Systemic use:** Phase III: age-related macular degeneration Phases II/III: cutaneous cancer, AIDS-related Kaposis Phase I: prostate cancer
Lutetium-texaphyrin (Lu-tex)	LUTRIN ANTRIN OPTRIN	Pharmacyclics Inc.	732	**Systemic use:** Phase II: locally recurrent breast cancer (LUTRIN™) Phase I: photoangioplasty of peripheral arterial diseases (ANTRIN™). Phase I: BCC, melanoma (LUTRIN™) Phase I: AMD (OPTRIN™)
Boronated porphyrin	BOPP	Pacific Pharmaceuticals Inc.	628	**Systemic use:** Phase I: brain cancer
Mono-N-Aspartyl chlorin e6 (NPe6)		Meiji Seika Kaisha Ltd.	664	**Systemic use:** Phase I: cutaneous cancer Phase I: early stage lung cancer
Hypericin	VIMRxyn	VIMRX Pharmaceuticals Inc.	600–1000	**Systemic use:** Phase I/II: glioblastoma Phase I: antiviral agent for HIV-infected diseases **Topical use:** Phase I: basal cell carcinoma, squamous cell carcinoma, warts and psoriasis

Sulfonated aluminium phthalocyanine (AlPcS)	Photosense	Russia	675	**Systemic use:** Phase I/II: cutaneous cancer, etc.
Porphycene (ATMPn)		Glaxo-Wellcome Inc.	630	**Topical use:** Phase I/II: psoriasis
5-aminolevulinic acid (ALA)	Levulan	DUSA Pharmaceuticals, Inc.	632	**Topical use:** Filing by US FDA for actinic keratosis Phase I/II: basal cell carcinoma, Bowen's disease, psoriasis Phase I/II: hair removal Phase I/II: acne **Intravesical instillation:** Phase I/II: bladder cancer photodetection
ALA methylester	Metvix	Photocure ASA Galderma	632	**Topical use:** Approved by EEC: actinic keratosis Approved by EEC: basal cell carcinoma
ALA hexylester	Hexvix	Photocure	632	**Intravesical instillation:** Phase III: detection of bladder cancer

1.7 The action mechanisms of PDT at the cellular level

The historical development of the understanding of these mechanisms has been dealt with in numerous papers, as reviewed elsewhere [60,65]. As a rough generalization of the current status it can be stated that cationic sensitizers localize in both the nucleus and mitochondria, lipophilic ones tend to stick to membrane structures, and water-soluble drugs are often found in lysosomes. However, not only the lipid/water partition coefficient is of importance but also other factors such as molecular weight and charge distribution (symmetry/asymmetry). In some cases light exposure leads to a relocalization of the sensitizers [74]. The mechanism of action is clearly linked to the intracellular localization of the sensitizers. This is due to the fact that the lifetime of the main active photoproduct, $^1O_2$, is short in cells, less than $0.05 \, \mu s$ [75]. Thus, $^1O_2$ can diffuse less than $0.02 \, \mu m$ from the site of production [75]. This is certainly the main reason why PDT has such a low mutagenic potential: most PDT sensitizers are localized outside the nucleus. The quantum yield of photoinactivation of cells is different for different sensitizers. Lipophilic sensitizers tend to have a larger quantum yield than the water-soluble ones [76]. PDT has a strong effect on cell division [77], which is probably mainly due to microtubule damage. Under certain conditions this kind of damage may contribute significantly to PDT induced cell death.

1.8 The action mechanisms of PDT at the tissue level

PDT acts via the $^1O_2$ pathway in vivo as well as in vitro. Strictly speaking, this has never been proven since nobody has been able to detect $^1O_2$ in vivo and it is for obvious reasons difficult to apply $^1O_2$-scavengers in sufficient concentrations in vivo. However, the oxygen dependency of the PDT effect is the same in vivo as in in vitro with a doubling of the exposure necessary to obtain a given effect at an oxygen concentration of about 1% (as compared to the in vivo concentration which is of the order of 5%, above which the PDT effect is maximal) [78,79]. The oxygen dependency is similar for PDT and ionizing radiation and the low oxygen concentration in the region of many tumors may limit the efficiency of PDT. The PDT-induced reactions themselves consume oxygen and it has been shown that low fluence rates or fractionation of the exposure may improve the PDT effects [80]. Kessel et al. proposed that lipophilic and liposome- or lipoprotein-bound dyes sensitized tumor cells in vivo directly, whereas water-soluble dyes mainly sensitized the vascular system in the tumor to photoinactivation [81]. The significance of vascular damage was first convincingly demonstrated by Henderson et al. [82]. Canti et al. were the first to observe and demonstrate immunological effects of PDT [83]. This research has been carried forward by Korbelik and others, and it is now agreed that the immunological effects of PDT are significant and play a role in tumor destruction as well as in the prevention of tumor recurrence [84].

Apoptosis is a popular field of cancer research. This mechanism of cell death has also been demonstrated for PDT [85]. Kessel has proposed that mitochondrial damage may be an important step in PDT-induced apoptosis [86].

Most sensitizers are photodegraded (bleached) during PDT. We have shown that bleaching may be taken advantage of to improve the tumor selectivity of PDT [87]. Furthermore, bleaching is induced by $^1O_2$ and the rate of bleaching can be used for dosimetric purposes.

1.9 Photochemical internalization

Some sensitizers, such as the water-soluble TPPS and AlPcS, localize in lysosomes, which disrupt during PDT [88,89]. The sensitizers are then relocalized in the cells and may even enter the nucleus [89]. Lysosomal damage leads to leakage of the contents of the lysosomes into the cytoplasm. One of the hottest and most promising applications of PDT may turn out to be photochemical internalization: the use of PDT to liberate molecules taken up by endosomes and lysosomes (toxins, DNA fragments etc.) into the cytoplasm. Several extremely promising applications of this technique have been proposed and demonstrated [90] since lysosomal disruption in vivo had been demonstrated in 1996 [91].

References

1. H. von Tappeiner, A. Jodlbauer (1904). Ueber die wirkung der photodynamischen (fluorescierenden) stoffe auf protozoen und enzyme. *Arch. Klin. Med.*, **80**, 427–487.
2. H. von Tappeiner, A. Jodlbauer (1970). *Die Sensibilisierende Wirkung Fluorescierender Substanzen*, Vogel, Leipzig.
3. H.F. Blum (1941). *Photodynamic Action and Diseases Caused by Light*, Rhinehold, New York (reprinted with an updated appendix, Hafner, New York, 1964).
4. J. Moan, K. Berg (1992). Photochemotherapy of cancer: experimental research. *Photochem. Photobiol.*, **55**, 931–948.
5. J.D. Spikes (1989). Photosensitization. In: K.S. Smith (Ed.), *The Science of Photobiology. Photosensitization* (pp. 79–110). Plenum Press, New York, London.
6. N.J. Turro (1978). *Modern Molecular Photochemistry* (Chapter 14). The Benjamin/ Cummings Publishing Co., London, Amsterdam, Sydney.
7. J.D. Spikes (1985). The historical development of ideas on applications of photosensitised reactions in health sciences. In: R.V. Bergasson, G. Jori, E.J. Land, T.G. Truscott (Eds), *Primary Photoprocesses in Biology and Medicine* (pp. 209–227). Plenum Press, New York.
8. J.F. Carvin (1815). *Des Bienfaits de L'insolation*. Paris.
9. M.F. Holick, J.A. MacLaughlin, J.A. Parrish, R.R. Anderson (1982). The photochemistry and photobiology of vitamin D_3. In: J.D. Regan, J.A. Parrish (Eds), *The Science of Photomedicine* (pp. 195–218). Plenum Press, New York, London.
10. N.R. Finsen (1901). *Phototherapy*. London.
11. R.J. Cremer, P.W. Perryman, D.H. Richards (1958). Influence of light on the hyper-bilirubinemia in infants. *Lancet*, **1**, 1094–1097.
12. S.S. Campbell, P.J. Murphy (1998). Circadian clock resetting in humans by extraocular light exposure. *Photochem. Photobiol.*, **67**, 57S.
13. J.A. Parrish, T.B. Fitzpatrick, L. Taneubaum, M.A. Pathac (1974). Photochemotherapy of psoriasis with oral methoxalen and longwave ultraviolet light. *N. Engl. J. Med.*, **291**, 1207–1211.

14. R. Edelson, C. Berger, F. Gasparro, B. Jegasothy, P.S. Held, B. Wintroub, E. Vonderheid, R. Knobler, K. Wolff, G. Plewig (1987). Treatment of cutaneous T-cell lymphoma by extracorporeal photochemotherapy. *New Engl. J. Med.*, **316**, 297–303.

15. H. von Tappeiner, A. Jesionek (1903). Therapeutische Versuche mit fluorescierenden Stoffen. *Munch. Med. Wochenschr.*, **47**, 2042–2044.

16. A. Jesionek, H. von Tappeiner (1905) Zur behandlung der Hautcarcinome mit fluorescierenden Stoffen. *Arch. Klin. Med.*, **82**, 223.

17. H. von Tappeiner (1900). Ueber die Wirkung fluorescierenden Stoffe auf Infusiorien nach Versuchen von O. Raab. *Munch Med. Wochenschr.*, **47**, 5.

18. O. Raab (1900). Ueber die Wirkung Fluorescierenden Stoffe auf Infusorien. *Z. Biol.*, **39**, 524–546.

19. H. von Tappeiner, A. Jodlbauer (1904). Über die Wirkung der photodynamischen (fluorescierenden) Stoffe auf Protozoen und Enzyme. *Dtsch. Arch. Klin. Med.*, **80**, 427–487.

20. H. von Tappeiner, A. Jodlbauer (1907). Die Sensibilisierende Wirkung Fluorescierender Substanzer. *Untersuchungen Uber die Photodynamische Erscheinung.* FCW Vogel. Leipzig.

21. J. Prime (1900). *Les Accidentes Toxiques par L'eosinate de Sodium.* Jouve & Boyer. Paris.

22. H. Scherer (1841). Chemical-physiological investigations. *Ann. D. Chem. Pharm.*, **40**, 1–64.

23. J.L. Thudichum (1867). *Tenth Report of the Medical Officer of the Privy Council* (pp. 152–233). HM Stationery Office, London.

24. F. Hoppe-Seyler (1871). *The Hematins* (pp. 124–528). Tübinger Med. Chem. Untersuchungen.

25. W. Hausmann (1911). Die sensibilisierende Wirkung des hamatoporphyrins. *Biochem. Z.*, **30**, 276–316.

26. H. Pfeifer (1911). Der Nachweis photodynamischer Wirkungen fluorescierenden Stoffe am lebenden Warmbluter. In: E. Abderhaldan (Ed.), *Handbuch der Biochemischen*, Arbeitsmethoden, (pp. 563–571), Berlin.

27. F. Meyer-Betz (1913). Untersuchungen uber die Biologische (photodynamische) Wirkung des hamatoporphyrins und anderer Derivative des Blut-und Gallenfarbstoffs. *Dtsch. Arch. Klin. Med.*, **112**, 476–503.

28. H. Fischer, H. Hilmer, F. Lindner, B. Pützer (1925). Zur kenntnis der natürlichen porphyrine. *Z. Physiol. Chem.*, **150**, 44–101.

29. A. Policard (1924). Etude sur les aspects offerts par des tumeurs experimentales examinées a la lumière de Wood. *C. R. Soc. Biol.*, **91**, 1423–1428.

30. J. Körbler (1931). Untersuchung von krebsgewebe im fluoreszenzerregenden licht. *Strahlentherapie*, **41**, 510–518.

31. F.H.J. Figge (1945). The relationship of pyrrol compounds to carcinogenesis. *AAAR Research Conference on Cancer* (pp. 147). Science Press, Washington D.C.

32. G. F. Rubino, L. Rasetti (1966). Porphyrin metabolism in human neoplastic tissues. *Panimerva Med.*, **8**, 290–292.

33. B. Zawirska (1979). Comparative porphyrin content in tumors with contiguous non-neoplastic tissues. *Neoplasms*, **26**, 223–229.

34. Q. Peng, K. Berg, J. Moan, M. Kongshaug, J.M. Nesland (1997). 5-aminolevulinic acid-based photodynamic therapy: principles and experimental research. *Photochem. Photobiol.*, **65**, 235–251.

35. H. Auler, G. Banzer (1942). Untersuchungen über die Rolle der Porphyrine bei geschwulstkranken Menschen und Tieren. *Z. Krebsforsch.*, **53**, 65–68.

36. F.H.J. Figge, G.S. Weiland, L.O.J. Manganiello (1948). Cancer detection and therapy: affinity of neoplastic, embryonic and traumatised tissues for porphyrins and metalloporphyrins. *Proc. Soc. Exp. Biol. Med.*, **68**, 640–641.

37. L.O. Manganiello, F.H.J. Figge (1951). Cancer detection and therapy II: Methods of preparation and biological effects of metallo-porphyrins. *Bull. Sch. Med. Univ. Maryland*, **36**, 3–7.

38. D.S. Rasmussen-Taxdal, G.E. Ward, F.H. Figge (1955). Fluorescence of human lymphatic and cancer tissues following high doses of intravenous hematoporphyrin. *Cancer*, **8**, 78–81.

39. R.L. Lipson, E.J. Baldes (1960). The photodynamic properties of a particular haemato-porphyrin derivative. *Arch. Dermatol.*, **82**, 508–516.

40. R.L. Lipson, E.J. Baldes, A.M. Olsen (1961). The use of a derivative of haematopor-phyrin in tumor detection. *J. Natl. Cancer Inst.*, **26**, 1–11.

41. R.L. Lipson, E.J. Baldes, A.M. Olsen (1961). Hematoporphyrin derivative: A new aid for endoscopic detection of malignant disease. *J. Thorac. Cardiovasc. Surg.*, **42**, 623–629.

42. R.L. Lipson, E.J. Baldes, A.M. Olsen (1964). Further evaluation of the use of haemato-porphyrin derivative as a new aid for the endoscopic detection of malignant disease. *Dis. Chest.*, **46**, 676–679.

43. R.L. Lipson, J.H. Pratt, E.J. Baldes (1964). Haematoporphyrin derivative for the detec-tion of cervical cancer. *Obstet. Gynecol.*, **24**, 78.

44. R.L. Lipson, E.J. Baldes, M.J. Gray (1967). Hematoporphyrin derivative for detection and management of cancer. *Cancer*, **20**, 2255–2257.

45. S. Schwartz, K. Absolon, H. Vermund (1955). Some relationship of porphyrins, X-rays and tumors. *Univ. Minn. Med. Bull.*, **27** (Oct. 15) 1–37.

46. L. Ohen, S. Schwartz (1966). Modification of radiosensitivity by porphyrins. II. Transplanted rhabdomyosarcoma in mice. *Cancer Res.*, **26**, 1769–1773.

47. J. Moan, E.O. Pettersen (1981). X-irradiation of human cells in culture in the presence of hematoporphyrin. *Int. J. Radiat. Biol.*, **40**, 107–109.

48. T.J. Dougherty (1987). Photosensitizers: therapy and detection of malignant tumors. *Photochem. Photobiol.*, **45**, 879–889.

49. J. Moan, S. Sommer (1983). Uptake of the components of hematoporphyrin derivative by cells and tumors. *Cancer Lett.*, **21**, 167–174.

50. J. Moan, T. Christensen, J. Sommer (1982). The main photosensitizing components of hematoporphyrin derivative. *Cancer Lett.*, **15**, 161–166.

51. J. Moan, S. Sandberg, T. Christensen, S. Elander (1983). Hematoporphyrin derivative: chemical composition, photochemical and photosensitizing properties. In: D. Kessel, T.J. Dougherty (Eds), *Porphyrin photosensitization* (Vol. 1, pp. 165–179). Plenum Publishing Co.

52. J. Moan, C. Rimington, A. Western (1988). Hematoporphyrin ethers Vol. III Cellular uptake and photosensitizing properties. *Int. J. Biochem.*, **20**, 1401–1404.

53. J. Moan, Q. Peng, J.F. Evensen, K. Berg, A. Western, C. Rimington (1987). Photosensitizing efficiencies, tumor and cellular uptake of different photosensitizing drugs relevant for photodynamic therapy of cancer. *Photochem. Photobiol.*, **46**, 713–721.

54. D. Kessel, P. Thompson (1987). Purification and analysis of hematoporphyrin and hematoporphyrin derivative by gel exclusion and reverse phase chromatography. *Photochem. Photobiol.*, **46**, 1023–1026.

55. M.D. Daniel, J.S. Hill (1991). A history of photodynamic therapy. *Aust. N.Z. J. Surg.*, **61**, 340–348.

56. T.J. Dougherty (1996). A brief history of clinical photodynamic therapy development at Roswell Park Cancer Institute. *J. Clin. Laser Med. Surg.*, **14**, 219–221.

57. J. Winkelman (1961). Intracellular localization of 'hematoporphyrin' in a transplanted tumor. *J. Natl. Cancer Inst.*, **27**, 1369–1377.

58. J. Winkelman (1962). The distribution of tetraphenylporphine-sulfonate in the tumor-bearing rat. *Cancer Res.*, **22**, 589–596.

59. E. Ben-Hur, I. Rosenthal (1985). Photosensitized inactivation of Chinese hamster cells by phthalocyanines. *Photochem. Photobiol.*, **42**, 129–133.

60. J. Moan (1986). Porphyrin photosensitization and phototherapy. *Photochem. Photobiol.*, **43**, 681–690.

61. B. Paquette, H. Ali, R. Langlois, J.E. van Lier (1988). Biological activities of phthalocyanines-VIII. Cellular distribution in V-79 Chinese hamster cells and phototoxicity of selectively sulfonated aluminium phthalocyanines. *Photochem. Photobiol.*, **47**, 215–220.

62. K. Berg, J.C. Bommer, J. Moan (1989). Evaluation of sulfonated aluminium phthalocyanines for use in photochemotherapy. A study on the relative efficiencies of photoinactivation. *Photochem. Photobiol.*, **49**, 587–594.

63. Q. Peng, J. Moan, J.M. Nesland, C. Rimington (1990). Aluminium phthalocyanines with asymmetrical lower sulfonation and with symmetrical higher sulfonation: a comparison of localizing and photosensitizing mechanism in human tumor LOX xenografts. *Int. J. Cancer*, **46**, 719–726.

64. Q. Peng, G.W. Farrants, K. Madslien, J.C. Bommer, J. Moan, H.E. Danielsen, J.M. Nesland (1990). Subcellular localization, redistribution and photobleaching of sulfonated aluminum phthalocyanines in a human melanoma cell line. *Int. J. Cancer*, **49**, 290–295.

65. J. Moan, K. Berg (1992). Photochemotherapy of cancer: Experimental research. *Photochem. Photobiol.*, **55**, 931–948.

66. M.C. Berenbaum, S.L. Acande, R. Bonnett, H. Kaur, S. Ioannov, R.D. White, U.J. Winfield (1986). Meso-tetra(hydroxyphenyl)porphyrin, a new class of potent tumor photosensitizers with favourable selectivity. *Br. J. Cancer*, **54**, 717–725.

67. T.J. Dougherty, C.J. Gomer, B.W. Henderson, G. Jori, D. Kessel, M. Korbelik, J. Moan, Q. Peng (1998). Photodynamic therapy. *J. Natl. Cancer Inst.*, **90**, 889–905.

68. Q. Peng, T. Warloe, K. Berg, J. Moan, M. Kongshaug, K.E. Giercksky, J.M. Nesland (1997). 5-Aminolevulinic acid-based photodynamic therapy: clinical research and future challenges. *Cancer*, **79**, 2282–2308.

69. Z. Malik, H. Lugaci (1987). Destruction of erythroleukemic cells by photoactivation of endogenous porphyrins. *Br. J. Cancer*, **56**, 589–595.

70. R.H. Pottier, Y.F.A. Chow, J.P. LaPlante, T.G. Truscott, J.C. Kennedy, L.A. Beiner (1986). Non-invasive technique for obtaining fluorescence excitation and emission spectra in vivo. *Photochem. Photobiol.*, **44**, 679–687.

71. J.C. Kennedy, R.H. Pottier, D.C. Pross (1990). Photodynamic therapy with endogenous protoporphyrin IX: basic principle and present clinical experience. *J. Photochem. Photobiol., B: Biol.*, **6**, 143–148.

72. Q. Peng, J.F. Evensen, C. Rimington, J. Moan (1987). A comparison of different photosensitizing dyes with respect to uptake by C3H-tumors and tissues of mice. *Cancer Lett.*, **36**, 1–10.

73. J. Moan, H.B. Steen, K. Fehren, T. Christensen (1981). Uptake of hematoporphyrin derivative and sensitizer photoinactivation of C3H cells with different oncogenic potential. *Cancer Lett.*, **14**, 291–296.

74. J. Moan, K. Berg, H. Anholt, K. Madslien (1994). Sulfonated aluminium phthalocyanines as sensitizers for photochemotherapy. Effects of small light doses on localization, dye fluorescence and photosensitivity in V-79 cells. *Int. J. Cancer*, **58**, 865–870.

75. J. Moan, K. Berg (1991). The photodegradation of porphyrins in cells can be used to estimate the lifetime of singlet oxygen. *Photochem. Photobiol.*, **53**, 543–553.

76. J. Moan, K. Berg, H.B. Steen, T. Warloe, K. Madslien (1992). Fluorescence and photodynamic effects of phthalocyanines and porphyrins in cells. In: B.W. Henderson, T.J. Dougherty (Eds), *Photodynamic Therapy* (pp.19–35). Marcel Dekker Inc., New York, Basel, Hong Kong.

77. K. Berg, J. Moan, J.C. Bommer, J.W. Winkelman (1990). Cellular inhibition of micro-tubule assembly by photoactivated sulphonated meso-tetraphenylporphines. *Int. J. Radiat. Biol.*, **58**, 475–487.
78. J. Moan, S. Sommer (1985). Oxygen dependence of the photosensitizing effect of hematoporphyrin derivative in NHIK 3025 cells. *Cancer Res.*, **45**, 1608–1610.
79. B.W. Henderson, V. Fingar (1987). Relationship of tumor hypoxia and response to photo-dynamic treatment in an experimental mouse tumor. *Cancer Res.*, **47**, 3110–3114.
80. T.J. Dougherty, G. Lawrence, J.H. Kaufman, D. Boyle, K.R. Weishaupt, A. Goldfarb (1979). Photoradiation in the treatment of recurrent breast carcinoma, *J. Natl. Cancer Inst.*, **62**, 231–237.
81. D. Kessel, P. Thompson, K. Saatio, K.D. Nantwi (1987). Tumor localization and photo-sensitization by sulfonated derivatives of tetraphenylporphine. *Photochem. Photobiol.*, **45**, 787–790.
82. B.W. Henderson, S.M. Waldow, T.S. Mang, W.R. Potter, P.B. Malone, T.J. Dougherty (1985). Tumor destruction and kinetics of tumor cell death in two experimental mouse tumors following photodynamic therapy. *Cancer Res.*, **45**, 572–576.
83. G. Canti, O. Marelli, L. Ricci, A. Nicolin (1981). Hematoporphyrin treated murine lymphocytes: in vitro inhibition of DNA synthesis and light-mediated inactivation of cells responsible for GVHR. *Photochem. Photobiol.*, **34**, 589–594.
84. M. Korbelik (1996). Induction of tumor immunity by photodynamic therapy. *J. Clin. Laser Med. Surg.*, **14**, 329–334.
85. M.L. Agarwall, M.E. Clay, E.J. Harvey, H.H. Evans, A.R. Antune, N.L. Oleinick (1991). Photodynamic therapy induces rapid cell death by apoptosis in L5178Y mouse lymphoma cells. *Cancer Res.*, **51**, 5993–5996.
86. D. Kessel, Y. Luo (1998). Mitochondrial photodamage and PDT-induced apoptosis. *J. Photochem. Photobiol., B: Biol.*, **42**, 89–95.
87. J. Moan (1985). Porphyrin photosensitization of cells. In: G. Jori, C. Perria (Eds), *Photodynamic Therapy of Tumors and other Diseases* (pp. 102–112). Libreria Progetto, Padova.
88. J. Moan, K. Berg, E. Kvam, A. Western, Z. Malik, A. Ruck, H. Schneckenburger (1989). *Intracellular localization of Photosensitizers. Photosensitizing Compounds: their Chemistry, Biology and Clinical Use* (Ciba Foundation Symposium 146, pp. 95–111). Wiley, Chichester.
89. K. Berg, A. Western, J.C. Bommer, J. Moan (1990). Intracellular localization of sulfonated meso-tetraphenylporphines in a human carcinoma cell line. *Photochem. Photobiol.*, **52**, 481–487.
90. K. Berg, P.K. Selbo, L. Prasmickaite, T.E. Tjelle, K. Sandvig, J. Moan, G. Gaudernack, Ø. Fodstad, S. Kjølsrud, H. Anholt, G.H. Rodal, S. Rodal, A. Høgset (1999). Photochemical internalization: A novel technology for delivery of macromolecules into cytosol. *Cancer Res.*, **59**, 1180–1183.
91. J. Moan, V. Iani, L.W. Ma, Q. Peng (1996). Photodegradation of sensitizers in mouse skin during PCT. *Proc. SPIE*, **2625**, 187–193.

Chapter 2

Mechanisms of photodynamic therapy

Anne C.E. Moor, Bernhard Ortel and Tayyaba Hasan*

*Corresponding Author

Table of contents

2.1 Introduction

Photodynamic therapy (PDT) may be considered to be coming of age. There have been a number of regulatory approvals worldwide (10 countries in Europe, Japan, Canada and U.S.A.) for the treatment of specific cancers with Photofrin® (PF). In addition, at the end of 1999 Verteporfin® (Benzoporhyrin derivative monoacid A (BPD-MA)) was approved in Switzerland for the treatment of the wet form of age-related macular degeneration (AMD), the leading cause of blindness amongst the elderly people in the Western world. A panel for the U.S.A. Food and Drug Administration (FDA) has also approved it in the U.S.A. Similarly, δ-aminolevulinic acid (ALA)-induced protoporphyrin based PDT has been approved by the FDA for the treatment of actinic keratoses, a pre-malignant dermatological disorder.

In the last decade, three major events in the field of PDT have possibly contributed to this transition into the clinic: (1) appropriate industrial involvement to sponsor controlled clinical trials, (2) recognition of the need for a larger variety of photosensitizers with improved optical and target localizing properties and (3) a better, and a more sophisticated understanding of biological mechanisms involved in PDT. The consequence of all this, and in particular of the mechanistic understanding, has been a dramatic increase in the potential applications of PDT such that it may now be considered a platform technology with a broad reach into many disease areas. In fact, while the initial focus of PDT was the treatment of cancer, there has been a rapid increase in studies investigating the use of PDT for the treatment of non-cancer diseases. Table 1 provides a list of some of these applications, which are in various stages of pre-clinical and clinical developments.

The goal of this chapter is to briefly introduce PDT, to summarize the salient features of mechanistic studies, and to discuss some strategies to further improve the selectivity of PDT. Different aspects, such as the photochemical and molecular/cellular processes, will be reviewed, as well as the mechanisms by which PDT is thought to achieve tumor cure in vivo. A separate section is dedicated to ALA-based PDT and its use in diagnosis and therapy. No attempt has been made to make this an exhaustive review of any of these aspects and this is essentially a subjective summary and analysis. Excellent reviews on PDT exist and are referred to for further reading [1–4].

2.2 Photodynamic therapy: light, photosensitizer and oxygen

PDT is based on different types of photochemical reactions, which are initiated by a photosensitizer, excited by the appropriate wavelength of light. In most cases, oxygen is required to obtain an effective biological response, which may be initiated by singlet oxygen ($^1O_2$) or free radicals.

2.2.1 Light

Most relevant to PDT is visible light, which covers the limited range of 400–700 nm of all electromagnetic radiation. However, in practice, the range of light used in

Table 1. New applications for PDT

Cardiology/vascular	intimal hyperplasia
	atherosclerosis
Dermatology	actinic keratosis
	psoriasis
	hair removal
Gynecology	endometrial destruction
	endometriosis
Microbiology	periodontal disease
	infection control
Ophthalmology	ocular neovascularizations
	– age related macular degeneration
	– corneal neovascularization
Orthopedics	rheumatoid arthritis
	osteoarthritis
Tissue repair	wound repair
	tissue welding/bonding
Transplantation biology	organ rejection
Blood banking	sterilization of blood products

PDT is mainly 600–900 nm, since endogenous molecules, such as hemoglobin, have a strong absorption below 600 nm and therefore capture most of the incoming photons [5]. The 900 nm upper limit is due to the energy content of the photons at higher wavelengths, which is not sufficient to induce the generation of singlet oxygen. Use of photosensitizers which absorb light at higher wavelengths between these limits has been an important focus of research, since light at longer wavelengths can penetrate deeper into the tissue. The effective penetration depth, δ_{eff}, of a given wavelength of light is a function of the optical properties, such as absorption and scattering, of the tissue being illuminated. The fluence (light dose) in a tissue is related to the depth, d, as: $e^{-d/\delta_{eff}}$. Typically, the effective penetration depth is about 2 to 3 mm at 630 nm and increases to 5 to 6 mm at longer wavelengths (700–800 nm) [6].

Absorption of a photon by a photosensitizer leads to the electronic excitation of the molecule. From this excited singlet state, called S_1, the molecule can undergo different transitions to other energy levels, as depicted in Figure 1. The electronic states are represented by the singlet states S_0 to S_2 and the triplet states T_1 and T_2. From the energized S_1 state the molecule may initiate photochemistry (depending on the chemical structure) or intersystem cross to the first triplet state, T_1, which is in general a longer-lived species. Alternatively, the excited S_1 molecule may also relax back to S_0 by a radiationless decay and generate heat or may reemit radiation as fluorescence. T_1, which is not only longer lived but also chemically more reactive than S_1, mediates most biologically relevant photochemistry. Two different photodynamic reactions have been defined [7–9]. From T_1, energy can be transferred to ground state oxygen ($^3O_2$) to generate $^1O_2$ (energy transfer or Type II reaction) or an electron can be transferred from the photosensitizer molecule to generate free radicals (electron transfer or Type I reactions). T_1 can also be potentially relaxed to S_0 by a radiationless decay or by a radiative decay similar to phosphorescence.

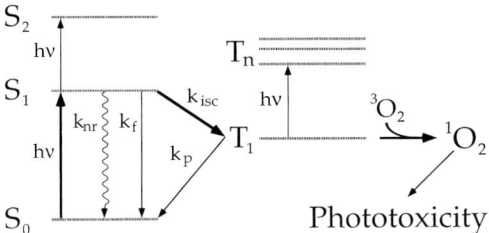

Figure 1. A simplified energy level diagram for the excitation by photons (hv) of a molecule. S_0, S_1 and S_2 represent singlet electronic states of the molecule. T_1 and T_n indicate the first and higher triplet states of the molecule respectively. The molecule can relax back to the ground state S_0 from either S_1 or T_1 radiatively or nonradiatively. k_{nr}, k_{isc}, k_f, and k_p represent rate constants for nonradiative decay, intersystem crossing, fluorescence, and phosphorescence, respectively.

The excitation to produce $^1O_2$ requires at least 20 kcal mol^{-1}, which places limits on the wavelength of absorption of the photosensitizer. Singlet oxygen can mediate photo-oxidative reactions with different cellular components, such as DNA, proteins and lipids, which leads to damage of cellular structures. Recently there has been increased interest in the use of two photon photochemistry, where the initial T_1 excited state is further excited to an upper triplet state by a second incoming photon matched to the wavelength of T_1 absorption (Figure 1). These excited upper triplet states can initiate non-oxygen dependent photochemistry, which is of importance for treatment of less vascularized/oxygenated regions. In this concept, the second photon, however, has to be delivered within a period shorter than the relaxation time of the T_1 state. This was shown in an elegant study by Andreoni et al. [10] in which hematoporphyrin (HP) photosensitized tryptophan oxidation was studied using pulsed irradiation. A significant increase in photosensitization was observed when a second pulse matched to the T_1 absorption was applied rapidly after the initial excitation. In contrast, when "long" delays (longer than the lifetime of the excited state) in the delivery of the second photon or a wavelength not matching the excited state of HP were used, no enhanced tryptophan destruction was observed. Similarly, in a study by Smith et al. it was shown that, using rose bengal, two color laser excitations could effectively generate upper triplet states, which could inhibit enzyme activity and induce cell death in an oxygen independent fashion [11]. In a different approach, short pulsed high intensity laser was used to excite photosensitizer molecules at wavelengths matched to the ground state of the photosensitizer used (BPD-MA). The hypothesis tested was that by virtue of the light photon density, there would be a rapid excitation of molecules leading to a transient "bleaching", which might allow light to penetrate deeper into tissue. In a subcutaneous rat prostate tumor model, no differences in the depth of necrosis was found between pulsed and continuous wave illumination [12], which was in contrast to previous reports [13,14]. The different observations in the various studies were attributed to the lower concentrations of BPD used in the study of Pogue et al. [12] so that shielding by the photosensitizer was not a limiting factor.

2.2.2 *Photosensitizers*

Most of the clinical experience has been obtained with PF, the first photosensitizer approved for the treatment of cancer. PF is a somewhat purified form of hematoporphyrin derivative (HPD), a mixture of porphyrins formed by the acid treatment of HP. Because of its nature as a mixture of porphyrins, PF is only partially chemically characterized. It has only low absorption in the 600–900 nm region, making it an inefficient photosensitizer at the wavelengths of therapeutic interest. Despite these drawbacks, PF has been shown to be effective in the treatment of a number of cancers, and the approvals include the treatment of esophageal, lung, bladder, gastric and cervical cancer. A significant side effect of PDT using PF is cutaneous phototoxicity, which can last 4 to 6 weeks. The disadvantages of PF have led to a search for the so-called 'second-'and now 'third-generation' photosensitizers [15,16], which is aimed at developing chemicals with (a) improved selectivity: i.e. high tumor (target) to normal tissue ratios (b) absorption at higher wavelengths, so that the deeper penetration of light can be used to treat larger target volumes and (c) higher molar absorption coefficients (molar absorptivity) at these wavelengths. Furthermore research has been directed towards delivery vehicles which are capable of directing the photosensitizers specifically to their target sites (see below) [17,18].

Over 30 compounds are presently being investigated in (pre-) clinical studies and most of these are tetrapyrrole compounds. They include porphyrins, chlorins, phthalocyanines, purpurins and texaphyrins. Figure 2 depicts their chemical structures and some photophysical properties. Of these second-generation photosensitizers, the most developed is BPD-MA. For clinical applications this molecule is liposomally formulated and has good absorbance at longer wavelengths (with an absorption maximum at 690 nm). It has been approved for the treatment of AMD and is in development for the treatment of various cancers and immune disorders. A different approach to PDT is the use of ALA, a precursor of endogenous porphyrins, to stimulate protoporphyrin IX (PpIX) synthesis. PpIX can act as a photosensitizer upon illumination with the appropriate wavelengths. This form of PDT, which has been approved for the treatment of actinic keratosis, will be discussed in more detail below.

2.2.3 *Oxygen*

Photodynamic reactions, as described above, require a photosensitizer, light and oxygen at the same localization to induce a response. The extent of this response is modulated by both the concentration of photosensitizer, as well as the number of captured photons (fluence). Several studies have shown that PDT efficacy is oxygen dependent [19–23]. This oxygen dependence is generally believed to be mediated by singlet oxygen, the reactive oxygen species responsible for most photodynamic processes in biological systems [24]. This assumption is based on extrapolation from solution chemistry; only most recently the detection of singlet oxygen in vivo has been possible [25]. Other reactive oxygen species such as hydroxyl radicals

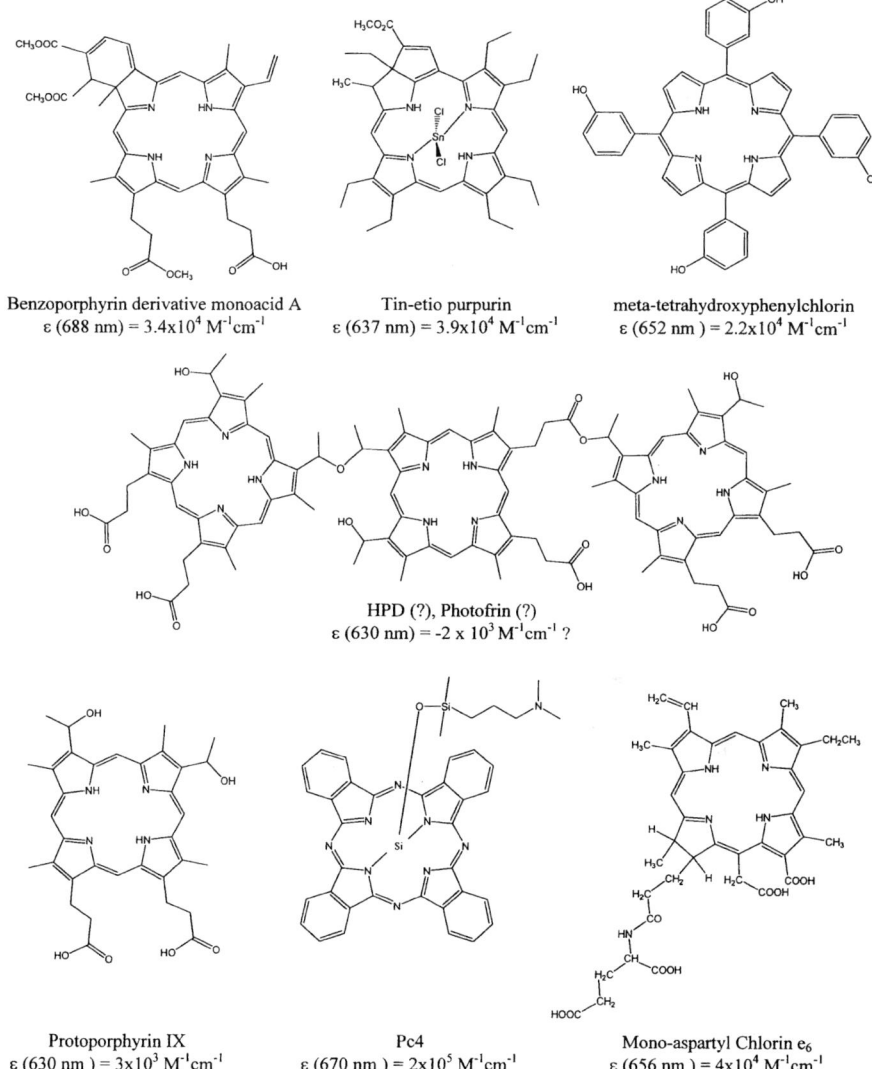

Figure 2. Chemical structures of selected photosensitizers. λ_{max}, the PDT-relevant maximum absorption wavelength, and ε, the corresponding molar absorption coefficient, are indicated. For PF/HPD, (?) indicates the uncertainty of the chemical structure for these entities.

and superoxide anions may well be equally important players [26]. In vivo, oxygen depletion has been determined by quantitation of the hypoxic cell fraction immediately after PDT [27] or by direct measurements of tissue oxygen tension using oxygen electrodes of various types [22,23,28].

The extent of the oxygen dependence of PDT effects is somewhat dependent on the nature of the photosensitizer. For example, it has been shown for PDT using PF that when the pO_2 is 5 kPa there is a maximum effect, while this effect is reduced to 50% at 1 kPa pO_2. On the other hand, chloroaluminium phthalocyanine (CASPc) based PDT is less dependent on oxygen tension: the oxygen levels have to be reduced to 0.33 kPa to reduce PDT effects to 50% [29]. As expected, based on the oxygen requirement for photodynamic reactions, it has been shown in vivo that, under anoxic conditions, PDT using PF has no effect [30]. The relationship between tumor blood flow or oxygen concentration and PDT is not a simple one, as demonstrated in the study by Fingar et al. [31]. In this study, the artificial oxygen carrier Fluosel-DA (20%) did not enhance PDT tumor destruction. In a different model system, nicotinamide, which homogenizes tumor flow and oxygen concentration, also failed to enhance PDT response in contrast to its effects in radiation therapy [32].

The oxygen dependence of PDT leads to decreased efficiency at higher fluence rates, similar to the observations in radiation therapy [33]. In various tumor models it was shown that, at typical fluence rates used in clinical PDT (~100–200 mW cm^{-2}), reduced tumor destruction was obtained than with lower fluence rates [20–23,34]. This lower effect in vivo has been attributed to oxygen depletion during the illumination caused by a high rate of oxygen consumption and an insufficient reperfusion of the tumor. In vitro, similar fluence rate effects were observed in erythrocytes and in solution chemistry, but it was concluded that oxygen depletion was not a likely explanation for the decreased efficiency, instead different intermediate pathways at high and low fluence rates contributed to the observed effects [35,36].

The inefficiency of PDT response at higher fluence rates can be partially circumvented using lower fluence rates or fractionated illumination, in analogy to radiation therapy [33]. Several preclinical studies have shown the increased efficiency of PDT by one of these two approaches [19,23,32,34,37,38]. This aspect of PDT has been pioneered and developed in most detail by Foster et al. [20,21,37–39] and has been confirmed by an expanded number of other investigators. In studies in an orthotopic rat bladder tumor model, Iinuma et al. [34] showed that PDT using BPD-MA was enhanced almost 1000-fold when a light fractionation regimen (λ = 690 nm) of 60 s on and 60 s off was used when compared to continuous illumination at 100 mW cm^{-2} (Figure 3). This effect was dependent on the interval of fractionation, probably due to an insufficient reperfusion at short intervals. In the same study, tumor cell cytotoxicity was much enhanced when the fluence rate was 30 mW cm^{-2} rather than 100 mW cm^{-2}. In more recent studies, Sitnik et al. [22,23] correlated for the first time directly the effects of fluence rate with oxygen tension in vivo. Median pO_2 before PDT ranged from 2.9 to 5.2 mmHg, and decreased rapidly in the first minute of illumination to values between 0.7 and 1.1 mmHg. During prolonged illumination (20–50 J cm^{-2}) at 30 mW cm^{-2} fluence rate pO_2 recovered, but pO_2 remained low at the 150 mW cm^{-2} fluence rate (median pO_2 of 1.7 mmHg). Tumor regrowth times were shown to be directly correlated with the recovery of oxygen levels within the tumor tissues. The above described in vivo studies have been supported by model studies of Foster et al. in which tumor cell spheroids were used to mathematically model the observed fluence rate effects [39].

These pre-clinical studies all show the oxygen dependency of PDT, and suggest that the currently clinically used fluence rates might be too high for an optimal PDT

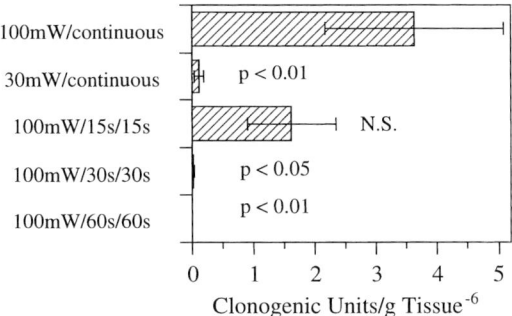

Figure 3. The effect of fluence rate and light fractionation on BPD-mediated PDT. BPD-MA was administered to rats with NBT II tumors implanted into the bladder wall. One hour later tumors were exposed to a total fluence of 30 J/cm^{-2} of 690 nm irradiation under the following conditions: 100 mWcm^{-2} continuous; 100 mW/cm^{-2} fractionated 15 s on/15 s off; 100 mWcm^{-2} 30 s on/30 s off; 100 mWcm^{-2} 60 s on/60 s off. Tumors were disaggregated 24 h later and tumor cells were plated for colony formation assay. Colonies (50 cells or more) were counted 9 days later after fixing with methanol and staining with crystal violet. The Wilcoxon rank sum test was used to compare the number of clonogenic cells with data at 100 mWcm^{-2} and continuous wave irradiations. NS: not significant. (Source: Iinuma et al. [32]. Reproduced with permission.)

response. Lowering the fluence rate, however, might not be practical in terms of treatment times. Furthermore, the fractionation schedule has to be determined for each model and photosensitizer used, since opposite results have been reported for longer dark intervals (>1 h) [40,41]. Factors which determine the efficiency of fractionation schedules are sublethal repair during dark periods, photobleaching of the photosensitizer and reoxygenation of the tumor [21,42,43]. Finally, the lack of oxygen in the tumor is not only due to oxygen depletion by PDT, but is also inherent to some tumor types. If the formation of neovessels stays behind tumor growth, hypoxic regions can develop in the tumor. In these areas, PDT will be less effective because of the limited oxygen supply [27] or because of a low sensitizer concentration.

In summary, the insights into photochemical processes have indicated the oxygen dependency of PDT. The introduction of two-photon chemistry, which gives rise to non-oxygen dependent photochemistry, can be of importance for the treatment of solid tumors, in which hypoxic regions pose a problem for treatment with more conventional PDT. In addition, it has been shown in vivo, by direct oxygen measurements, that oxygen depletion by PDT can strongly decrease the efficacy of the treatment. Based on these studies, treatment protocols which include low fluence rates or fractionated protocols are being evaluated for their effect on oxygen depletion and PDT efficacy.

2.3 Selectivity of photodynamic treatment

Efficient biological responses to PDT require three components and this renders PDT a complicated treatment modality. There is, however, some opportunity for an increased selectivity of the treatment, because of the following aspects: (a) localization of the photosensitizer at the treatment site and (b) illumination of a specified volume of tissue. Both can be modulated to further improve the efficiency of the treatment. Selective localization of the photosensitizer in target tissue is dependent on its biochemical properties and can be modulated by the use of various delivery vehicles, which will be discussed below [17,18]. A specific subcellular localization may also be obtainable by choosing the right photosensitizer properties. This is of importance since the cellular effect of PDT is dependent on the site where the primary damage in the cell occurs, which may determine the mode of cell death [44,45]. Both electron transfer and energy transfer reactions require the photosensitizer to be close to the target, since either there has to be a direct interaction or the mediator ($^1O_2$) should be quenched efficiently [46]. Therefore, the cellular structures close to both a high sensitizer and a high oxygen concentration will be preferentially damaged upon illumination. In addition to the right photosensitizer and delivery vehicle, the status of the cell itself is important for its reaction to PDT. Therefore, by modifying the status of the target cells, e.g. by induction of differentiation, additional specificity of the treatment might be achieved [47].

Selectivity of PDT effects may also be enhanced by appropriate manipulation of the illumination process. The spatial control of the illumination is an obvious advantage of PDT, but it is also an important source of the complexity of this modality [48,49]. Since most photosensitizers are non-toxic in the dark at the clinically used doses, only illuminated areas will be damaged by the treatment. The introduction of fiber-optic coupling to lasers has opened a range of opportunities to treat not only superficial lesions, but also more complex body sites, such as the abdominal cavity. An additional way to achieve specificity is the timing of the illumination. Since photosensitizer distribution is dependent on the time after delivery (initially more vascular, later more in tumor cells) [50,51], the time of illumination is an important tool in tailoring PDT for specific applications.

Finally, there appears to be a threshold for PDT effects: at low doses of light and/or photosensitizer tissue damage seems reparable [52]. Since this threshold value can be different for tumor and normal tissue it provides an opportunity for added selectivity.

2.3.1 Delivery vehicles: lipoproteins, liposomes and antibodies

At complex sites, the issue of photosensitizer selectivity and careful dosimetry becomes crucial so as to avoid unacceptable damage to normal tissues. Delivery vehicles might provide this selectivity by analogy with the idea of drug targeting [53]. In comparison with chemotherapeutics and toxins, delivery of photosensitizers is fundamentally different in two aspects: (a) most drugs are inactive while coupled to the delivery vehicle and need to be released by cellular biochemical processes to be active. In contrast, the photosensitizer can elicit a response even when it is coupled to the macromolecule used for delivery, although the efficiency might be compromised [17,18]. (b) Requirements for specificity are less stringent for macromolecularly bound photosensitizers than, for example, toxins. This is due to two above-discussed characteristics of PDT: the presence of a threshold, under which PDT induced damage is reparable, and the dual selectivity achieved by photosensitizer localization and light delivery. The use of vehicles can improve opportunities for the use of PDT in different ways: by increasing the concentration of photosensitizers at the target, by broadening the choice of photosensitizer since non-tumor localizing agents with appropriate photochemistry can be applied and finally by mitigating the requirement for accurate dosimetry, if photosensitizer delivery is effectively targeted. The drawbacks of the use of large molecules as carriers are not specific for photosensitizers. Drug targeting with these compounds requires complicated syntheses, is limited in crossing of transport barriers, and might lead to systemic toxicity.

Several lipoprotein formulations have been tested for delivery of photosensitizers to tumor tissue [54–56] and it has been shown that LDL is a particularly suitable delivery vehicle for different porphyrins. Uptake of LDL complex is mainly mediated via an active endocytotic pathway, involving LDL receptors on both tumor cells and endothelial cells [57–60]. Depending on the time after administration, the LDL complexed photosensitizers can be found either in the vasculature [61], or in the tumor cells themselves [57,62]. Covalent linkage of LDL to chlorin e6 resulted in conjugates which were specifically taken up by LDL-receptor carrying fibroblasts, and a retinoblastoma cell line, indicating its potential use in ocular tumors [54].

A different approach to targeting tumors has been based on the enhanced accumulation of porphyrins in tumor-associated macrophages. This enhancement is attributed to the association of most porphyrins with LDL in vivo and the recognition of LDL by the scavenger receptor on the macrophages [63]. In an attempt to target these macrophages specifically, photosensitizers have been coupled to maleylated bovine serum albumin, a known scavenger receptor ligand. Specific uptake and enhanced phototoxicity in vitro has been reported with these conjugates [64]. The high number of scavenger receptors on the macrophages in atherosclerotic plaques and the role of macrophages in the pathogenesis of the disease make these receptors useful targets in the photodynamic treatment of atherosclerosis [65]. Using acetylated LDL and BPD, Allison et al. showed specific uptake in atherosclerotic plaques in an in vivo model, but these complexes were not stable upon systemic administration [66]. In vitro studies with oxidized LDL and aluminum phthalocyanine (AlPc) showed specific uptake of these complexes by macrophages and no exchange with serum proteins, indicating a promising formulation for this specific application [67].

Alternatively, liposomes have been used as delivery vehicles for photosensitizers. The mechanisms by which liposomes can specifically deliver photosensitizers to tumors are not clear, but it is suggested that the photosensitizer is transferred from liposomes to lipoproteins and delivered in this form [57].

A more specific approach to photosensitizer delivery involves coupling them to monoclonal antibodies (Mab), directed against specific (tumor) antigens [18]. These antigens can either be exclusively expressed on tumor cells – but this is only true in rare cases – or be overexpressed as compared to normal tissue. In combination with the inherent selectivity of PDT (see above) this might prove sufficient for successful targeting. In contrast to Mab-toxin or Mab-radionuclide conjugates, photoimmuno-targeting requires conjugates with high photosensitizer-to-Mab ratios, which makes the syntheses complicated, especially since both photosensitizer and Mab need to retain their functions. Mab-photosensitizer conjugates can be synthesized in two ways: (a) a photosensitizer is linked directly to a Mab or (b) photosensitizers are linked to Mabs via polymers. The use of polymers was introduced to achieve high photosensitizer:Mab ratios without serious impairment of the binding capabilities of the Mab. Several polymers have been used in the synthesis of Mab-photosensitizer conjugates, including dextrans [68,69], polyglutamic acid (PGA) [70–72], poly(vinyl alcohols) (PVAs) [73,74], poly[N-(2-hydroxypropyl)methacrylamide] [75–77] and poly-L-lysines [78,79]. The development of methods for site-specific conjugation on the antibody has contributed to the maintenance of the antigen-binding properties of the Mab [68–72,80].

Directly coupled Mab-photosensitizer conjugates were used in the first photoim-munotherapy study by Mew et al. (D. Mew, C.K. Wat, G.H. Towers, J.G. Levy (1983). Photoimmunotherapy: treatment of animal tumors with tumor-specific monoclonal antibody-hematoporphyrin conjugates. *J. Immunol.*, **130**, 1473-1477). The conjugates were shown to be slightly more phototoxic in vivo than the parent photosensitizer. More recently, a direct coupling method was described using a Mab directed against the epidermal growth factor receptor and *meta*-tetra(hydroxyphenyl)chlorin (mTHPC) and BPD-MA [81, M.D. Savellano and T. Hasan (2003). Targeting cells that over express the epidermal growth factor receptor with polyethylene glycolated BPD verteporfin photosensitizer immunoconjugates. *Photochem. photobiol.*, **77**, 431-439.] Increased tumor selectivity of the conjugates in vitro and in vivo was reported, compared with the free photosensitizers. Further evaluation of these conjugates in PDT in terms of cure rates is still needed. A different use of a directly coupled Mab-photosensitizer conjugate was reported by Steele et al. [82] who targeted T-suppressor cells DBA/2J in mice. Upon illumination, tumor regression was correlated with increased activity of specific cytotoxic T lymphocytes against the target tumor cells. The advantage of this approach is the targeting of easily available cells, in contrast with targeting tumor cells, which especially in solid tumors might be hard to reach for large Mab-antibody conjugates.

Several studies have shown the use of photoimmunoconjugates in PDT where large numbers of photosensitizers were linked to Mabs via polymers. Tumor cells, both T-cell leukemia cells and bladder carcinoma cells, were killed using the appropriate Mabs coupled via either dextran or PGA to chlorin e6-monoethylene diaminemonoamide (CMA) [68,70,71]. Alternatively, a reaction scheme was described using PVA to couple BPD-MA in a non-specific way to a Mab. Good affinity, specificity and phototoxicity of the conjugate were reported probably because of the minimal number of sites on the Mab involved in the linkage [73]. These initial studies suffer from insufficient conjugate characterization and purification.

Later studies, using PGA and dextran intermediaries showed clear, site specific, covalent linkage of the photosensitizer CMA on the carbohydrate moiety of the heavy chain of the antibody [69,72]. In vitro, these conjugates showed light- and photosensitizer-dose-dependent killing of target melanoma cells [69] and ovarian cancer cells both from a cell line and from human ovarian cancer patients [72]. In addition, using an in vivo model for ovarian cancer in mice, a prolonged survival after photoimmunotherapy was shown [80, 83]. In follow-up studies of the effect of photoimmunotargeting on human ovarian cancer cells ex vivo, Duska et al. [84]

demonstrated that combination treatment of cisplatin (CDDP) and photoimmuno-targeting produced a 7-fold enhanced cytotoxicity over CDDP treatment alone. Interestingly this enhancement was synergistic and greater for CDDP resistant cells (up to 13-fold). These and similar observations with other PDT agents [85–88] demonstrate the possibility of using PDT in the destruction of tumor cells that have developed resistance to chemotherapy agents.

Clinical applications of photoimmunotherapy have been sparse, probably because of the complexity of the approach. It was reported that photoimmuno-conjugates of Mabs recognizing CA125 on human ovarian cancer cells were used in humans [89,90]. In this study, selective photocytotoxicity to target cells in vitro and in vivo in a tumor bearing nude rat model was shown and 3 patients with advanced ovarian cancer were treated. The group (Hahn and co-workers) at the University of Pennsylvania continues to develop the treatment for cancer in the peritoneal cavity.

In conclusion, localization of the photosensitizer is a key factor in the specificity of PDT. This holds true not only for the specific uptake of the photosensitizer in the tumor, or other diseased tissue, but also for the subcellular localization, which is an important determinant for the cellular response to PDT (see below). The development of specific delivery vehicles has contributed to both aspects of photosensitizer localization. The existing investigations of Mab-photosensitizer conjugates are both challenging and promising. Better characterized and purified conjugates are needed, along with careful pharmacokinetic information in vivo in appropriate animal models. Since the conjugates available presently are large in size, a problem might be their availability and penetration into the tumors. Studies into the use of Mab-immunoconjugates synthesized with antibody fragments, synthetic Mabs and fragments, single chain and chimeric antibodies are therefore underway [91]. Molecular charge [78,92] may be critical in establishing the route of delivery for optimal selectivity. Similarly, the presence of enzyme-cleavable linkages [75] could further enhance the efficacy of photoimmunoconjugates.

An innovative approach in drug targeting called photochemical internalization is not aimed at delivering the photosensitizer to the desired (cellular) localization, but uses photosensitizers and light to free endocytosed macromolecules from the lysosomes; this serves to circumvent the problem of inactivity of bound and localized toxins. In this approach, macromolecules are administered to the target cells, and are taken up via endocytosis. Photosensitizers which have a localized effect on the lysosomal membrane can disrupt them upon sublethal doses of illumination, thereby releasing the content of the lysosomes in the cytosol. Applications that have been suggested for this technology include gene therapy, vaccination and cancer treatment [93].

2.4 Cellular mechanisms of PDT

In this and the next section, the biological mechanisms, emanating from photodynamic treatment will be discussed. In recent years, significant effort has been directed towards understanding the molecular and cellular mechansims of PDT (reviewed in [4]). Most of these studies have been performed in vitro, and further studies are necessary to confirm these findings in in vivo models and to determine their relevance for tumor cure. Cells can react in different ways to PDT: upon lethal doses they can undergo either necrosis or apoptosis, and in the case of sublethal damage, the cells can elicit a rescue response [94]. The responses are dependent primarily on the localization of the photo-sensitizers in the cell, since, as explained above, structures close to high photosensitizer and high oxygen concentration are damaged upon illumination. However, other factors, such as cell line [95–97] and PDT doses [98,99], play important roles as well.

2.4.1 Necrosis

Cell death in a necrotic fashion can be induced following organelle damage, such as membrane lipid peroxidation, disruption of lysosomal membrane, membrane enzyme inhibition [100] or damage to nuclear components [101]. In contrast to apoptosis as described below, necrosis is a less controlled way of cell death, which does not seem to involve complex signaling cascades. In PDT using photosensitizers which localize in the lysosomes, cell death is possibly due to release of lysosomal enzymes and other toxic moieties (reviewed in [102]). There is, however, a possibility of lysosomally localized photosensitizer relocating to mitochondria within the first few seconds of illumination, where they may be considerably more phototoxic [103].

2.4.2 Apoptosis

In recent years a number of studies have addressed the induction of apoptosis by PDT in various models. PDT induces apoptosis via several pathways, and its most striking aspect is the very rapid induction of the apoptotic process. Several studies have shown that photosensitizers which localize preferentially in mitochondria are very rapid inducers of apoptosis, in contrast to photosensitizers localized in lysosomes and plasma membranes [44,104–106]. Apoptosis induction by mitochon-drial based photosensitizers is an extremely rapid process: cells can enter the execu-tion phase of apoptosis within 30 min after illumination in the presence of the silicon phthalocyanine Pc 4 [107] and the mechanisms behind this rapid induction have been partly elucidated. BPD-MA is an example of a photosensitizer which shows a primarily mitochondrial localization and induces apoptosis efficiently upon illumi-nation (Figure 4). In general, mitochondrial localized photosensitizers induce apoptosis in accordance with the hypothesis of Liu et al. [108] which proposes the release of cytochrome c from mitochondria as being a critical signal for the induc-tion of apoptosis. Following PDT of various cell lines, a very early step is the loss of cytochrome c into the cytosol [109–111]. In addition, a rapid loss of mitochon-drial membrane potential is observed upon PDT in some cases and is attributed to the opening of the so-called mitochondrial permeability transition pore [112] . The loss of cytochrome c after PDT results in a sharp increase of caspase 3 (cysteine proteases acting on aspartic acid) activity [110] via complex formation with dATP, apoptosis-activating factor-1 (APAF-1) and pro-caspase 9. The formation of this complex induces cleavage of pro-caspase 9, yielding the active caspase 9. This is

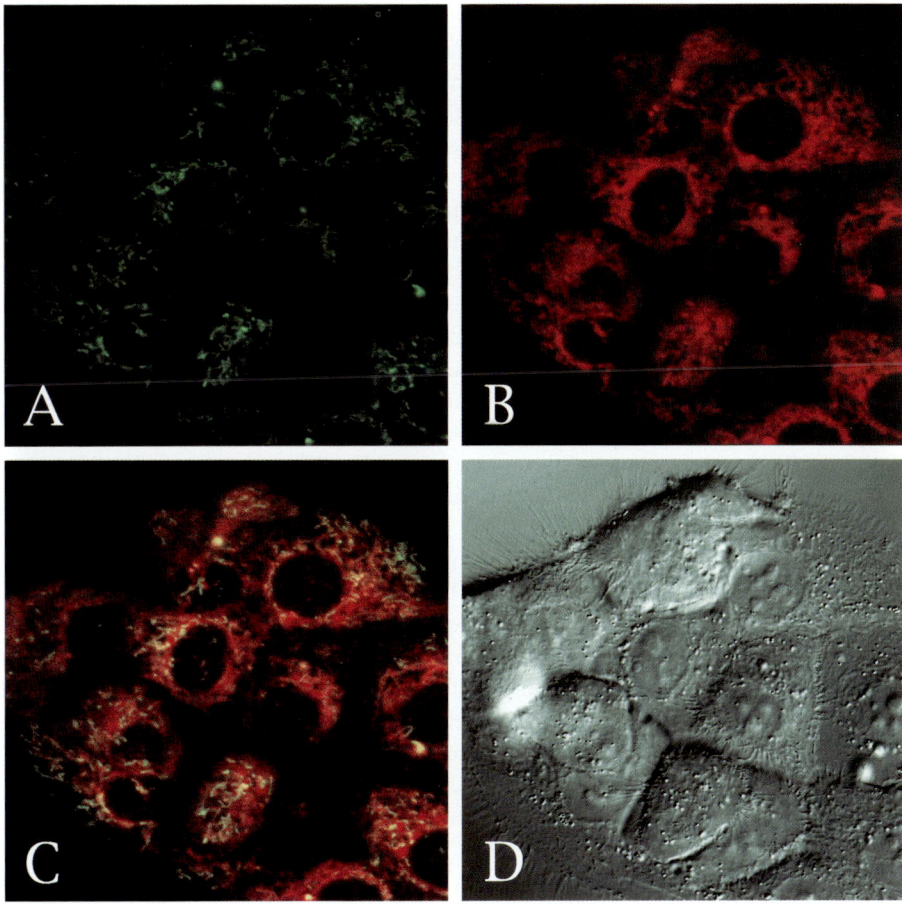

Figure 4. Subcellular localization of the photosensitizer BPD-MA. OVCAR-5 cells were incubated in 92 nM BPD-MA for 3 h and 10 nM rhodamine 123, a mitochondrial probe, for 20 min. Imaging was performed using confocal laser scanning microscopy (CLSM). (A) Exclusively mitochondrial green fluorescence of rhodamine 123; (B) red BPD-MA fluorescence; (C) overlay of A + B, where yellow indicates co-localization; (D) DIC transmission image. The colocalization in (C) indicates that BPD-MA localizes to mitochondria, but also stains other subcellular structures.

the start of a cascade of protease reactions, in which the cleaving of pro-caspase 3 leads to the activated form, caspase 3, a key player in the execution phase of apoptosis [113,114]. Caspase 3 is involved in the cleavage of a number of proteins [115], including DNA fragmentation factor (DFF) and poly (ADP-ribose) polymerase (PARP). The latter is involved in the final phases of the apoptotic process. Photodynamic treatment with the photosensitizers Pc 4, BPD-MA, and AlPc has been shown to induce cleavage of PARP in different cell lines [98,106,107]. In addition, DFF activation was shown after PDT [109]. Although for PDT using Pc 4 the pathway of mitochondrial induced apoptosis has been well characterized, this does not mean that other cascades are not involved. Ceramide formed after activation of sphingomyelinase has been shown to play a role in this respect [116–118], as well as several kinases, such as the stress activated kinases SAPK/JNK and p38/HOG1 [119,120] and the non-receptor tyrosine kinase Etk/bmx [121]. Further studies should determine how these pathways interact with the above-described mitochondrial pathway.

In many tumor cells, anti-apoptotic control mechanisms are present, which make the cells less sensitive to cytotoxic agents. Under certain conditions, PDT appears unimpaired by these mechanisms. For example, bcl-2, a protein found in the outer membrane of the mitochondria, is known to be an anti-apoptotic moiety. The over-expression of this protein has been associated with chemotherapy and radiation resistance [122,123]. Consistent with these observations, it was reported that in CHO cells the presence of Bcl-2 partly protects against apoptosis induction by photodynamic treatment with Pc 4 [124]. This could be due to the known anti-oxidant effect of Bcl-2 [125] but also to its ability to interfere with calcium homeostasis, which has been shown to play a role in photodynamically induced cell death [126]. However, it is more likely that Bcl-2 is involved in the inhibition of the cytochrome c release after PDT, known to be an important mechanism of modulation of apoptosis by Bcl-2 [127,128]. Similarly, it was shown that PDT with BPD-MA was less effective in apoptosis induction in HL-60 cells overexpressing Bcl-2 [106]. In these cells the activation of caspase 3 and 6 was also diminished, indicating again their key role in PDT induced apoptosis. In accordance with these results, it was shown that blocking of Bcl-2 using retrovirus transfection with antisense Bcl-2 increases the sensitivity of MGC803 cells to PDT induced apoptosis [129]. However, a reversal of the conventional inverse relationship between Bcl-2 expression and apoptosis induction was shown recently in an interesting study conducted by Kim et al. [130]. Using AlPc as sensitizer an enhanced sensitivity of a Bcl-2 transfected breast cancer cell line was demonstrated. This unexpected result was explained by the simultaneous increase in Bax, a pro-apoptotic Bcl-2 family member (Figure 5). It was postulated that Bcl-2

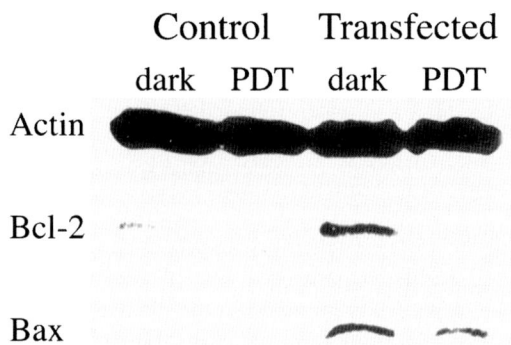

Figure 5. Western blot analysis of actin, Bcl-2, and Bax in control cells and in cells directly after PDT was carried out at 10°C. MCF10A cells, either transfected with Bcl-2 or not, were photosensitized with AlPc and then irradiated with a light dose of 100 mJcm^{-2}. Reprinted with permission from Kim et al. [130]

might be preferentially damaged by PDT thereby increasing the Bax:Bcl-2 ratio which subsequently leads to enhanced apoptosis. These observations are of significance in cancer therapy since, as mentioned above, overexpression of Bcl-2 is often involved in resistance mechanisms against chemotherapeutic agents [122].

A different signal transduction cascade has been shown to play a role in Pc 4 based PDT induced growth arrest [131–133]. These studies, which involve relatively high doses of PDT, show the induction of the WAF1/CIP1/p21 protein. This cyclin kinase inhibitor inhibits the function of cyclin D1 and D2 and their catalytic subunits cyclin-dependent kinase 2 (cdk2) and cdk6. This leads to growth arrest of the cell by accumulation in the G0/G1 phase of the cell cycle. NO might be the inducer of WAF1/CIP1/p21 and the subsequent growth arrest but this has not been confirmed [133]. In a follow up study, it was shown that PDT, using Pc 4, can cause an hypophosphorylation of retinoblastoma protein (Rb), and inhibit free E2F [132]. E2F is a family of transcription factors, which regulate the G1-S transition in the cell cycle, and its inhibition causes arrest of the cells in the G0/G1 phase. This is the final step in the cascade involved in cell cycle regulation that is affected by PDT. Similar findings were reported by Fisher et al. [134] who showed that PDT with PF caused hypophosphorylation of Rb and subsequent cell cycle arrest. They used both wild type and cells transfected with the viral protein E6, which inhibit the function of p53. Growth arrest was independent of the p53 status of the cells, but the apoptotic response was clearly diminished in the cells without functional p53. However, despite the inhibition of apoptosis in these p53-deficient cells, the clonogenic survival following PDT was similar for cells with p53 wild type or cells with abrogated p53 function. This opens up opportunities for the treatment of tumors with apoptosis resistant cells, while a mutation in p53 which also occurs in 50% of human tumors does not seem to influence the sensitivity to PDT.

2.4.3 Responses to sublethal PDT

Apart from a necrotic or apoptotic response, cells can also undergo a rescue response after PDT, dependent on PDT dose, cell type, and photosensitizer. Several stress proteins involved in cell rescue have been shown to be regulated upon PDT: members of the family of heat shock proteins [111,135–137], glucose regulated proteins [138–140] and heme oxygenase [141]. In addition, phospholipase A, prostaglandin E2 and cAMP have been implicated in cellular rescue responses after PDT [142–144]. However, both phospholipase A and C have also been shown to play a role in the induction of apoptosis [145]. PDT, at sublethal doses, not only can induce a rescue response, but it also regulates gene and protein expression which is involved in other cellular functions [146]. Various transcription factors, such as AP-1 [147] and NFκB [148], are activated by PDT. In turn, these transcription factors control, among other proteins, the expression of various cytokines, which indeed are induced by PDT [147,149,150] . These cytokines play an important role by the induction of anti-tumor immunity, which will be discussed below. In addition to soluble mediators, PDT has also been shown to regulate adhesion molecules on cells [151–153], which may be relevant to long-term effects, such as tumor metastasis. Alterations in the expression of surface receptors such as MHC class I and II [154, 155] have been reported and PDT-associated immune response has been attributed to these alterations.

In summary, these studies into the working mechanisms of PDT at a cellular level have shown that (a) PDT with mitochondria-based photosensitizers can induce extremely rapid apoptosis, (b) it can bypass some of the normal control mechanisms, such as bcl-2 and (c) although PDT induced apoptosis is dependent on p53 status, overall cell killing is not, which is of importance since many human tumors have non-functional p53. The majority of these studies have been performed in vitro and further studies should indicate whether the same responses occur in vivo. In addition, the role of apoptosis in the overall tumor response is not clear and recent data, using cell lines deficient in apoptotic machinery, indicate that these cells are still sensitive to PDT, probably by using a necrotic pathway [134]. This sensitivity could possibly be used in the treatment of tumors which are chemoresistant, and indeed several studies have already shown the effective use of PDT in this area [84,85,88,156].

2.5 Mechanisms of PDT in vivo

PDT can generally induce tumor destruction in vivo in three different ways: vascular destruction, direct tumor cell destruction and elicitation of an anti-tumor immune response. The relative contribution of each depends on the localization of the photosensitizer within the tumor tissue, which is partly dependent on the time between photosensitizer administration and illumination and the properties of the tumor such as the degree of vascularity and its immune cell content. Depending on their characteristics, photosensitizers have the ability to induce any of these responses. For example, Henderson and Dougherty [29] showed that bacterio-chlorophyll a has a direct cell kill potential of $\approx 50\%$ at the end of illumination and does not induce vascular damage until 3 to 4 h after. On the other hand, in the same model using PF-based PDT, vascular shutdown begins almost immediately after the initiation of light exposure. Although the different mechanisms will be discussed separately below, it has to be kept in mind that a clean separation is not possible, as pointed out by several studies using RIF cells [157,158]. RIF cells in which PDT resistance had been induced in vitro were implanted in mice and subjected to PF-mediated PDT under conditions in which a shutdown of the vasculature was expected to be the dominant mode of tumor destruction. Since resistance to PDT was induced within the tumor cells, it was expected that in vivo the tumor response to PDT would not be dependent on their resistance. Surprisingly, the resistance of the RIF cells to PDT was maintained in vivo, suggesting that either direct tumor cell kill was a major component in the observed effects, or that induction of PDT resistance also made these cells hypoxia resistant.

2.5.1 Vascular effects of PDT

Most clinical protocols are based on vascular damage as the dominant mechanism of tumor death. PF is the major example of a photosensitizer which targets the vasculature, but other porphyrins are also capable of induction of vascular damage. Macroscopically, the vascular PDT response is characterized by acute erythema, edema, blanching, and sometimes necrosis. Microscopically, the PDT treated tumor tissue is characterized by endothelial cell damage [29,61], platelet aggregation [61], vasoconstriction, and hemorrhage. The vascular damage is believed to be initiated by release of factors such as eicosanoids, in particular thromboxane [159], histamines, and tumor necrosis factor-α (TNF-α) [150]. As is the case with other modalities, extrapolation of in vitro observations to in vivo mechanisms is not always possible. For example, while PF-mediated PDT in vivo causes platelet aggregation, this could not be supported by in vitro studies in which photosensitization led to an inhibition of platelet aggregation [160].

BPD-MA is an example of a porphyrin, which in pre-clinical studies was shown to target the tumor vasculature. Intra-ocular tumors implanted in rabbit eyes were used as a model for neovasculature and the very efficient destruction of these tumors could be attributed primarily to the destruction of the tumor vasculature. The established choroidal vessels remained largely intact [61,161]. Based on these studies, this compound has been developed by QLT Phototherapeutics and Ciba-Vision as a first line treatment for age-related macular degenerateion (AMD) of the eye [162–166]. In 1999, approval for PDT using BPD-MA in the treatment of AMD was obtained in Switzerland and more recently in Canada, the USA and several other European countries. It is anticipated that the development of this agent or its analogues for oncological indications will be resumed.

2.5.2 Direct tumor cell kill

Direct cell destruction is expected to dominate when the photosensitizer content is high within the tumor cells at the time of illumination. The actual mechanisms of cell death have been discussed above and may be apoptotic or necrotic in nature or a combination. In contrast to the porphyrins, which derive their PDT effect in large part via destruction of the tumor vasculature, cationic photosensitizers are believed to be cellularly localized molecules and act at the tumor cell level. The basis for their preferential accumulation in tumor tissue is hypothesized to be the much steeper electrical potential across the mitochondrial membrane in tumor cells [68,167]. This steep gradient leads to a high accumulation of compounds with a delocalized positive charge. None of these hypotheses have been tested in a rigorous fashion for photosensitizers, and so must remain a conjecture at this point. Cincotta et al. showed high cure rates by PDT in two animal models of sarcoma using 5-ethyl-amino-9-diethylamino-benzo[a]phenothiazinium chloride [168]. Minimal damage to both the surrounding tissue and the irradiated vasculature was observed, indicating the selectivity of this compound. Cellular uptake of these compounds appears to occur within seconds. In an attempt to combine cellular and vascular effects, a benzophenothiazinium sensitizer and BPD-MA were used in PDT of EMT-6 tumors in Balb/c mice. The treatment produced a synergistic effect as compared to the single treatments [169].

2.5.3 Immunological effects of PDT

Several studies show PDT induced immune modulation, either immune stimulating, mediated by natural killer (NK) cells and macrophages, or immune suppressing. It has been suggested that the modulation of immune effects may play a role in PDT-induced destruction of tumors [4,82,94,149,150,170–173]. Several cell types have been implicated in the stimulation of an anti-tumor response by PDT. Both myeloid (neutrophils and macrophages) and lymphoid (T helper and cytotoxic T-cells) cells were shown to be important for the long-term cure of EMT-6 tumors by PF based PDT [174]. Enhanced natural killer cell activity following PDT was suggested to be important because of their possible capacity to lower the metastatic potential of surviving tumor cells [94,171]. Increased tumor immunity was demonstrated in mice treated with BPD-MA mediated PDT by using in vitro colony inhibition assays involving tumor cells and splenocytes [172]. Neutrophils have been shown to play an important role in initial tumor response to PDT using PF, partially mediated by G-CSF [175,176]. Macrophages have been reported to be involved in anti-tumor immunity via their TNF-α production [150,173,177,178]. This is supported by studies which show that tumor-associated macrophages accumulate up to 9 times the PF levels present in tumor cells [63]. In addition to their TNF-α production, macrophages might also play an indirect role in PDT induced cytotoxicity [173]. According to this hypothesis, sublethal PDT-induced damage causes tumor cells to expose lipid fragments. These fragments are then recognized as targets by macrophages and phagocytosed. This recognition of possibly reparable cells by macrophages is then responsible for tumor cell cytotoxicity. As described above, photoimmunotherapy can be used to modulate the anti-tumor immune response by specifically inactivating a certain subset of T-cells, in this case the suppressor cells [82]. The cellular mediated anti-tumor immunity might be induced by different cytokines, which have been described to be induced by PDT. Various studies have shown regulation of cytokines after PDT in vivo. Nseyo et al. [170] have reported high concentrations of interleukin (IL)-1β, IL-2, and TNF-α in the urine of patients treated with PDT for bladder cancer. The reason for the release of these cytokines and the role they may play in PDT are not well understood. Gollnick et al. [149] demonstrated in a BALB/c mouse model that PDT delivered to normal and tumor tissue in vivo causes marked changes in the expression of IL-6 and IL-10 but not TNF-α. IL-6 mRNA and protein were strongly enhanced in both the PDT-treated EMT6 tumor and exposed spleen and skin. The investigators concluded that the general inflammatory response to PDT might be mediated at least in part by IL-6. In contrast, IL-10 mRNA in the tumor decreased following PDT while it was induced in the normal skin of mice exposed to a PDT regime that strongly inhibits the contact hypersensitivity (CHS) response. The coincidence of the kinetics of IL-10 induction with the known kinetics of CHS inhibition in the normal skin suggests that the enhanced IL-10 expression is instrumental in the observed suppression of cell-mediated responses seen following PDT. In the tumor tissue, however, the decrease in IL-10 mRNA might contribute to an enhanced cellular response.

In addition to the above evidence of immune stimulation, immune suppression has also been reported following PDT with both PF and BPD-MA [179,180]. This observed immune system suppression is being investigated for novel applications such as organ transplantation [181] and the treatment of autoimmune diseases [182].

The studies of the in vivo mechanisms of tumor cure by PDT have led to one of the breakthroughs in the field. Based on the observations that BPD-MA in liposomal formulation targets the neovasculature in intraocular tumors, PDT for the treatment of AMD has been successfully developed. Further studies, including the fairly new field of the effects of PDT on the immune system, can lead to more applications for PDT, as already shown by the interest in using PDT in autoimmune disease and organ transplantation.

2.6 Mechanisms of ALA-based PDT

In a fundamentally different approach to PDT, the photosensitizer is not administered exogenously, but synthesized in living cells from a photochemically inactive precursor molecule. This is the case with ALA, the physiological precursor of porphyrins and heme [183]. Figure 6 shows a simplified scheme of those steps that are involved in the conversion of ALA into porphyrins and ultimately heme. This

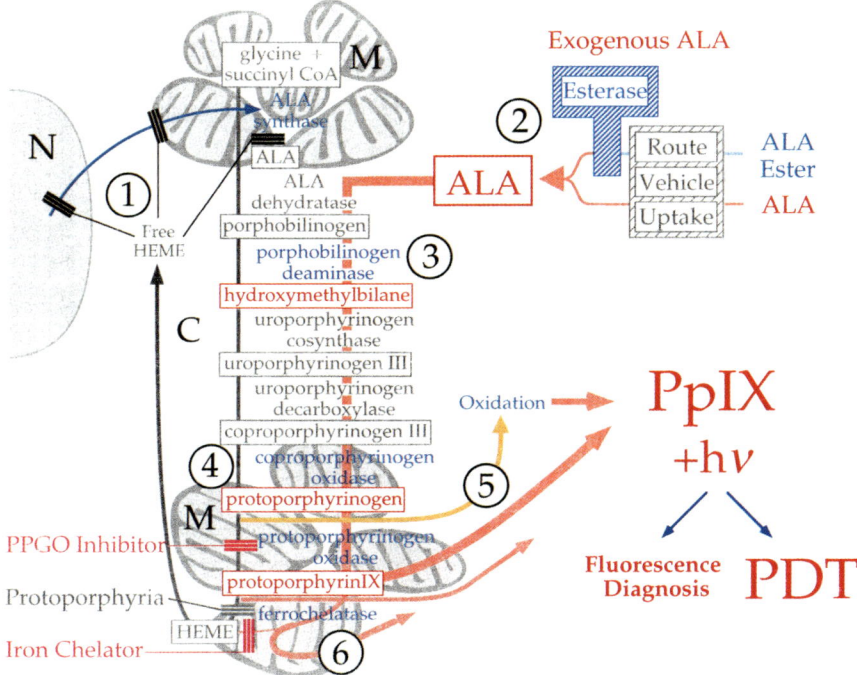

Figure 6. The biosynthetic pathway for the production of heme. The enzymes and metabolites are located in the mitochondria (M) and the cytosol (C). Heme negatively regulates ALA synthase at several points, three of which are shown in the figure. Exogenous ALA induces excess formation of heme precursors, including PpIX that can be utilized for photosensitization. (N) nucleus where transcriptional control occurs.

process is under normal conditions in regular cells controlled by cytoplasmic free heme via feedback inhibition of the initial enzyme, ALA synthase (Figure 6, #1). In the genetic disorder erythropoietic protoporphyria the production of heme is severely reduced because of a deficiency in ferrochelatase, the enzyme responsible for the conversion of PpIX into heme. The low heme levels result in overproduction of heme precursors and accumulation of PpIX, which leads to clinically relevant cutaneous photosensitivity [184]. In the case of ALA-based PDT, the addition of exogenous ALA makes the cellular control mechanisms of ALA production irrelevant and creates a situation similar to protoporphyria: the conversion of PpIX into heme by ferrochelatase becomes rate limiting, and PpIX accumulates [185] (Figure 6, #6).

The photoactivation of exogenous ALA-induced PpIX can be exploited for both diagnostic and therapeutic purposes [186]. The basic principle of this approach had been considered earlier for therapeutic use by several researchers, but it only gained broader interest after Kennedy et al. reported on the treatment of non-melanoma skin cancer [187]. The accessibility of the skin to light treatment and the efficacy of topical ALA preparations made the initial choice of dermatological applications an obvious one [186,188]. Once ALA-based PDT was found to be safe and efficient in human skin cancer therapy, clinical applications for PDT and fluorescence detection expanded rapidly to other organs. However, the clinical results in skin cancer were not completely satisfying because of incomplete responses and recurrences [189–193]. This sub-optimal efficiency resulted in research that aimed at improving ALA-based PDT. Consequently, mechanistic studies tried to gain a better understanding of those steps that lead from ALA administration to photoactivation of PpIX. These studies used two general approaches: (a) the analysis of target cell properties that determine heme synthetic capacity and (b) the optimization of ALA availability at the target cells. Selectivity in ALA-based PDT may be increased by an altered target cell status, such as in transformed or differentiated cells [47,194] or by an altered availability of ALA, such as with the use of ALA esters [195,196] (see below).

2.6.1 Cellular determinants of PpIX production

The cellular levels of iron and ferrochelatase are important determinants of PpIX yield under exogenous ALA stimulation. The use of iron chelators is based on the relative inefficiency of ferrochelatase compared to other enzymes, which causes PpIX build-up when heme precursors are produced at an increased rate. In some cancer cell lines lower levels of ferrochelatase have been found than in normal cells and this may contribute to tumor selectivity in certain cancers [197]. In an attempt to further reduce or totally abrogate heme formation, exogenous chelating agents were used to remove iron. It was shown in vitro that iron chelation caused both increased PpIX formation and improved PDT efficacy and in vivo applications in animals and humans have confirmed the concept [191,198,199]. In lymphocytes that express the transferrin receptor (CD71), which is interpreted as an indication of low intracellular iron levels, higher PpIX concentrations were reached under ALA stimulation [200]. In addition, an analysis of iron availability at the molecular level indicated that the iron regulatory protein might be useful as an indicator of the susceptibility of tissues to ALA-based PDT [201].

Ferrochelatase is the enzyme which is rate limiting in heme formation, thereby passively increasing the PpIX contents of the cells. However, PpIX formation itself requires the activity of several other enzymes. The weakest link in this metabolic chain limits its overall synthetic capacity and thus determines photodynamic efficacy. The mitochondria play an important role in heme formation and a correlation between mitochondrial content of cells and their capacity to form PpIX under exogenous ALA exposure has been described by Gibson et al. [202]. Studies from Hilf and co-workers have correlated increased PpIX-forming capacity with an increased activity of porphobilinogen deaminase (PBGD, Figure 6, #3), suggesting that PBGD conversion of porphobilinogen into hydroxymethylbilane acts as a rate-limiting step [202–204]. These data are supported by the fact that cellular PBGD levels are lower than those of other heme enzymes [205]. In addition, the pathologic metabolite excretion pattern in acute porphyrias supports rate limitation by PBGD. Another enzyme, protoporphyrinogen oxidase (PPO) has also been targeted for increasing PpIX formation. Although this may seem counter-intuitive, PPO inhibitors have been successful in increasing cellular PpIX levels. This approach is based on the concept that PPO inhibition results in accumulation of protoporphyrinogen and its subsequent non-enzymatic oxidation to PpIX. This approach was efficient in vitro and in animals, but some of the enzyme inhibitors are highly toxic [206].

In an attempt to better understand factors determining the selectivity of ALA-based PDT, Li et al. [207] compared PpIX accumulation in a number of fibroblast-derived cells and cell lines. Their findings confirmed earlier reports [208–210], which showed that transformed cell lines had higher PpIX levels than primary cells or 'normal' cell lines. Similarly, the transfection with oncogenes of fibroblast-derived cell lines resulted in greater ALA-induced PpIX quantities [207]. This report indicates that the oncogene-dependent malignant phenotype is associated with an increased heme synthetic capacity, which in some settings may be the basis for the tumor selectivity of ALA-based PDT. However, it should be noted that these and other studies have observed great variations of cellular PpIX content in different ALA-exposed cell lines, independently of species or phylogenetic origin [208].

An increased rate of PpIX accumulation in tumor cell lines is in agreement with the concept that more rapidly proliferating cells have the capacity to produce larger amounts of PpIX under exogenous ALA exposure [208,210] . This is also supported by the increased PpIX formation in CD71-positive lymphocytes, since CD71 expression is a marker of proliferation [200], and the increased PpIX levels in antigen-activated lymphocytes [211]. However, more rapid growth is not always associated with more efficient PpIX production. It has been demonstrated that certain growth-arrested, differentiated cells produce more PpIX than their proliferating, undifferentiated counterparts [47,194]. The increased PpIX production in differentiated cells has been accompanied by elevated mRNA levels of coproporphyrin oxidase but not ALA dehydratase.

For murine keratinocytes, growth arrest and differentiation in vitro lead to increased PpIX accumulation, and this finding has been validated in normal mouse skin in vivo. Figure 7 shows a fluorescence image of the skin of a newborn ALA-treated mouse. The levels of the epidermis closer to the surface, which contain the

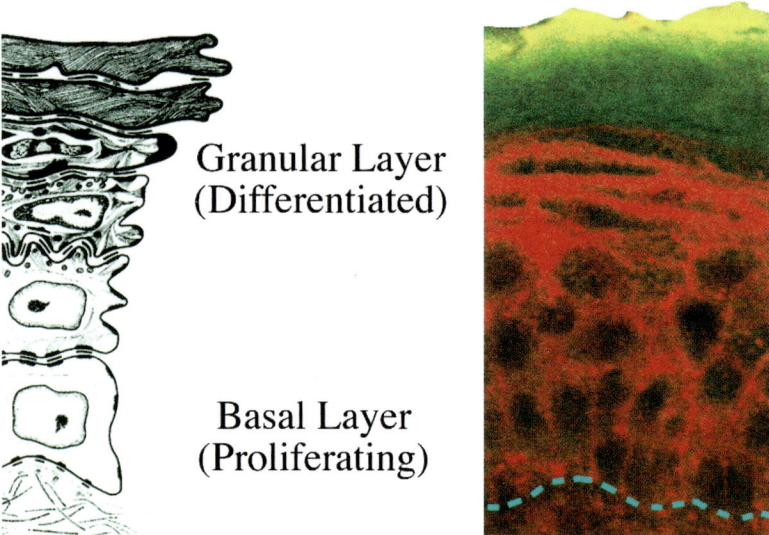

Figure 7. Fluorescence image of the epidermis of a neonatal mouse that received 500 mg kg^{-1} ALA intraperitoneally for 4 h. Vertical sections of the skin were imaged by CLSM. The fluorescence image is aligned with a schematic representation of the epidermal stratification. The green dotted line represents the basal membrane. Red fluorescence can be seen in the membranes of the viable keratinocytes. There is stronger fluorescence in the more differentiated (higher) keratinocyte layers than in the basal layer.

differentiated cells, show a stronger fluorescence than the basal, undifferentiated keratinocyte layer. It is remarkable that the cell membranes show a strong fluorescence, since it has been postulated that, in accordance with the physiological subcellular distribution of heme precursors, PpIX would primarily localize to mitochondria [212,213]. However, it seems that at higher concentrations PpIX redistributes within the cells and reaches high levels in cellular membranes. Figure 8 shows primary keratinocytes in vitro during proliferation and in a growth-arrested, differentiated state. It appears that the cell membrane staining becomes more prominent in differentiated cells. These data demonstrate that ALA-based PDT is still a wide open field with a multitude of opportunities to better understand the underlying biology and the chance to ever improve the clinical applications.

2.6.2 Optimized ALA availability in target cells

The availability of ALA at the cellular level has been improved by addition of, for example, DMSO (dimethyl sulfoxide) to topical formulations for increased tissue penetration [214,215] and by application of iontopheresis to induce reproducible levels of PpIX in normal skin [216]. It has been shown that ALA uses renal and intestinal peptide transporters for intracellular uptake [217]. High ALA-induced PpIX levels in intestinal epithelia suggest clinical relevance for this transport mechanism.

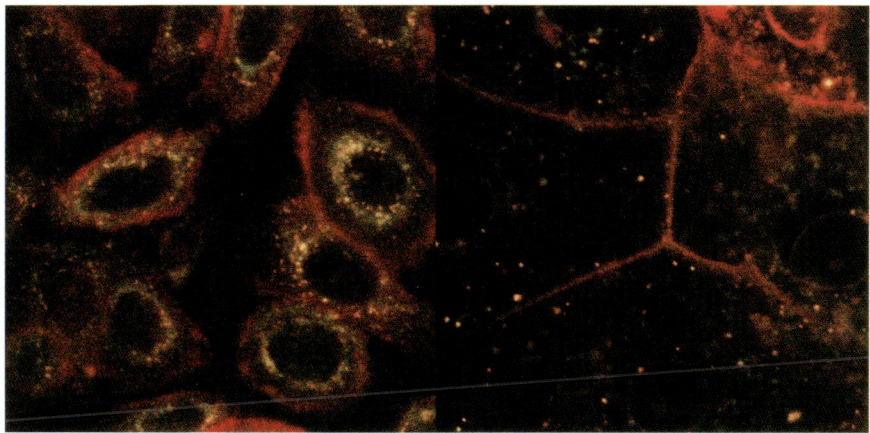

Figure 8. Fluorescence images (CLSM) of primary mouse keratinocytes after 4 h incubation with 0.1 mM ALA. On the left, proliferating keratinocytes show red fluorescence of the cell membranes. Perinuclear yellow autofluorescence indicates lysosomes, and green autofluorescence indicates mitochondria. The right-hand panel shows fluorescence of large, flat, differentiated keratinocytes with accentuated PpIX fluorescence in the cellular and nuclear membranes.

The use of ALA derivatives for improved topical ALA delivery has been investigated, since ALA uptake in cells is relatively low. ALA esters are being studied for this purpose, and add a step to the pro-drug PDT approach [218,219]. After uptake in target cells, ALA esters need to be hydrolyzed to release ALA, which then enters the heme synthesis pathway (Figure 6, #2). Since ALA needs to be freed by hydrolysis, this approach conceptually might be hampered by insufficient cellular esterases, which may become rate limiting for PpIX formation. Esterification alters the physicochemical properties and thus the pharmaco-dynamics of the parent molecule. Alkyl esters of ALA have a changed cellular uptake which depends on the length of the alkyl side chain [220]. Lower concentrations of the hexyl ALA ester are needed as compared to free ALA to induce similar PpIX levels with an improved selectivity [221]. Although some esters are less efficient in inducing maximal PpIX levels they still may increase selectivity. Several esters are at different levels of pre-clinical and clinical investigations [196,198,219,221].

2.7 Summary

In summary, the field of photodynamic therapy is at an exciting juncture with approval for its use as palliation and cure, and with an improved understanding of basic mechanisms of the underlying processes. Numerous challenges remain, including selectivity of photosensitizers and light delivery and dosimetry in complex sites. As the paradigm for the application of PDT shifts from mere palliation to cure, it becomes imperative that orthotopic animal models reflecting the physiology of the disease be incorporated into PDT investigations. It is reasonable to expect that there will be growth in this area, at least in the short term; the overall longevity and impact of PDT in science and medicine remain to be established.

Acknowledgements

Support to the authors was provided by National Institutes of Health Grants R01 AR40352-03 and P01 CA84203, and Office of Naval Research Contract N00014-94-1-0927 during the preparation of this chapter.

References

1. T.J. Dougherty (1987). Photosensitizers: therapy and detection of malignant tumors. *Photochem. Photobiol.*, **45**, 879–889.
2. B.W. Henderson, T.J. Dougherty (1992). *Photodynamic Therapy: Basic Principles and Clinical Applications.* Marcel Dekker, New York.
3. D. Kessel (1990). *Photodynamic Therapy of Neoplastic Diseases.* CRC Press, Boca Raton, FL.
4. T.J. Dougherty, C.J. Gomer, B.W. Henderson, G. Jori, D. Kessel, M. Korbelik, J. Moan, Q. Peng (1998). Photodynamic therapy. *J. Natl Cancer Inst.*, **90**, 889–905.
5. J.A. Parrish, R.R. Anderson, F. Urbach, D. Pitts (1978). *Optical properties of the Skin and Eyes. UV-A: Biological Effects of Ultraviolet Radiation with Emphasis on Human Responses to Longwave Ultraviolet.* Plenum, New York.
6. L.O. Svaasand, R. Ellingsen (1983). Optical properties of human brain. *Photochem. Photobiol.*, **38**, 293–299.
7. K. Gollnick (1968). Type II photooxygenation reactions in solution. *Adv. Photochem.*, **6**, 1–122.
8. G.O. Schenk (1963). Photosensitization. *Ind. Eng. Chem.*, **55**, 40–43.
9. C.S. Foote (1991). Definition of type I and type II photosensitized oxidation [editorial]. *Photochem. Photobiol.*, **54**, 659.
10. A. Andreoni (1987). Two-step photoactivation of hematoporphyrin by excimer-pumped dye-laser pulses. *J. Photochem. Photobiol. B*, **1**, 181–193.
11. G. Smith, W.G. McGimpsey, M.C. Lynch, I.E. Kochevar, R.W. Redmond (1994). An efficient oxygen independent two-photon photosensitization mechanism. *Photochem. Photobiol.*, **59**, 135–139.
12. B.W. Pogue, T. Momma, H.C. Wu, T. Hasan (1999). Transient absorption changes in vivo during photodynamic therapy with pulsed-laser light. *Br. J. Cancer*, **80**, 344–351.
13. P.C. Rausch, F. Rolfs, M.R. Winkler, A. Kottysch, A. Schauer, W. Steiner (1993). Pulsed versus continuous wave excitation mechanisms in photodynamic therapy of differently graded squamous cell carcinomas in tumor-implanted nude mice. *Eur. Arch. Otorhinolaryngol.*, **250**, 82–87.
14. T. Okunaka, H. Kato, C. Konaka, H. Sakai, H. Kawabe, K. Aizawa (1992). A comparison between argon-dye and excimer-dye laser for photodynamic effect in transplanted mouse tumor. *Jpn J. Cancer Res.*, **83**, 226–231.
15. C.J. Gomer (1991). Preclinical examination of first and second generation photosensitizers used in photodynamic therapy. *Photochem. Photobiol.*, **54**, 1093–1107.

16. R.W. Boyle, D. Dolphin (1996). Structure and biodistribution relationships of photodynamic sensitizers. *Photochem. Photobiol.*, **64**, 469–485.
17. L. Strong, D.M. Yarmush, M.L. Yarmush (1994). Antibody-targeted photolysis. Photophysical, biochemical, and pharmacokinetic properties of antibacterial conjugates. *Ann. N. Y. Acad. Sci.*, **745**, 297–320.
18. T. Hasan (1992). Photosensitizer delivery mediated by macromolecular carrier systems. In: T.J. Dougherty, B.W. Henderson (Eds), *Photodynamic Therapy: Basic Principles and Clinical Applications* (pp. 187–200). Marcel Dekker, New York.
19. W.M. Star, H.P. Marijnissen, A.E. van den Berg-Blok, J.A. Versteeg, K.A. Franken, H.S. Reinhold (1986). Destruction of rat mammary tumor and normal tissue microcirculation by hematoporphyrin derivative photoradiation observed in vivo in sandwich observation chambers. *Cancer Res.*, **46**, 2532–2540.
20. S.L. Gibson, K.R. van der Meid, R.S. Murant, R. Hilf (1990). Increased efficacy of photodynamic therapy of R3230AC mammary adenocarcinoma by intratumoral injection of Photofrin II. *Br. J. Cancer*, **61**, 553–557.
21. T.H. Foster, R.S. Murant, R.G. Bryant, R.S. Knox, S.L. Gibson, R. Hilf (1991). Oxygen consumption and diffusion effects in photodynamic therapy. *Radiat. Res.*, **126**, 296–303.
22. T.M. Sitnik, J.A. Hampton, B.W. Henderson (1998). Reduction of tumor oxygenation during and after photodynamic therapy in vivo: effects of fluence rate. *Br. J. Cancer*, **77**, 1386–1394.
23. T.M. Sitnik, B.W. Henderson (1998). The effect of fluence rate on tumor and normal tissue responses to photodynamic therapy. *Photochem. Photobiol.*, **67**, 462–466.
24. T.J. Dougherty (1986). Photosensitization of malignant tumors. *Semin. Surg. Oncol.*, **2**, 24–37.
25. M. Niedre, M.S. Patterson, B.C. Wilson (2002). Direct near-infrared luminescence detection of singlet oxygen generated by photodynamic therapy in cells in vitro and tissues in vivo. *Photochem. Photobiol.*, **75**, 382–391.
26. G.R. Buettner, L.W. Oberley (1990). The apparent production of superoxide and hydroxyl radicals by hematoporphyrin and light as seen by spin-trapping. *FEBS Lett.*, **121**, 161–164.
27. B.W. Henderson, V.H. Fingar (1989). Oxygen limitation of direct tumor cell kill during photodynamic treatment of a murine tumor model. *Photochem. Photobiol.*, **49**, 299–304.
28. B.J. Tromberg, A. Orenstein, S. Kimel, S.J. Barker, J. Hyatt, J.S. Nelson, M.W. Berns (1990). In vivo tumor oxygen tension measurements for the evaluation of the efficiency of photodynamic therapy. *Photochem. Photobiol.*, **52**, 375–385.
29. B.W. Henderson, T.J. Dougherty (1992). How does photodynamic therapy work? *Photochem. Photobiol.*, **55**, 145–157.
30. B.W. Henderson (1990). Probing the effects of photodynamic therapy through in vivo-in vitro methods. In: D. Kessel (Ed.), *Photodynamic Therapy of Neoplastic Disease* (Vol. 1, pp. 169–188). CRC Press, Boca Raton, FL.
31. V.H. Fingar, T.S. Mang, B.W. Henderson (1988). Modification of photodynamic therapy-induced hypoxia by fluosol-DA (20%) and carbogen breathing in mice. *Cancer Res.*, **48**, pp. 3350–3354.
32. S. Iinuma, G. Wagnieres, K. Schomacker, M. Bamberg, T. Hasan (1995). The importance of fluence rate in photodynamic therapy with ALA-induced PPIX and BPD-MA in a rat bladder tumor model. In: T.J. Dougherty (Ed.), *Optical Methods for Tumor Treatment and Detection: Mechanisms and Techniques in PDT IV* (Proc SPIE., Vol. 2392, pp. 136–140).
33. R.B. Veenhuizen, F.A. Stewart (1995). The importance of fluence rate in photodynamic therapy: is there a parallel with ionizing radiation dose-rate effects? *Radiother. Oncol.*, **37**, 131–135.

34. S. Iinuma, K.T. Schomacker, G. Wagnieres, M. Rajadhyaksha, M. Bamberg, T. Momma, T. Hasan (1999). In vivo fluence rate and fractionation effects on tumor response and photobleaching in photodynamic therapy with two photosensitizers in an orthotopic rat tumor model. *Cancer Res.*, **59**, 6164–6170.

35. E. Ben-Hur, N.E. Geacintov, B. Studamire, M.E. Kenney, B. Horowitz (1995). The effect of irradiance on virus sterilization and photodynamic damage in red blood cells sensitized by phthalocyanines. *Photochem. Photobiol.*, **61**, 190–195.

36. A.C. Moor, J.W. Lagerberg, K. Tijssen, S. Foley, T.G. Truscott, I.E. Kochevar, A. Brand, T. M. Dubbelman, J. VanSteveninck (1997). In vitro fluence rate effects in photodynamic reactions with AlPcS4 as sensitizer. *Photochem. Photobiol.*, **66**, 860–865.

37. S.L. Gibson, K.R. VanDerMeid, R.S. Murant, R.F. Raubertas, R. Hilf (1990). Effects of various photoradiation regimens on the antitumor efficacy of photodynamic therapy for R3230AC mammary carcinomas. *Cancer Res.*, **50**, 7236–7241.

38. S.L. Gibson, T.H. Foster, R.H. Feins, R.F. Raubertas, M.A. Fallon, R. Hilf (1994). Effects of photodynamic therapy on xenografts of human mesothelioma and rat mammary carcinoma in nude mice. *Br. J. Cancer*, **69**, 473–481.

39. T.H. Foster, D.F. Hartley, M.G. Nichols, R. Hilf (1993). Fluence rate effects in photo-dynamic therapy of multicell tumor spheroids. *Cancer Res.*, **53**, 1249–1254.

40. G. Graschew, M. Shopova, A. Anastassova, A. Chakarova (1988). Light dose fractiona-tion versus single dose irradiation in photodynamic therapy of tumors. *Lasers Med. Sci.*, **3**, 173–177.

41. I.P. van Geel, H. Oppelaar, J.P. Marijnissen, F.A. Stewart (1996). Influence of fractiona-tion and fluence rate in photodynamic therapy with Photofrin or mTHPC. *Radiat. Res.*, **145**, 602–609.

42. T.S. Mang, T.J. Dougherty, W.R. Potter, D.G. Boyle, S. Somer, J. Moan (1987). Photobleaching of porphyrins used in photodynamic therapy and implications for therapy. *Photochem. Photobiol.*, **45**, 501–506.

43. H. Anholt, J. Moan (1992). Fractionated treatment of CaD2 tumors in mice sensitized with aluminium phthlocyanine tetrasulfonate. *Cancer Lett.*, **61**, 263–267.

44. D. Kessel, Y. Luo, Y. Deng, C.K. Chang (1997). The role of subcellular localization in initiation of apoptosis by photodynamic therapy. *Photochem. Photobiol.*, **65**, 422–426.

45. B.B. Noodt, K. Berg, T. Stokke, Q. Peng, J.M. Nesland (1999). Different apoptotic pathways are induced from various intracellular sites by tetraphenylporphyrins and light. *Br. J. Cancer*, **79**, 72–81.

46. Q. Peng, J. Moan, J.M. Nesland (1996). Correlation of subcellular and intratumoral photosensitizer localization with ultrastructural features after photodynamic therapy. *Ultrastruct. Pathol.*, **20**, 109–129.

47. B. Ortel, N. Chen, J. Brissette, G.P. Dotto, E. Maytin, T. Hasan (1998). Differentiation-specific increase in ALA-induced protoporphyrin IX accumulation in primary mouse keratinocytes. *Br. J. Cancer*, **77**, 1744–1751.

48. B.C. Wilson (1989). Photodynamic therapy: light delivery and dosage for second-generation photosensitizers. In G. Bock and S. Harnett (Eds), *Photosensitizing Compounds: Their Chemistry, Biology and Clinical use* (pp. 60–77). John Wiley, Chichester, NY.

49. L. Lilge, K. Molpus, T. Hasan, B.C. Wilson (1998). Light dosimetry for intraperitoneal photodynamic therapy in a murine xenograft model of human epithelial ovarian carci-noma. *Photochem. Photobiol.*, **68**, 281–288.

50. J.S. Nelson, L.H. Liaw, A. Orenstein, W.G. Roberts, M.W. Berns (1988). Mechanism of tumor destruction following photodynamic therapy with hematoporphyrin derivative, chlorin, and phthalocyanine. *J. Natl. Cancer Inst.*, **80**, 1599–1605.

51. M.W. Reed, A.P. Mullins, G.L. Anderson, F.N. Miller, T.J. Wieman (1989). The effect of photodynamic therapy on tumor oxygenation. *Surgery*, **106**, 94–99.

52. M.S. Patterson, B.C. Wilson, R. Graff (1990). In vivo tests of the concept of photodynamic threshold dose in normal rat liver photosensitized by aluminum chlorosulphonated phthalocyanine. *Photochem. Photobiol.*, **51**, 343–349.

53. P. Ehrlich (1906). *Collected Studies on Immunity*. Wiley, New York.

54. U. Schmidt-Erfurth, H. Diddens, R. Birngruber, T. Hasan (1997). Photodynamic targeting of human retinoblastoma cells using covalent low-density lipoprotein conjugates. *Br. J. Cancer*, **75**, 54–61.

55. B.A. Allison, P.H. Pritchard, J.G. Levy (1994). Evidence for low-density lipoprotein receptor-mediated uptake of benzoporphyrin derivative. *Br. J. Cancer*, **69**, 833–839.

56. A. Barel, G. Jori, A. Perin, P. Romandini, A. Pagnan, S. Biffanti (1986). Role of high-, low- and very low-density lipoproteins in the transport and tumor-delivery of hematoporphyrin in vivo. *Cancer Lett.*, **32**, 145–150.

57. C.N. Zhou, C. Milanesi, G. Jori (1988). An ultrastructural comparative evaluation of tumors photosensitized by porphyrins administered in aqueous solution, bound to liposomes or to lipoproteins. *Photochem. Photobiol.*, **48**, 487–492.

58. C. Candide, J.P. Reyftmann, R. Santus, J.C. Maziere, P. Morliere, S. Goldstein (1988). Modification of epsilon-amino group of lysines, cholesterol oxidation and oxidized lipid-apoprotein cross-link formation by porphyrin-photosensitized oxidation of human low density lipoproteins. *Photochem. Photobiol.*, **48**, 137–146.

59. P. Morliere, E. Kohen, J.P. Reyftmann, R. Santus, C. Kohen, J.C. Maziere, S. Goldstein, W.F. Mangel, L. Dubertret (1987). Photosensitization by porphyrins delivered to L cell fibroblasts by human serum low density lipoproteins. A microspectrofluorometric study. *Photochem. Photobiol.*, **46**, 183–191.

60. C.M. West, D.C. West, S. Kumar, J.V. Moore (1990). A comparison of the sensitivity to photodynamic treatment of endothelial and tumor cells in different proliferative states. *Int. J. Radiat. Biol.*, **58**, 145–156.

61. U. Schmidt-Erfurth, W. Bauman, E. Gragoudas, T.J. Flotte, N.A. Michaud, R. Birngruber, T. Hasan (1994). Photodynamic therapy of experimental choroidal melanoma using lipoprotein-delivered benzoporphyrin. *Ophthalmology*, **101**, 89–99.

62. C. Milanesi, C. Zhou, R. Biolo, G. Jori (1990). Zn(II)-phthalocyanine as a photodynamic agent for tumors. II. Studies on the mechanism of photosensitised tumor necrosis. *Br. J. Cancer*, **61**, 846–850.

63. M.R. Hamblin, E.L. Newman (1994). On the mechanism of the tumor-localising effect in photodynamic therapy. *J. Photochem. Photobiol. B*, **23**, 3–8.

64. N. Brasseur, R. Langlois, C. La Madeleine, R. Ouellet, J.E. van Lier (1999). Receptor-mediated targeting of phthalocyanines to macrophages via covalent coupling to native or maleylated bovine serum albumin. *Photochem. Photobiol.*, **69**, 345–352.

65. A. Matsumoto, M. Naito, H. Itakura, S. Ikemoto, H. Asaoka, I. Hayakawa, H. Kanamori, H. Aburatani, F. Takaku, H. Suzuki, et al. (1990). Human macrophage scavenger receptors: primary structure, expression, and localization in atherosclerotic lesions. *Proc. Natl. Acad. Sci. U.S.A.*, **87**, 9133–9137.

66. B.A. Allison, M.T. Crespo, A.K. Jain, A.M. Richter, Y.N. Hsiang, J.G. Levy (1997). Delivery of benzoporphyrin derivative, a photosensitizer, into atherosclerotic plaque of Watanabe heritable hyperlipidemic rabbits and balloon-injured New Zealand rabbits. *Photochem. Photobiol.*, **65**, 877–883.

67. H.E. de Vries, A.C. Moor, T.M. Dubbelman, T.J. van Berkel, J. Kuiper (1999). Oxidized low-density lipoprotein as a delivery system for photosensitizers: implications for photodynamic therapy of atherosclerosis. *J. Pharmacol. Exp. Ther.*, **289**, 528–534.

68. A.R. Oseroff, D. Ohuoha, T. Hasan, J.C. Bommer, M.L. Yarmush (1986). Antibody-targeted photolysis: selective photodestruction of human T-cell leukemia cells using monoclonal antibody-chlorin e6 conjugates. *Proc. Natl. Acad. Sci. U.S.A.*, **83**, 8744–8748.

69. S.L. Rakestraw, R.G. Tompkins, M.L. Yarmush (1986). Antibody-targeted photolysis: in vitro studies with Sn(IV) chlorin e6 covalently bound to monoclonal antibodies using a modified dextran carrier. *Proc. Natl. Acad. Sci. U.S.A.*, **87**, 4217–4221.

70. T. Hasan, C.W. Lin, A. Lin (1989). Laser-induced selective cytotoxicity using monoclonal antibody-chromophore conjugates. *Prog. Clin. Biol. Res.*, **288**, 471–477.

71. T. Hasan, A. Lin, D. Yarmush (1989). Monoclonal antibody-chromophore conjugates as selective phototoxins. *J. Control. Release*, **10**, 107–117.

72. B.A. Goff, M. Bamberg, T. Hasan (1991). Photoimmunotherapy of human ovarian carcinoma cells ex vivo. *Cancer Res.*, **51**, 4762–4767.

73. F.N. Jiang, S. Jiang, D. Liu, A. Richter, J.G. Levy (1990). Development of technology for linking photosensitizers to a model monoclonal antibody. *J. Immunol. Methods*, **134**, 139–149.

74. F.N. Jiang, D.J. Liu, H. Neyndorff, M. Chester, S.Y. Jiang, J.G. Levy (1991). Photodynamic killing of human squamous cell carcinoma cells using a monoclonal antibody-photosensitizer conjugate. *J. Natl. Cancer Inst.*, **83**, 1218–1225.

75. V. Omelyanenko, C. Gentry, P. Kopeckova, J. Kopecek (1998). HPMA copolymer-anticancer drug-OV-TL16 antibody conjugates. II. Processing in epithelial ovarian carcinoma cells in vitro. *Int. J. Cancer*, **75**, 600–608.

76. V. Omelyanenko, P. Kopeckova, C. Gentry, J.G. Shiah, J. Kopecek (1996). HPMA copolymer-anticancer drug-OV-TL16 antibody conjugates. 1. influence of the method of synthesis on the binding affinity to OVCAR-3 ovarian carcinoma cells in vitro. *J. Drug Target*, **3**, 357–373.

77. N.L. Krinick, Y. Sun, D. Joyner, J.D. Spikes, R.C. Straight, J. Kopecek (1994). A polymeric drug delivery system for the simultaneous delivery of drugs activatable by enzymes and/or light. *J. Biomater. Sci. Polym. Ed.*, **5**, 303–324.

78. M.R. Hamblin, J.L. Miller, T. Hasan (1996). Effect of charge on the interaction of site-specific photoimmunoconjugates with human ovarian cancer cells. *Cancer Res.*, **56**, 5205–5210.

79. N.S. Soukos, M.R. Hamblin, T. Hasan (1997). The effect of charge on cellular uptake and phototoxicity of polylysine chlorin(e6) conjugates. *Photochem. Photobiol.*, **65**, 723–729.

80. B.A. Goff, U. Hermanto, J. Rumbaugh, J. Blake, M. Bamberg, T. Hasan (1994). Photoimmunotherapy and biodistribution with an OC125-chlorin immunoconjugate in an in vivo murine ovarian cancer model. *Br. J. Cancer*, **70**, 474–480.

81. M.B. Vrouenraets, G.W. Visser, F.A. Stewart, M. Stigter, H. Oppelaar, P.E. Postmus, G.B. Snow, G.A. van Dongen (1999). Development of meta-tetrahydroxyphenylchlorin-monoclonal antibody conjugates for photoimmunotherapy. *Cancer Res.*, **59**, 1505–1513.

82. J.K. Steele, D. Liu, A.T. Stammers, S. Whitney, J.G. Levy (1988). Suppressor deletion therapy: selective elimination of T suppressor cells in vivo using a hematoporphyrin conjugated monoclonal antibody permits animals to reject syngeneic tumor cells. *Cancer Immunol. Immunother.*, **26**, 125–131.

83. B.A. Goff, J. Blake, M.P. Bamberg, T. Hasan (1996). Treatment of ovarian cancer with photodynamic therapy and immunoconjugates in a murine ovarian cancer model. *Br. J. Cancer*, **74**, 1194–1198.

84. L. Duska, M. Hamblin, J. Miller, T. Hasan (1999). Combination photoimmunotherapy and cisplatin: effects on human ovarian cancer ex vivo. *J. Natl. Cancer Inst.*, **91**, 1557–1563.

85. P.F. Brophy, S.M. Keller (1992). Adriamycin enhanced in vitro and in vivo photo-dynamic therapy of mesothelioma. *J. Surg. Res.*, **52**, 631–634.

86. G. Canti, D. Lattuada, A. Nicolin, P. Taroni, G. Valentini, R. Cubeddu (1994). Antitumor immunity induced by photodynamic therapy with aluminum disulfonated phthalocyanines and laser light. *Anticancer Drugs*, **5**, 443–447.

87. L.W. Ma, K. Berg, H.E. Danielsen, O. Kaalhus, V. Iani, J. Moan (1996). Enhanced antitumor effect of photodynamic therapy by microtubule inhibitors. *Cancer Lett.*, **109**, 129–139.

88. M.Y. Nahabedian, R.A. Cohen, M.F. Contino, T.M. Terem, W.H. Wright, M.W. Berns, A.G. Wile (1988). Combination cytotoxic chemotherapy with cisplatin or doxorubicin and photodynamic therapy in murine tumors. *J. Natl. Cancer Inst.*, **80**, 739–743.

89. S. Schmidt, U. Wagner, P. Oehr, D. Krebs (1992). Clinical use of photodynamic therapy in gynecologic tumor patients–antibody-targeted photodynamic laser therapy as a new oncologic treatment procedure. *Zentralbl Gynäkol.*, **114**, 307–311.

90. S. Schmidt, U. Wagner, B. Schultes, P. Oehr, W. Decleer, W. Ertmer, H. Lubaschowski, H.J. Biersack, D. Krebs (1992). Photodynamic laser therapy with antibody-bound dyes. A new procedure in therapy of gynecologic malignancies. *Fortschr. Med.*, **110**, 298–301.

91. M. Birchler, F. Viti, L. Zardi, B. Spiess, D. Neri (1999). Selective targeting and photo-coagulation of ocular angiogenesis mediated by a phage-derived human antibody fragment. *Nat Biotechnol.*, **17**, 984–988.

92. L.R. Duska, M.R. Hamblin, M.P. Bamberg, T. Hasan (1997). Biodistribution of charged F(ab')2 photoimmunoconjugates in a xenograft model of ovarian cancer. *Br. J. Cancer*, **75**, 837–844.

93. K. Berg, P.K. Selbo, L. Prasmickaite, T.E. Tjelle, K. Sandvig, J. Moan, G. Gaudernack, O. Fodstad, S. Kjolsrud, H. Anholt, G.H. Rodal, S.K. Rodal, A. Hogset (1999). Photochemical internalization: a novel technology for delivery of macromolecules into cytosol. *Cancer Res.*, **59**, 1180–1183.

94. C.J. Gomer, A. Ferrario, N. Hayashi, N. Rucker, B.C. Szirth, A.L. Murphree (1988). Molecular, cellular, and tissue responses following photodynamic therapy. *Lasers Surg. Med.*, **8**, 450–463.

95. M.A. Laukka, K.K. Wang, J.A. Bonner (1994). Apoptosis occurs in lymphoma cells but not in hepatoma cells following ionizing radiation and photodynamic therapy. *Dig. Dis. Sci.*, **39**, 2467–2475.

96. X.Y. He, R.A. Sikes, S. Thomsen, L.W. Chung, S.L. Jacques (1994). Photodynamic therapy with photofrin II induces programmed cell death in carcinoma cell lines. *Photochem. Photobiol.*, **59**, 468–473.

97. B.B. Noodt, K. Berg, T. Stokke, Q. Peng, J.M. Nesland (1996). Apoptosis and necrosis induced with light and 5-aminolevulinic acid-derived protoporphyrin IX. *Br. J. Cancer*, **74**, 22–29.

98. Y. Luo, D. Kessel (1997). Initiation of apoptosis versus necrosis by photodynamic therapy with chloroaluminum phthalocyanine. *Photochem. Photobiol.*, **66**, 479–483.

99. E. Ben-Hur, J. Oetjen, B. Horowitz (1997). Silicon phthalocyanine Pc 4 and red light causes apoptosis in HIV-infected cells. *Photochem. Photobiol.*, **65**, 456–460.

100. T.M.A.R. Dubbelman (1985). Porphyrin-photosensitized modification of subcellular structures. In: G. Jori, C. Perria (Ed.), *Photodynamic Therapy of Tumors and Other Diseases* (pp. 93–95). Libreria Progetto Editore, Padova.

101. B.B. Noodt, E. Kvam, H.B. Steen, J. Moan (1993). Primary DNA damage, HPRT mutation and cell inactivation photoinduced with various sensitizers in V79 cells. *Photochem. Photobiol.*, **58**, 541–547.

102. K. Berg, J. Moan (1997). Lysosomes and microtubules as targets for photo-chemotherapy of cancer. *Photochem. Photobiol.*, **65**, 403–409.

103. S.R. Wood, J.A. Holroyd, S.B. Brown (1997). The subcellular localization of Zn(II) phthalocyanines and their redistribution on exposure to light. *Photochem. Photobiol.*, **65**, 397–402.

104. D. Kessel, Y. Luo (1998). Mitochondrial photodamage and PDT-induced apoptosis. *J. Photochem. Photobiol. B*, **42**, 89–95.

105. J. Dahle, H.B. Steen, J. Moan (1999). The mode of cell death induced by photodynamic treatment depends on cell density. *Photochem. Photobiol.*, **70**, 363–367.

106. D.J. Granville, J.G. Levy, D.W. Hunt (1997). Photodynamic therapy induces caspase-3 activation in HL-60 cells. *Cell Death Diff.*, **4**, 623–629.

107. J. He, C.M. Whitacre, L.Y. Xue, N.A. Berger, N.L. Oleinick (1998). Protease activation and cleavage of poly(ADP-ribose) polymerase: an integral part of apoptosis in response to photodynamic treatment. *Cancer Res.*, **58**, 940–946.

108. X. Liu, C.N. Kim, J. Yang, R. Jemmerson, X. Wang (1996). Induction of apoptotic program in cell-free extracts: requirement for dATP and cytochrome c. *Cell*, **86**, 147–157.

109. D.J. Granville, C.M. Carthy, H. Jiang, G.C. Shore, B.M. McManus, D.W. Hunt (1998). Rapid cytochrome c release, activation of caspases 3, 6, 7 and 8 followed by Bap31 cleavage in HeLa cells treated with photodynamic therapy. *FEBS Lett.*, **437**, 5–10.

110. M.E. Varnes, S.M. Chiu, L.Y. Xue, N.L. Oleinick (1999) Photodynamic therapy-induced apoptosis in lymphoma cells: translocation of cytochrome c causes inhibition of respiration as well as caspase activation. *Biochem. Biophys. Res. Commun.*, **255**, 673–679.

111. D. Kessel, Y. Luo (1999). Photodynamic therapy: a mitochondrial inducer of apoptosis. *Cell Death Diff.*, **6**, 28–35.

112. D.R. Green, J.C. Reed (1998). Mitochondria and apoptosis. *Science*, **281**, 1309–1312.

113. N.A. Thornberry, Y. Lazebnik (1998). Caspases: enemies within. *Science*, **281**, 1312–1316.

114. G. Nunez, M.A. Benedict, Y. Hu, N. Inohara (1998). Caspases: the proteases of the apoptotic pathway. *Oncogene*, **17**, 3237–3245.

115. A.G. Porter, R.U. Janicke (1999). Emerging roles of caspase-3 in apoptosis. *Cell Death Diff.*, **6**, 99–104.

116. D. Separovic, J. He, N.L. Oleinick (1997). Ceramide generation in response to photo-dynamic treatment of L5178Y mouse lymphoma cells. *Cancer Res.*, **57**, 1717–1721.

117. D. Separovic, K.J. Mann, N.L. Oleinick (1997). Association of ceramide accumulation with photodynamic treatment-induced cell death. *Photochem. Photobiol.*, **68**, 101–109.

118. D. Separovic, J.J. Pink, N.A. Oleinick, M. Kester, D.A. Boothman, M. McLoughlin, L.A. Pena, A. Haimovitz-Friedman (1999). Niemann-Pick human lymphoblasts are resistant to phthalocyanine 4-photodynamic therapy-induced apoptosis. *Biochem. Biophys. Res. Commun.*, **258**, 506–512.

119. Z. Assefa, A. Vantieghem, W. Declercq, P. Vandenabeele, J.R. Vandenheede, W. Merlevede, P. de Witte, P. Agostinis (1999). The activation of the c-Jun N-terminal kinase and p38 mitogen-activated protein kinase signaling pathways protects HeLa cells from apoptosis following photodynamic therapy with hypericin. *J. Biol. Chem.*, **274**, 8788–8796.

120. L. Xue, J. He, N.L. Oleinick (1999). Promotion of photodynamic therapy-induced apoptosis by stress kinases. *Cell Death Diff.*, **6**, 855–864.

121. L.Y. Xue, Y. Qiu, J. He, H.J. Kung, N.L. Oleinick (1999). Etk/Bmx, a PH-domain containing tyrosine kinase, protects prostate cancer cells from apoptosis induced by photodynamic therapy or thapsigargin. *Oncogene*, **18**, 3391–3398.

122. J.P. van Brussel, G.H. Mickisch (1998). Circumvention of multidrug resistance in genitourinary tumors. *Int. J. Urol.*, **5**, 1–15.

123. J.C. Reed, T. Miyashita, S. Takayama, H.G. Wang, T. Sato, S. Krajewski, C. Aime-Sempe, S. Bodrug, S. Kitada, M. Hanada (1996). BCL-2 family proteins: regulators of cell death involved in the pathogenesis of cancer and resistance to therapy. *J. Cell. Biochem.*, **60**, 23–32.

124. J. He, M.L. Agarwal, H.E. Larkin, L.R. Friedman, L.Y. Xue, N.L. Oleinick (1996). The induction of partial resistance to photodynamic therapy by the protooncogene BCL-2. *Photochem. Photobiol.*, **64**, 845–852.

125. C.M. Payne, C. Bernstein, H. Bernstein (1995). Apoptosis overview emphasizing the role of oxidative stress, DNA damage and signal-transduction pathways. *Leuk. Lymphoma*, **19**, 43–93.

126. E. Ben-Hur, T.M. Dubbelman (1993). Cytoplasmic free calcium changes as a trigger mechanism in the response of cells to photosensitization. *Photochem. Photobiol.*, **58**, 890–894.

127. J. Yang, X. Liu, K. Bhalla, C.N. Kim, A.M. Ibrado, J. Cai, T.I. Peng, D.P. Jones, X. Wang (1997). Prevention of apoptosis by Bcl-2: release of cytochrome c from mitochondria blocked. *Science*, **275**, 1129–1132.

128. R.M. Kluck, E. Bossy-Wetzel, D.R. Green, D.D. Newmeyer (1997). The release of cytochrome c from mitochondria: a primary site for Bcl-2 regulation of apoptosis. *Science*, **275**, 1132–1136.

129. W.G. Zhang, L.P. Ma, S.W. Wang, Z.Y. Zhang, G.D. Cao (1999). Antisense bcl-2 retrovirus vector increases the sensitivity of a human gastric adenocarcinoma cell line to photodynamic therapy. *Photochem. Photobiol.*, **69**, 582–586.

130. H.R. Kim, Y. Luo, G. Li, D. Kessel (1999). Enhanced apoptotic response to photodynamic therapy after bcl-2 transfection. *Cancer Res.*, **59**, 3429–3432.

131. N. Ahmad, D.K. Feyes, R. Agarwal, H. Mukhtar (1998). Photodynamic therapy results in induction of WAF1/CIP1/P21 leading to cell cycle arrest and apoptosis. *Proc. Natl. Acad. Sci. U.S.A.*, **95**, 6977–6982.

132. N. Ahmad, S. Gupta, H. Mukhtar (1999). Involvement of retinoblastoma (Rb) and E2F transcription factors during photodynamic therapy of human epidermoid carcinoma cells A431. *Oncogene*, **18**, 1891–1896.

133. S. Gupta, N. Ahmad, H. Mukhtar (1998). Involvement of nitric oxide during phthalocyanine (Pc4) photodynamic therapy-mediated apoptosis. *Cancer Res.*, **58**, 1785–1788.

134. A.M. Fisher, A. Ferrario, N. Rucker, S. Zhang, C.J. Gomer (1999). Photodynamic therapy sensitivity is not altered in human tumor cells after abrogation of p53 function. *Cancer Res.*, **59**, 331–335.

135. L.Y. Xue, J. He, N.L. Oleinick (1997). Rapid tyrosine phosphorylation of HS1 in the response of mouse lymphoma L5178Y-R cells to photodynamic treatment sensitized by the phthalocyanine Pc 4. *Photochem. Photobiol.*, **66**, 105–113.

136. C.J. Gomer, S.W. Ryter, A. Ferrario, N. Rucker, S. Wong, A.M. Fisher (1996). Photodynamic therapy-mediated oxidative stress can induce expression of heat shock proteins. *Cancer Res.*, **56**, 2355–2360.

137. A.K. Verrico, J.V. Moore (1997). Expression of the collagen-related heat shock protein HSP47 in fibroblasts treated with hyperthermia or photodynamic therapy. *Br. J. Cancer*, **76**, 719–724.

138. C.J. Gomer, A. Ferrario, N. Rucker, S. Wong, A.S. Lee (1991). Glucose regulated protein induction and cellular resistance to oxidative stress mediated by porphyrin photosensitization. *Cancer Res.*, **51**, 6574–6579.

139. J. Morgan, J.E. Whitaker, A.R. Oseroff (1998). GRP78 induction by calcium ionophore potentiates photodynamic therapy using the mitochondrial targeting dye victoria blue BO. *Photochem. Photobiol.*, **67**, 155–164.

140. L.Y. Xue, M.L. Agarwal, M.E. Varnes (1995). Elevation of GRP-78 and loss of HSP-70 following photodynamic treatment of V79 cells: sensitization by nigericin. *Photochem. Photobiol.*, **62**, 135–143.

141. C.J. Gomer, M. Luna, A. Ferrario, N. Rucker (1991). Increased transcription and translation of heme oxygenase in Chinese hamster fibroblasts following photodynamic stress or Photofrin II incubation. *Photochem. Photobiol.*, **53**, 275–279.

142. L.C. Penning, M.H. Rasch, E. Ben-Hur, T.M. Dubbelman, A.C. Havelaar, J. Van der Zee, J. Van Steveninck (1992). A role for the transient increase of cytoplasmic free calcium in cell rescue after photodynamic treatment. *Biochim. Biophys. Acta*, **1107**, 255–260.

143. L.C. Penning, M.J. Keirse, J. Van Steveninck, T.M. Dubbelman (1993). Ca(2+)-mediated prostaglandin E2 induction reduces haematoporphyrin-derivative-induced cytotoxicity of T24 human bladder transitional carcinoma cells in vitro. *Biochem. J.*, **292**, 237–240.

144. L.C. Penning, J. Van Steveninck, T.M. Dubbelman (1993). HPD-induced photodynamic changes in intracellular cyclic AMP levels in human bladder transitional carcinoma cells, clone T24. *Biochem. Biophys. Res. Commun.*, **194**, 1084–1089.

145. M.L. Agarwal, H.E. Larkin, S.I. Zaidi, H. Mukhtar, N.L. Oleinick (1993). Phospholipase activation triggers apoptosis in photosensitized mouse lymphoma cells. *Cancer Res.*, **53**, 5897–5902.

146. B. Piret, S. Legrand-Poels, C. Sappey, J. Piette (1995). NF-kappa B transcription factor and human immunodeficiency virus type 1 (HIV-1) activation by methylene blue photosensitization. *Eur. J. Biochem.*, **228**, 447–455.

147. G. Kick, G. Messer, A. Goetz, G. Plewig, P. Kind (1995). Photodynamic therapy induces expression of interleukin 6 by activation of AP-1 but not NF-kappa B DNA binding. *Cancer Res.*, **55**, 2373–2379.

148. S. Legrand-Poels, S. Schoonbroodt, J.Y. Matroule, J. Piette (1998). NF-kappa B: an important transcription factor in photobiology. *J. Photochem. Photobiol. B*, **45**, 1–8.

149. S.O. Gollnick, X. Liu, B. Owczarczak, D.A. Musser, B.W. Henderson (1997). Altered expression of interleukin 6 and interleukin 10 as a result of photodynamic therapy in vivo. *Cancer Res.*, **57**, 3904–3909.

150. S. Evans, W. Matthews, R. Perry, D. Fraker, J. Norton, H.I. Pass (1990). Effect of photodynamic therapy on tumor necrosis factor production by murine macrophages. *J. Natl. Cancer Inst.*, **82**, 34–39.

151. N. Rousset, V. Vonarx, S. Elèouet, J. Carrè, E. Kerninon, Y. Layat, T. Patrice (1999). Effects of photodynamic therapy on adhesion molecules and metastasis. *J. Photochem. Photobiol. B Biol.*, **52**, 65–73.

152. J.M. Runnels, N. Chen, B. Ortel, D. Kato, T. Hasan (1999). BPD-MA-mediated photosensitization in vitro and in vivo: cellular adhesion and beta1 integrin expression in ovarian cancer cells. *Br. J. Cancer*, **80**, 946–953.

153. P. Margaron, R. Sorrenti, J.G. Levy (1997). Photodynamic therapy inhibits cell adhesion without altering integrin expression. *Biochim. Biophys. Acta*, **1359**, 200–210.

154. D.J. Blom, H.J. Schuitmaker, I. de Waard-Siebinga, T.M. Dubbelman, M.J. Jager (1997). Decreased expression of HLA class I on ocular melanoma cells following in vitro photodynamic therapy. *Cancer Lett.*, **112**, 239–243.

155. D.E. King, H. Jiang, G.O. Simkin, M.O. Obochi, J.G. Levy, D.W. Hunt (1999). Photodynamic alteration of the surface receptor expression pattern of murine splenic dendritic cells. *Scand. J. Immunol.*, **49**, 184–192.

156. G. Canti, A. Nicolin, R. Cubeddu, P. Taroni, G. Bandieramonte, G. Valentini (1998). Antitumor efficacy of the combination of photodynamic therapy and chemotherapy in murine tumors. *Cancer Lett.*, **125**, 39–44.

157. C.J. Gomer, A. Ferrario, A. Fisher, M. Luna, N. Rucker, S. Wong (1995) Molecular studies associated with photodynamic therapy mediated oxidative stress. Presented at the 15th Annual Meeting of the American Society for Laser Medicine and Surgery.

158. K. Adams, A.J. Rainbow, B.C. Wilson, G. Singh (1999). In vivo resistance to photofrin-mediated photodynamic therapy in radiation-induced fibrosarcoma cells resistant to in vitro Photofrin-mediated photodynamic therapy. *J. Photochem. Photobiol. B*, **49**, 136–141.

159. V.H. Fingar, T.J. Wieman (1990). Studies on the mechanism of photodynamic therapy induced tumor destruction. In: T.J. Dougherty (Ed.), PDT Mechanisms II, (Proc. SPIE. Vol. 1203, pp. 168–177).

160. B.W. Henderson, B. Owczarczak, J. Sweeney, T. Gessner (1992). Effects of photodynamic treatment of platelets or endothelial cells in vitro on platelet aggregation. *Photochem. Photobiol.*, **56**, 513–521.

161. U. Schmidt-Erfurth, T.J. Flotte, E.S. Gragoudas, K. Schomacker, R. Birngruber, T. Hasan (1996). Benzoporphyrin-lipoprotein-mediated photodestruction of intraocular tumors. *Exp. Eye Res.*, **62**, 1–10.

162. U. Schmidt-Erfurth, J. Miller, M. Sickenberg, A. Bunse, H. Laqua, E. Gragoudas, L. Zografos, R. Birngruber, H. van den Bergh, A. Strong, U. Manjuris, M. Fsadni, A.M. Lane, B. Piguet, N.M. Bressler (1998). Photodynamic therapy of subfoveal choroidal neovascularization: clinical and angiographic examples. *Graefes. Arch. Clin. Exp. Ophthalmol.*, **236**, 365–374.

163. U. Schmidt-Erfurth (1998). Photodynamic therapy. Minimally invasive treatment of choroidal neovascularization. *Ophthalmologe*, **95**, 725–731.

164. U. Schmidt-Erfurth, T. Hasan, K. Schomacker, T. Flotte, R. Birngruber (1995). In vivo uptake of liposomal benzoporphyrin derivative and photothrombosis in experimental corneal neovascularization. *Lasers Surg. Med.*, **17**, 178–188.

165. U. Schmidt-Erfurth, J.W. Miller, M. Sickenberg, H. Laqua, I. Barbazetto, E.S. Gragoudas, L. Zografos, B. Piguet, C.J. Pournaras, G. Donati, A.M. Lane, R. Birngruber, H. van den Berg, H.A. Strong, U. Manjuris, T. Gray, M. Fsadni, N.M. Bressler (1999). Photodynamic therapy with verteporfin for choroidal neovascularization caused by age-related macular degeneration: results of retreatments in a phase 1 and 2 study. *Arch. Ophthalmol.*, **117**, 1177–1187.

166. J.W. Miller, U. Schmidt-Erfurth, M. Sickenberg, C.J. Pournaras, H. Laqua, I. Barbazetto, L. Zografos, B. Piguet, G. Donati, A.M. Lane, R. Birngruber, H. van den Berg, A. Strong, U. Manjuris, T. Gray, M. Fsadni, N.M. Bressler, E.S. Gragoudas (1999). Photodynamic therapy with verteporfin for choroidal neovascularization caused by age-related macular degeneration: results of a single treatment in a phase 1 and 2 study. *Arch. Ophthalmol.*, **117**, 1161–1173.

167. I.C. Summerhayes, T.J. Lampidis, S.D. Bernal, J.J. Nadakavukaren, K.K. Nadakavukaren, E.L. Shepherd, L.B. Chen (1982). Unusual retention of rhodamine 123 by mitochondria in muscle and carcinoma cells. *Proc. Natl. Acad. Sci. U.S.A.*, **79**, 5292–5296.

168. L. Cincotta, J.W. Foley, T. MacEachern, E. Lampros, A.H. Cincotta (1994). Novel photodynamic effects of a benzophenothiazine on two different murine sarcomas. *Cancer Res.*, **54**, 1249–1258.

169. L. Cincotta, D. Szeto, E. Lampros, T. Hasan, A.H. Cincotta (1996). Benzophenothiazine and benzoporphyrin derivative combination phototherapy effectively eradicates large murine sarcomas. *Photochem. Photobiol.*, **63**, 229–237.

170. U.O. Nseyo, R.K. Whalen, M.R. Duncan, B. Berman, S.L. Lundahl (1990). Urinary cytokines following photodynamic therapy for bladder cancer. A preliminary report. *Urology*, **36**, 167–171.

171. C.J. Gomer, A. Ferrario, A.L. Murphree (1987). The effect of localized porphyrin photodynamic therapy on the induction of tumor metastasis. *Br. J. Cancer*, **56**, 27–32.

172. P.M. Logan, J. Newton, A. Richter, S. Yip, J.G. Levy (1990). Immunological effects of photodynamic therapy. In: T. J. Dougherty (Ed.), *PDT Mechanisms II* (Proc. SPIE. Vol. 1203, pp. 153–158).

173. M. Korbelik, G. Krosl (1994). Enhanced macrophage cytotoxicity against tumor cells treated with photodynamic therapy. *Photochem. Photobiol.*, **60**, 497–502.

174. M. Korbelik, I. Cecic (1999). Contribution of myeloid and lymphoid host cells to the curative outcome of mouse sarcoma treatment by photodynamic therapy. *Cancer Lett.*, **137**, 91–98.

175. W.J. de Vree, M.C. Essers, H.S. de Bruijn, W.M. Star, J.F. Koster, W. Sluiter (1996). Evidence for an important role of neutrophils in the efficacy of photodynamic therapy in vivo. *Cancer Res.*, **56**, 2908–2911.

176. W.J. de Vree, M.C. Essers, J.F. Koster, W. Sluiter (1997). Role of interleukin 1 and granulocyte colony-stimulating factor in photofrin-based photodynamic therapy of rat rhabdomyosarcoma tumors. *Cancer Res.*, **57**, 2555–2558.

177. M. Korbelik, G.J. Dougherty (1999). Photodynamic therapy-mediated immune response against subcutaneous mouse tumors. *Cancer Res.*, **59**, 1941–1946.

178. M. Korbelik, V.R. Naraparaju, N. Yamamoto (1997). Macrophage-directed immunotherapy as adjuvant to photodynamic therapy of cancer. *Br. J. Cancer*, **75**, 202–207.

179. D.H. Lynch, S. Haddad, V.J. King, M.J. Ott, R.C. Straight, C.J. Jolles (1989). Systemic immunosuppression induced by photodynamic therapy (PDT) is adoptively transferred by macrophages. *Photochem. Photobiol.*, **49**, 453–458.

180. G. Simkin, M. Obochi, D.W.C. Hunt, A.H. Chan, J.G. Levy (1995). Effect of photodynamic therapy using benzoporphyrin derivative on the cutaneous immune response. In: T.J. Dougherty (Ed.), *Optical Methods for Tumor Treatment and Detection: Mechanisms and Techniques in PDT IV* (Proc. SPIE. Vol. 2392, pp. 23–33).

181. M.O.K. Obochi, L.G. Ratkay, J.G. Levy (1997). Prolonged skin allograft survival after photodynamic therapy associated with modification of donor skin antigenicity. *Transplantation*, **63**, 810–817.

182. L.G. Ratkay, R.K. Chowdhary, A. Iamaroon, A.M. Richter, H.C. Neyndorff, E.C. Keystone, J.D. Waterfield, J.G. Levy (1998). Amelioration of antigen-induced arthritis in rabbits by induction of apoptosis of inflammatory cells with local application of transdermal photodynamic therapy. *Arthritis. Rheum.*, **41**, 525–534.

183. Z. Malik, H. Lugaci (1987). Destruction of erythroleukaemic cells by photoactivation of endogenous porphyrins. *Br. J. Cancer*, **56**, 589–595.

184. M.R. Moore, K.E.L. McColl, C. Rimington, A. Goldberg (1987). Erythropoietic protoporphyria. In: M.R. Moore, K.E.L. McColl, C. Rimington, A. Goldberg (Eds), *Disorders of Porphyrin Metabolism* (pp. 201–211). Plenum Medical Book Company, New York, N.Y.

185. Z. Malik, B. Ehrenberg, A. Faraggi (1989). Inactivation of erythrocytic, lymphocytic and myelocytic leukemic cells by photoexcitation of endogenous porphyrins. *J. Photochem. Photobiol. B Biol.*, **4**, 195–205.

186. Q. Peng, K. Berg, J. Moan, M. Kongshaug, J.M. Nesland (1997). 5-Aminolevulinic acid-based photodynamic therapy: principles and experimental research. [Review] [234 refs]. *Photochem. Photobiol.*, **65**, 235–251.

187. J.C. Kennedy, R.H. Pottier, D.C. Pross (1990). Photodynamic therapy with endogenous protoporphyrin IX: Basic principles and present clinical experience. *J. Photochem. Photobiol. B*, **6**, 143–148.
188. J.C. Kennedy, R.H. Pottier (1992). Endogenous protoporphyrin IX, a clinically useful photosensitizer for photodynamic therapy. *J. Photochem. Photobiol. B*, **14**, 275–292.
189. P. Wolf, E. Rieger, H. Kerl (1993). Topical photodynamic therapy with endogenous porphyrins after application of 5-aminolevulinic acid. An alternative treatment modality for solar keratoses, superficial squamous cell carcinomas, and basal cell carcinomas? *J. Am. Acad. Dermatol.*, **28**, 17–21.
190. R. Fink-Puches, H.P. Soyer, A. Hofer, H. Kerl, P. Wolf (1998). Long-term follow-up and histological changes of superficial nonmelanoma skin cancers treated with topical delta-aminolevulinic acid photodynamic therapy. *Arch. Dermatol.*, **134**, 821–826.
191. S. Fijan, H. Honigsmann, B. Ortel (1995). Photodynamic therapy of epithelial skin tumors using delta-aminolaevulinic acid and desferrioxamine. *Br. J. Dermatol.*, **133**, 282–288.
192. A.F. Hurlimann, G. Hanggi, R.G. Panizzon (1998). Photodynamic therapy of superficial basal cell carcinomas using topical 5-aminolevulinic acid in a nanocolloid lotion. *Dermatology*, **197**, 248–254.
193. K. Svanberg, T. Andersson, D. Killander, I. Wang, U. Stenram, S. Andersson-Engels, R. Berg, J. Johansson, S. Svanberg (1994). Photodynamic therapy of non-melanoma malignant tumors of the skin using topical delta-aminolevulinic acid sensitization and laser irradiation. *B. J. Dermatol.*, **130**, 743–751.
194. G. Li, M.R. Szewczuk, R.H. Pottier, J.C. Kennedy (1999). Effect of mammalian cell differentiation on response to exogenous 5-aminolevulinic acid. *Photochem. Photobiol.*, **69**, 231–235.
195. C. Fritsch, B. Homey, W. Stahl, P. Lehmann, T. Ruzicka, H. Sies (1998). Preferential relative porphyrin enrichment in solar keratoses upon topical application of delta-aminolevulinic acid methylester. *Photochem. Photobiol.*, **68**, 218–221.
196. J. Kloek, W. Akkermans, G.M. Beijersbergen van Henegouwen (1998). Derivatives of 5-aminolevulinic acid for photodynamic therapy: enzymatic conversion into protopor-phyrin. *Photochem. Photobiol.*, **67**, 150–154.
197. R. Van Hillegersberg, J.W. Van den Berg, W.J. Kort, O.T. Terpstra, J.H. Wilson (1992). Selective accumulation of endogenously produced porphyrins in a liver metastasis model in rats. *Gastroenterology*, **103**, 647–651.
198. A. Casas, A.M. Batlle, A.R. Butler, D. Robertson, E.H. Brown, A. MacRobert, P.A. Riley (1999). Comparative effect of ALA derivatives on protoporphyrin IX production in human and rat skin organ cultures. *Br. J. Cancer*, **80**, 1525–1532.
199. S.C. Chang, A.J. MacRobert, J.B. Porter, S.G. Bown (1997). The efficacy of an iron chelator (CP94) in increasing cellular protoporphyrin IX following intravesical 5-aminolevulinic acid administration: an in vivo study. *J. Photochem. Photobiol. B Biol.*, **38**, 114–122.
200. K. Rittenhouse-Diakun, H. Van Leengoed, J. Morgan, E. Hryhorenko, G. Paszkiewicz, J.E. Whitaker, A.R. Oseroff (1995). The role of transferrin receptor (CD71) in photo-dynamic therapy of activated and malignant lymphocytes using the heme precursor delta-aminolevulinic acid (ALA). *Photochem. Photobiol.*, **61**, 523–528.
201. C. Pourzand, O. Reelfs, E. Kvam, R.M. Tyrrell (1999). The iron regulatory protein can determine the effectiveness of 5-aminolevulinic acid in inducing protoporphyrin IX in human primary skin fibroblasts. *J. Invest. Dermatol.*, **112**, 419–425.

202. S.L. Gibson, M.L. Nguyen, J.J. Havens, A. Barbarin, R. Hilf (1999). Relationship of delta-aminolevulinic acid-induced protoporphyrin IX levels to mitochondrial content in neoplastic cells in vitro. *Biochem. Biophys. Res. Commun.*, **265**, 315–321.

203. S.L. Gibson, L.T. Anderson, J.J. Havens, R. Hilf (1999). Effect of estrogenic perturbations on delta-aminolevulinic acid-induced porphobilinogen deaminase and protoporphyrin IX levels in rat Harderian glands, liver, and R3230AC tumors. *Biochem. Pharmacol.*, **58**, 1821–1829.

204. S.L. Gibson, D.J. Cupriks, J.J. Havens, M.L. Nguyen, R. Hilf (1998). A regulatory role for porphobilinogen deaminase (PBGD) in delta-aminolaevulinic acid (delta-ALA)-induced photosensitization? *Br. J. Cancer*, **77**, 235–243.

205. M.R. Moore, K.E.L. McColl, C. Rimington, A. Goldberg (1987). *Disorders of Porphyrin Metabolism*. Plenum Medical Book Company, New York, N.Y.

206. V.H. Fingar, T.J. Wieman, K.S. McMahon, P.S. Haydon, B.P. Halling, D.A. Yuhas, J.W. Winkelman (1997). Photodynamic therapy using a protoporphyrinogen oxidase inhibitor. *Cancer Res.*, **57**, 4551–4556.

207. G. Li, M.R. Szewczuk, L. Raptis, J.G. Johnson, G.E. Weagle, R.H. Pottier, J.C. Kennedy (1999). Rodent fibroblast model for studies of response of malignant cells to exogenous 5-aminolevulinic acid. *Br. J. Cancer*, **80**, 676–684.

208. S. Iinuma, S.S. Farshi, B. Ortel, T. Hasan (1994). A mechanistic study of cellular photodestruction with 5-aminolaevulinic acid-induced porphyrin. *Br. J. Cancer*, **70**, 21–28.

209. M.T. Wyss-Desserich, C.H. Sun, P. Wyss, C.S. Kurlawalla, U. Haller, M.W. Berns, Y. Tadir (1996). Accumulation of 5-aminolevulinic acid-induced protoporphyrin IX in normal and neoplastic human endometrial epithelial cells. *Biochem. Biophys. Res. Commun.*, **224**, 819–824.

210. L. Wyld, J.L. Burn, M.W. Reed, N.J. Brown (1997). Factors affecting aminolaevulinic acid-induced generation of protoporphyrin IX. *Br. J. Cancer*, **76**, 705–712.

211. E.A. Hryhorenko, A.R. Oseroff, J. Morgan, K. Rittenhouse-Diakun (1999). Deletion of alloantigen-activated cells by aminolevulinic acid-based photodynamic therapy. *Photochem. Photobiol.*, **69**, 560–565.

212. Z. Malik, M. Dishi, Y. Garini (1996). Fourier transform multipixel spectroscopy and spectral imaging of protoporphyrin in single melanoma cells. *Photochem. Photobiol.*, **63**, 608–614.

213. K.P. Uberriegler, E. Banieghbal, B. Krammer (1995). Subcellular damage kinetics within co-cultivated WI38 and VA13-transformed WI38 human fibroblasts following 5-aminolevulinic acid-induced protoporphyrin IX formation. *Photochem. Photobiol.*, **62**, 1052–1057.

214. Z. Malik, G. Kostenich, L. Roitman, B. Ehrenberg, A. Orenstein (1995). Topical application of 5-aminolevulinic acid, DMSO and EDTA: protoporphyrin IX accumulation in skin and tumors of mice. *J. Photochem. Photobiol. B*, **28**, 213–218.

215. Y. Harth, B. Hirshowitz, B. Kaplan (1998). Modified topical photodynamic therapy of superficial skin tumors, utilizing aminolevulinic acid, penetration enhancers, red light, and hyperthermia. *Dermatol. Surg.*, **24**, 723–726.

216. L.E. Rhodes, M.M. Tsoukas, R.R. Anderson, N. Kollias (1997). Iontophoretic delivery of ALA provides a quantitative model for ALA pharmacokinetics and PpIX phototoxicity in human skin. *J. Invest. Dermatol.*, **108**, 87–91.

217. F. Doring, J. Walter, J. Will, M. Focking, M. Boll, S. Amasheh, W. Clauss, H. Daniel (1998). Delta-aminolevulinic acid transport by intestinal and renal peptide transporters and its physiological and clinical implications. *J. Clin. Invest.*, **101**, 2761–2767.

218. J. Kloek, H. van Beijersbergen (1996). Prodrugs of 5-aminolevulinic acid for photo-dynamic therapy. *Photochem. Photobiol.*, **64**, 994–1000.
219. R. Washbrook, P.A. Riley (1997). Comparison of delta-aminolevulinic acid and its methyl ester as an inducer of porphyrin synthesis in cultured cells. *Br. J. Cancer*, **75**, 1417–1420.
220. J.M. Gaullier, K. Berg, Q. Peng, H. Anholt, P.K. Selbo, L.W. Ma, J. Moan (1997). Use of 5-aminolevulinic acid esters to improve photodynamic therapy on cells in culture. *Cancer Res.*, **57**, 1481–1486.
221. N. Lange, P. Jichlinski, M. Zellweger, M. Forrer, A. Marti, L. Guillou, P. Kucera, G. Wagnieres, H. van den Bergh (1999). Photodetection of early human bladder cancer based on the fluorescence of 5-aminolevulinic acid hexylester-induced protoporphyrin IX: a pilot study. *Br. J. Cancer*, **80**, 185–193.

Chapter 3

Sensitizers in photodynamic therapy

Nathalie Rousset, Ludovic Bourré and Sonia Thibaud

Table of contents

Abstract

Photodynamic therapy (PDT), by exposure of tissues to a specific wavelength of laser light, induces the photoactivation of a sensitizer that is relatively retained by cancer cells and produces cytotoxic oxygen species. Regardless of the sensitizer used, the main indications concern limited in-depth cancers or precancerous lesions treated for curative purposes. However, PDT is also actually emerging as a treatment for non-oncological indications, announcing the beginning of a new PDT era. PDT with Photofrin® has already been authorized for certain applications in Japan, the USA and France, and powerful second-generation sensitizers are now being approved (Levulan®, Visudyne®) or considered for approval. Porphyrins and their analogues have been extensively reviewed and new classes of photosensitizers have appeared and proven their efficacy.

3.1 Introduction

Photodynamic therapy (PDT) has been widely investigated in recent decades and is emerging as a promising therapy for a wide variety of malignant tumors. A major objective for cancer treatment is the selective destruction of malignant cells without damage to normal tissues and functions. Although the mechanisms involved in PDT cancer cell destruction have not been thoroughly defined, recent studies have helped to elucidate the effects of PDT at molecular, cellular and tissue levels. PDT is based on photoactivation by a given wavelength of a sensitizer relatively retained by cancer cells, which produces highly reactive oxygen species [1,2] that can destroy tumor cells. The subcellular localisation of sensitizers (lysosomes, mitochondria, and/or cell membranes) depends essentially on their nature and the cell line studied [3,4]. Intracellular sites of photodamage do not necessarily correspond to the sensitizer binding site in cells [5], but relate to the PDT parameters involved, i.e. sensitizer concentration, incubation time, exposure time and laser power [6]. The first photosensitizer used was a porphyrin derivative agent named hematoporphyrin derivative (HPD) and has become Photofrin® after purification. The approval of the FDA obtained in France (April 1996) after the USA (1995) and Japan (1994) promoted the synthesis and development of second generation photosensitizers.

3.2 First generation photosensitizers

3.2.1 Haematoporphyrin derivative

Porphyrins are heteroaromatic compounds characterised by a tetrapyrrolic structure that consists of four pentagonal pyrroles linked by four methylene bridges, the porphine structure (Figure 1).

Even if this porphine structure can be oxidised or reduced, it is usually considerably stable. It is characterised by an absorption spectrum having a specific band

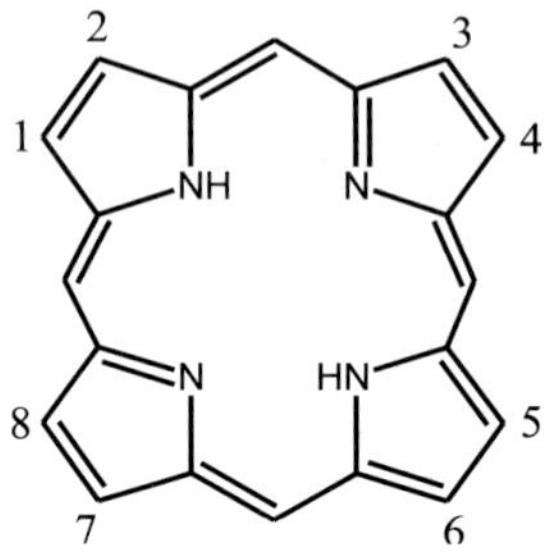

Figure 1. The porphine structure.

around 400 nm (Soret band) and usually four further absorption bands around 500–600 nm.

Porphyrins are essential as they constitute the haemoglobin, the myoglobin and other essential proteins in animal species [7].

Haematoporphyrin derivative is synthesised from porcine hemine by treating a hematoporphyrin chlorhydrate with acetic acid in presence of sulfuric acid used as a catalyst [8]. The resulting mixture is then filtered, neutralised and dried. The final step of HPD preparation, involving the product solubilisation in a diluted alkaline solution (pH adjusted at 7.4), provokes a wide variety of unexpected reactions with the formation of dimers and even oligomers of superior degrees ($n < 7$).

HPD absorbs light between 350 and 630 nm. Its basic structure, represented in Figure 2, is made up of a porphine structure substituted by:

- 4 methyl radicals (CH_3) in positions 1, 3, 5 and 8.
- 2 propanoyl radicals (CH_2CH_2COOH) in positions 6 and 7.
- 2 hydroxyethyl radicals ($CHOHCH_3$) in positions 2 and 4.

HPD is a complex compound, heterogeneous and weakly reproducible. However, its tumor tropism has clearly been demonstrated [9,10]. The first clinical trials with HPD, and for the treatment of lung cancer, were reported by Hayata et al. in 1982 [11].

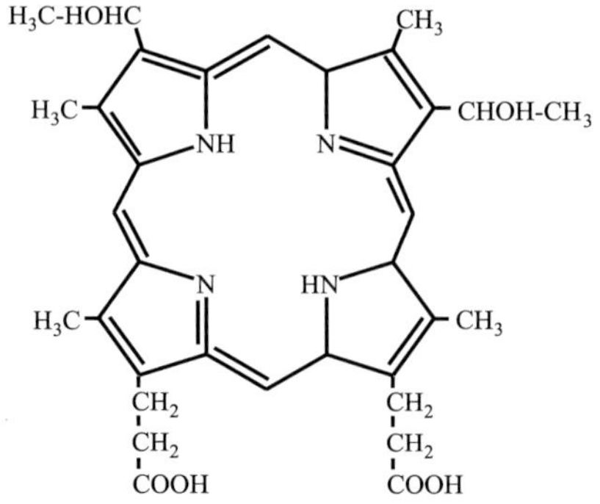

Figure 2. Haematoporphyrin Derivative (HPD).

3.2.2 Photofrin®

Photofrin® (Porfimer Sodium), a purified mixture of HPD, was developed by QLT PhotoTherapeutics which recently sold the worldwide rights of this compound to Axcan Pharma [12], including the marketing rights in the USA, which were obtained back from Sanofi-Synthelabo. While Photofrin® mainly contains the active fraction of HPD [13,14], there are several other components that have been characterised by different spectroscopic, pharmacokinetic and pharmacological properties, showing that Photofrin consists mainly of monomers (including HPD, the two isomers of hydroxyethylvinyldeuteroporphyrin, and Protoporphyrin IX), dimers of hematoporphyrin with an ether, an ester or a carbon–carbon bond (Figure 3), and some very large oligomers.

This first generation of photosensitizers present several deficiencies: the tumor selectivity is not as good as can be obtained with some of the newer sensitizers; the weak absorption in the near-infrared does not allow them to easily penetrate tissues (1 cm depth at most) [15,16]; they generally induce a strong and lasting

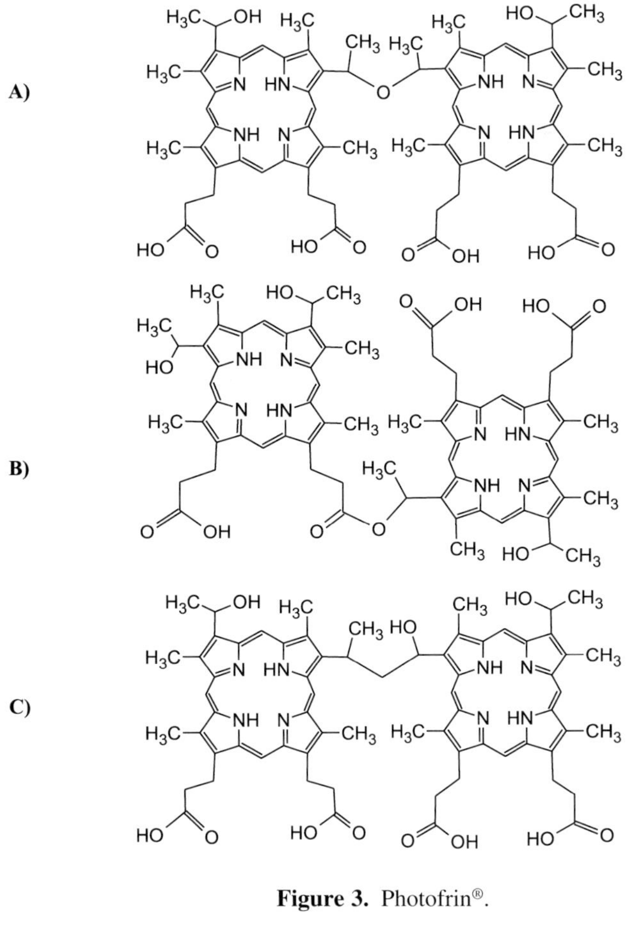

Figure 3. Photofrin®.

(several weeks) skin photosensitization. Photofrin®, however, is safe and several clinical trials have been performed with this agent. Photofrin® was the first sensitizer to reach the market and was approved for specific clinical procedures by the government regulatory groups in several countries [17]. It was approved in Holland in 1995 for, esophageal tumor, which causes dysphagia, and obstructing lesions caused by non-small cell lung cancer (NSCLC) or other cancers that spread to the lungs. It was also approved in 1995 in Japan for stomach tumor, oesophageal disease and cervical tumor [18]. Subsequent approvals followed in Canada for oesophageal and superficial bladder cancer [19,20], in Japan for early stage lung cancer, oesophageal tumor, gastric and cervical cancers, including cervical dysplasia [21], in Finland as a palliative treatment for advanced lung and oesophageal cancers [22] and in the USA for oesophageal tumor [23]. Photofrin® was also approved in Germany for early-stage endobronchial NSCLC in patients where surgery or radiotherapy is not indicated [24], in the USA as a palliative treatment for late-stage lung cancer [25], in the UK as a palliative treatment for oesophageal and late-stage lung cancers [26] and in France for lung and oesophageal cancers [27].

Clinical results indicated Photofrin® as a potential cure for some early-stage superficial lung cancers, Kaposi's sarcoma, Barrett's oesophagus (in conjunction with omeprazole) and a cost-effective first-line alternative to surgery [28–30]. Photofrin® PDT was also assessed as an effective palliation for inoperable dysphagia in which there was minimal extrinsic compression [31]. A recent pilot study also showed good results with combined PDT/hyperbaric oxygen (HBO) treatments in dealing with oesophageal and cardiac cancer [32]. Some preclinical evaluations in progress have indicated Photofrin® as a potential cure in atherosclerosis [33].

3.3 Second generation photosensitizers derived from porphine

3.3.1 1,5-aminolevulinic acid (ALA) (Also see Chapter 4)

5-Aminolevulinic acid (ALA, Levulan®) originally developed by Queen's University, Kingston, Ontario, is licensed to, and being developed by, DUSA and

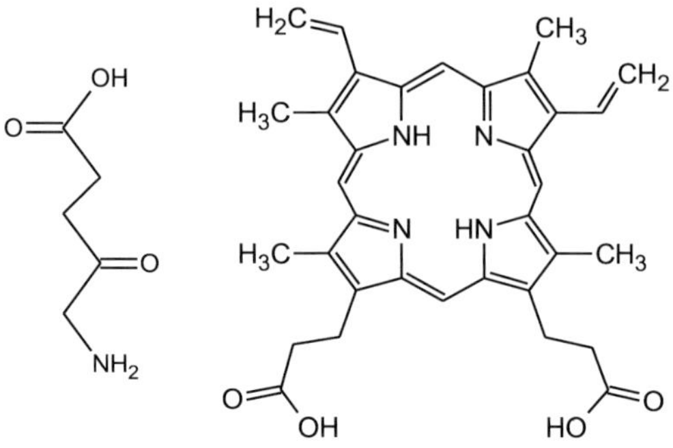

Figure 4. ALA (left) and PPIX (right).

Schering Plough. ALA is not exactly a photosensitizer but a prodrug enzymatically transformed into the protoporphyrin IX (PPIX) photosensitizer (Figure 4). ALA synthesis constitutes the first step of heme biosynthesis (Figure 5), the penultimate step being the PPIX formation, an effective sensitizer. The last step, the heme formation corresponding to iron incorporation in PPIX, takes place in mitochondria under the action of the enzyme ferrochelatase. Addition of exogenous ALA allows PPIX formation because of the limited ability of the ferrochelatase to transform PPIX into heme [34]. ALA is a hydrophilic molecule and, for this reason, it does not penetrate easily intact skin, cell membranes or biological barriers [35,36]. While PPIX has only a weak absorption at 630 nm, ALA has been used with success for skin basal carcinoma cells and GI adenocarcinoma as well as for tumor diagnosis [37,38]. ALA is particularly efficient in dermatology for the treatment and the diagnosis of neoplasic skin tissues [37]. Topical treatment is accompanied by excellent healing and cosmetic outcome.

Levulan® was approved in the USA for the treatment of actinic keratosis (February 2000). This molecule, which is not patented in Europe, will be developed by Schering AG via the Medac German company.

Moreover several ALA derivatives are actually being intensively evaluated and seem to have a promising future [36].

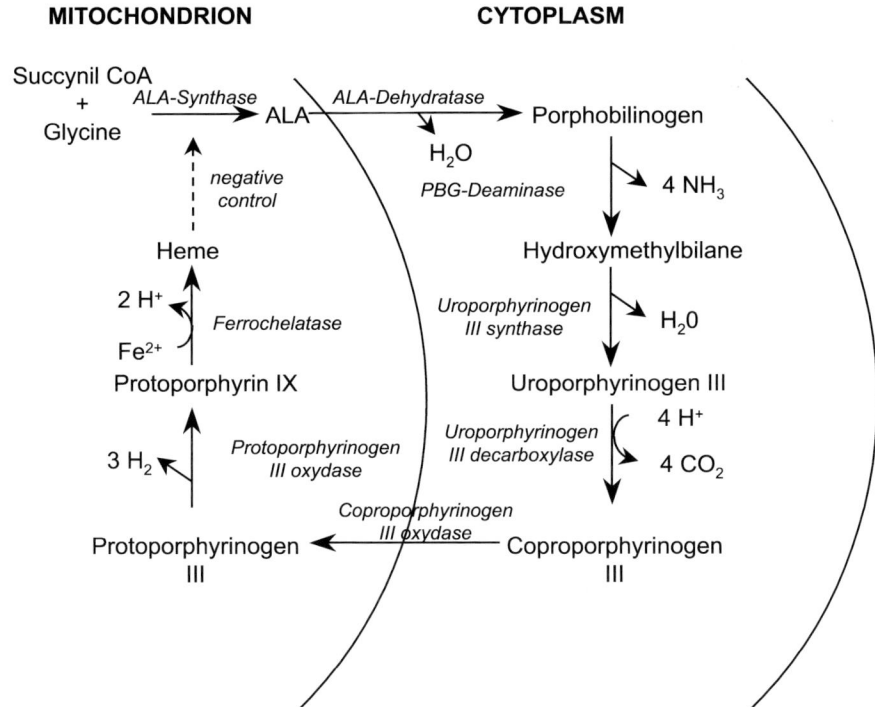

Figure 5. Heme biosynthesis.

3.3.2 Chlorins

Unlike porphyrins, chlorins strongly absorb in the red (between 640 and 700 nm) [39]. Chlorins have different origins as they can be derived from a chlorophyll by modification or through chemical synthesis.

3.3.2.1 Chlorin e6 and derivative
An example of chlorin derived from a modification of chlorophyll a is Chlorin e6 (Ce6) sometimes named phytochlorin. This hydrophilic sensitizer is mainly monomeric in a phosphate buffer solution, in which, at pH 7.4, it exhibits a main Soret peak at 402 nm and a strong peak in the red band at 654 nm. Its fluorescence is maximal at 675 nm with a life-time of 5.3 s (Figure 6) [40].

The Ce6 is localised in lisosomes where it induces damage after irradiation. The Ce6 derivative agents, including the mono-L-aspartyl chlorine e6 (Ace6), the diaspartyl Ce6, the monoseryl Ce6 and other derivative amino acids, which are more lipophilic, are better retained by the tumors and therefore induce a better tumor destruction than the Ce6 agent itself [41,42].

3.3.2.2 Purpurins
Purpurins have a strong absorption in the red, between 630 and 715 nm [39]. Purpurins are either base free or complexed with a metal (silver, nickel, tin, zinc etc.). Only the metallic purpurins with tin (Sn) or zinc (Zn) are effective. The best representative agent of this category is the purlytin SnET$_2$ (Figure 7) PhotoPoint, PNU-167524, tin ethyl etiopurpurin dichloride, developed by Miravant Medical Technologies for several ophthalmological, oncological and urological indications. This compound is in a Phase II/III clinical trial for basal cell carcinoma, cutaneous metastatic breast cancer with favourable preliminary results [43,44]. SnET$_2$ is also in a Phase I/II clinical trial for AIDS-related Kaposi's sarcoma [45], under clinical evaluation for Age-related Macular Degeneration or AMD [46] and prostate

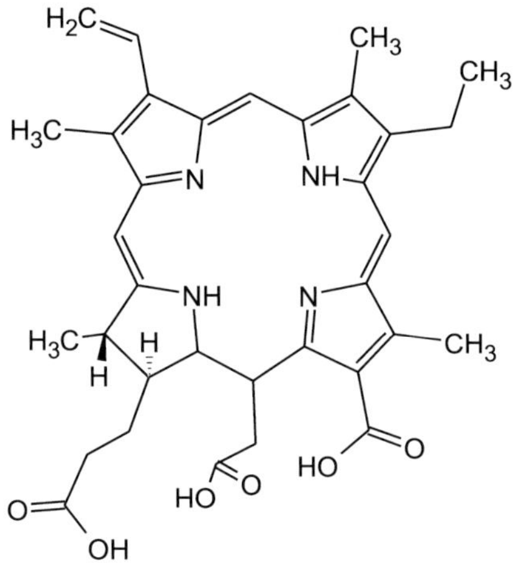

Figure 6. Chlorin e6.

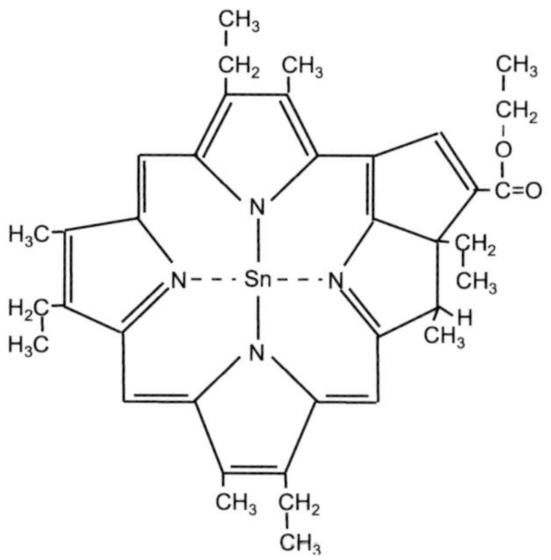

Figure 7. Purlytin SnET$_2$.

cancer [47] and has shown promising preclinical results for the prevention of restenosis after balloon angioplasty for coronary artery disease [48].

3.3.2.3 The benzoporphyrin derivative (BPD)

This chlorin (Verteporfin, Visudyne®) is synthesised from the protoporphyrin. It can exist as a monoacid or a diacid but the monoacid is considerably more active [49]. Its lipophilicity facilitates its association with the cell membrane [50]. The main advantage of BPD is its strong absorption at 690 nm, promoting the tissue penetration of the light.

The benzoporphyrin derivative monoacid ring A (BPD-MA) exists in the form of two isomers which differ by the position of their ester group but have similar effectiveness as photosensitizers (Figure 8) [51]. BPD (Visudyne®), developed by QLT PhotoTherapeutics in collaboration with Ciba Vision, was approved by the Food and Drug Administration (FDA) in the USA in November 1999 and in Switzerland in December 1999 for an ophthalmologic application, the damp form of the Age-related Macular Degeneration (AMD), and launched there rapidly afterwards. The vascular localisation of the sensitizer immediately after injection, along with a light irradiation

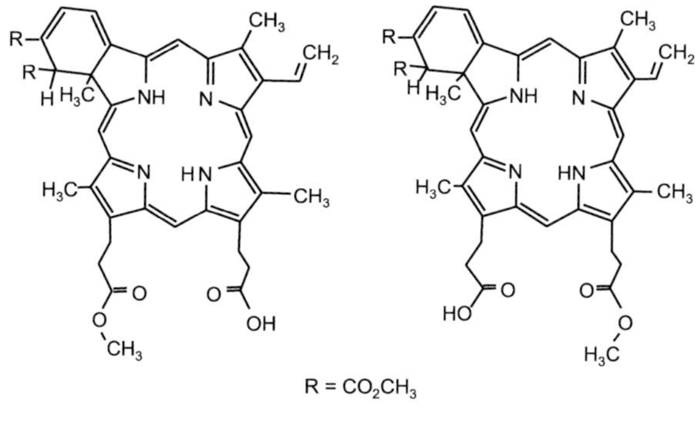

R = CO$_2$CH$_3$

Figure 8. BPD (Verteporfin or Visudyne®).

at 690 nm allows treatment of this disease [52]. Several Phase III clinical trials are ongoing in the cancer field [53]. A Phase IIIb trial, granted by the FDA, is also actually ongoing for CNV due to pathological myopia and idiopathic causes [54] and a Phase I trial showed encouraging results with psoriasis and psoriatic arthritis [55].

3.3.2.4 meta(Tetra)hydroxyphenylchlorin (m-THPC) and derivatives

m-THPC (temoporfin, Foscan, Figure 9) appears to be one of the most active photo-sensitizing agents, requiring very low drug and light doses. As this molecule is not water soluble, it needs a specific formulation to be given parenterally. This sensitizer strongly absorbs in the red band with a maximal absorption at 652 nm [56].

Foscan, was developed by Scotia which has been acquired in 2001 by Biolitec. Although its approval was refused by the FDA in 1999, Foscan was approved in the European Union, Norway and Iceland as a local therapy for the palliative treatment of patients with advanced head and neck cancer and has been successfully used in Zurich in a feasibility study on breast cancer and in London for the treatment of advanced pancreas cancer [57].

Several preclinical studies have been performed with conjugates of poly(ethylene) glycol (PEG) with m-THPC and encouraging results have been obtained concerning tumor selectivity increase compared to m-THPC PDT [58–60].

3.3.3 Phthalocyanines (Also see Chapter 5)

Phthalocyanines (Figure 10) are synthetic porphyrins where pyrrol groups are condensed with a benzenic group and where a nitrogenous bridge replaces a methine one. Because of their very high molar absorption coefficient between 675 nm and 700 nm, some phthalocyanines are very potent photosensitizers [61]. Phthalocyanines are often completed with diamagnetic metals such as aluminium or zinc which lengthen the triplet state lifetime and thus increase their phototoxicity [15]. Some preclinical evaluations with conjugated phthalocyanines showed increased efficiency in terms of tumor selectivity and phototoxicity [62,63]. A group of zinc phthalocyanines and corresponding conjugates has been patented for use in

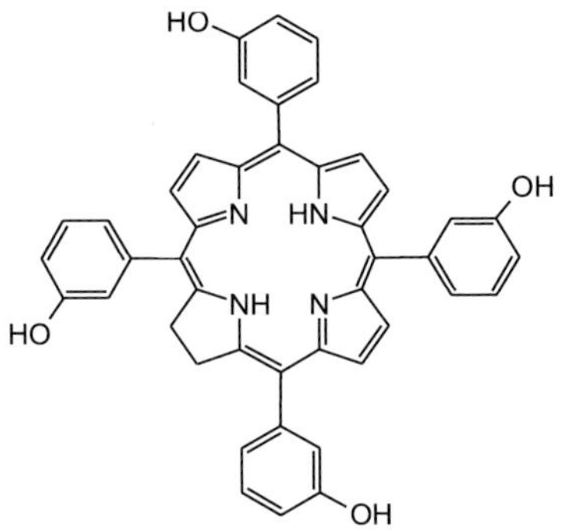

Figure 9. m-THPC™.

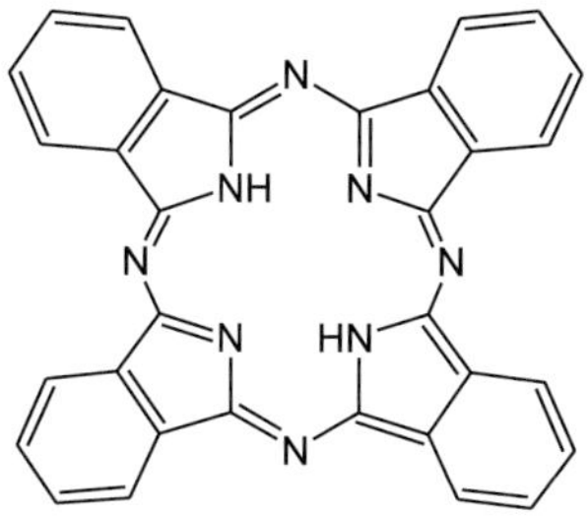

Figure 10. Phthalocyanine.

PDT in the treatment and the photodiagnosis of proliferative diseases and infections [64]. Fluorine-substituted zinc phthalocyanine has also been patented for use in PDT in the treatment of cancer and for the inactivation of viruses in blood [65]. Aluminium tetrasulfophthalocyanine (AlPcS4) conjugated to a monoclonal antibody directed against carcinoembryogenic antigen (CEA) has shown, in vivo, encouraging results concerning tumor selectivity and phototoxicity [66].

3.3.4 Texaphyrins

Texaphyrins are synthetic porphyrins, activated by deep tissue-penetrating red light, which are water soluble, tumor selective but rapidly cleared by the circulation [67,68]. The main sensitizer of the texaphyrins category is the Lutetium texaphyrin (Lu-Tex, Antrin, Lutrin, Optrin). This new photosensitizer, developed by Pharmacyclics in agreement with Alcon absorbs at 732 nm. This is a pure product with a wide absorption band centered on 732 nm (Figure 11) [69]. Phase II studies with Optrin are ongoing for the treatment of AMD and have shown good preliminary results with mild side effects and a strong compound fluorescence which could allow the optimization of the dose and the timing of irradiation in patients [70].

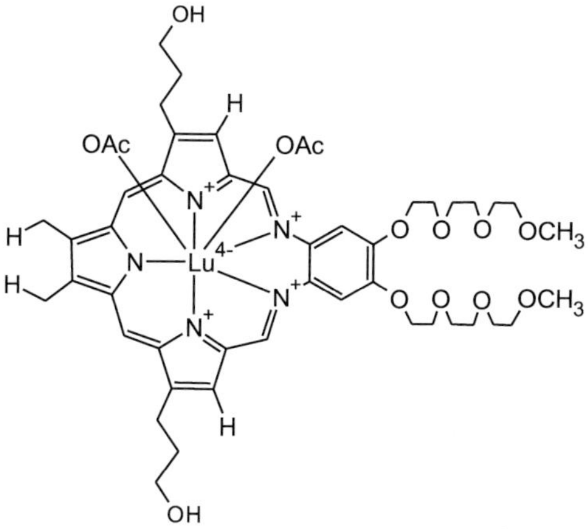

Figure 11. Lu-Tex.

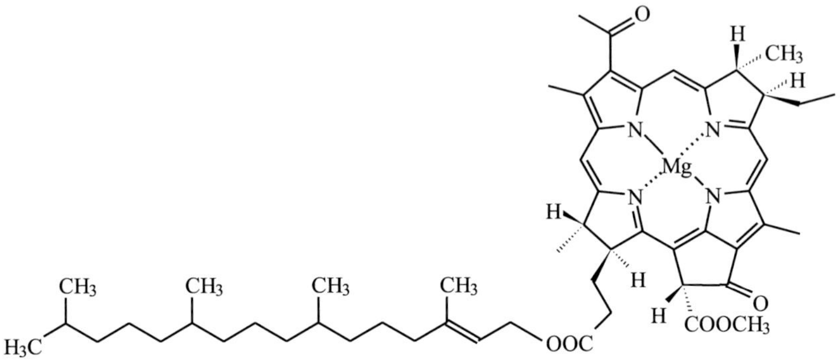

Figure 12. Bacteriochlorophyll a.

Optrin has demonstrated a reduction of atherosclerosis by eliminating inflammatory cells and is actually in Phase II studies for the primary treatment of peripheral
artery disease and for the prevention of restenosis following angioplasty [71].

Clinical results for Lutrin have been published for locally recurrent breast cancer
after unsuccessful radiotherapy or chemotherapy [72]. Lutrin has also been selected
by the National Cancer Institute for trials in primary cancers of the bladder, oesophagus, head and neck, and ovaries.

3.3.5 Bacteriochlorophyll a derivative agents

Bacteriochlorophyll a (Figure 12) is a natural pigment with an absorption band
around 780 nm. At this wavelength, the penetration depth of light is approximately
three times greater than that reached at 630 nm, the wavelength generally used in
clinical PDT with Photofrin® [73,74].

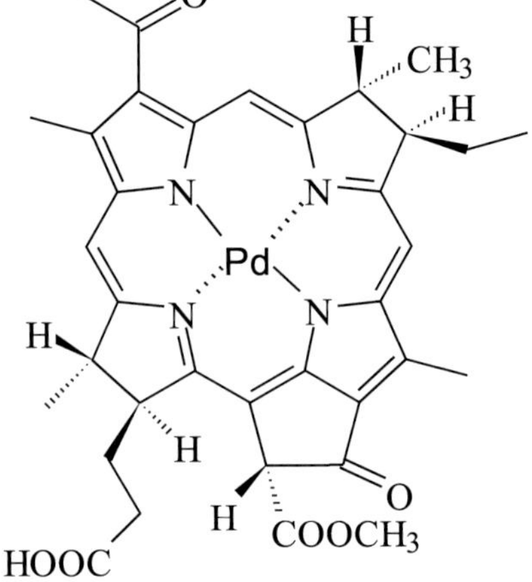

Figure 13. WST09 (Tookad®).

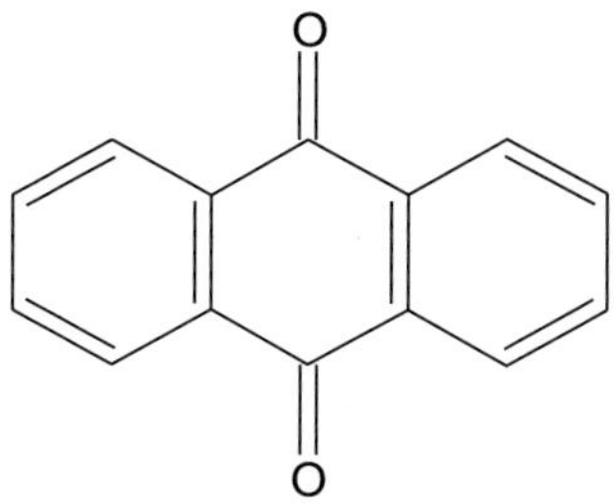

Figure 14. Anthraquinone.

Among bacteriochlorophyll a derivative agents, WST09 (Tookad®) is being developed by Negma Laboratories and synthesised from photosynthetic bacteria (Figure 13). Some preclinical in vivo experiments performed with bacteriochlorophyll a and its derivatives have already been published and have shown promising results [75–77].

3.4 Second generation photosensitizers not porphyrinic [78]

3.4.1 Quinones

3.4.1.1 The anthraquinones
The anthraquinones (Figure 14) are widely used as dyes, antibiotics, solar energy collectors and photosensitizers. They are also implicated in several biological processes such as respiration, photosynthesis and energy transport. Photobiological properties depend on several factors such as the nature, the number, the position of substituents, the intramolecular and intermolecular hydrogen bondings, the solvent, the concentration etc. The photophysical and photochemical properties of anthraquinones are influenced by their substituents, particularly the amino- and hydroxy- groups, and these can decrease the quantum yield of triplet state formation.

3.4.1.2 The perylenequinones
While mushrooms are regarded as the richest sources of perylenequinones (Figure 15) it is also possible to obtain them from some soils or insects. They generally have

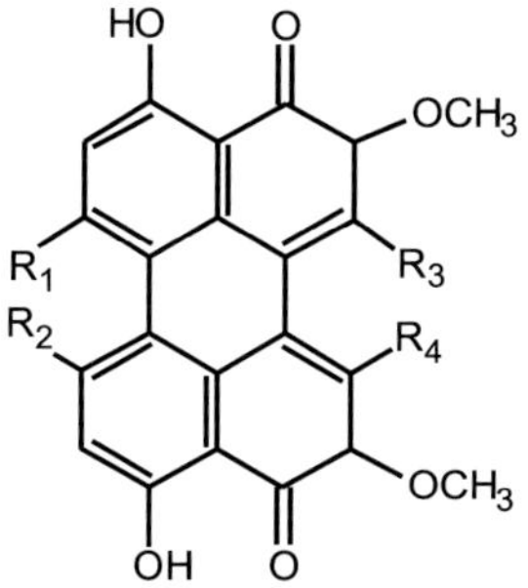

Figure 15. Perylenequinone.

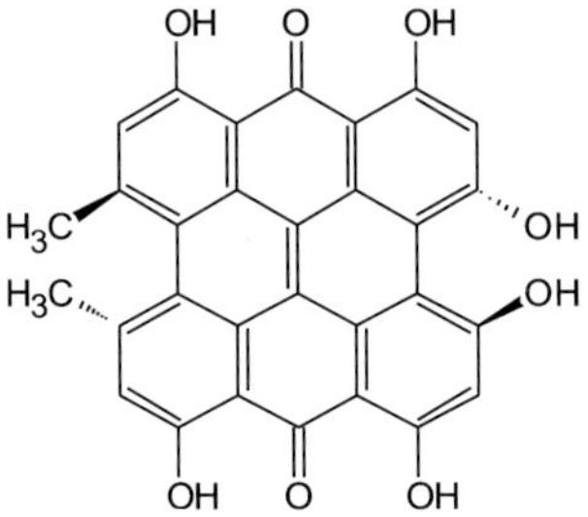

Figure 16. Hypericin.

three absorption bands in the visible and fluorescence spectrum with a main peak and a shoulder. They are effective generators of singlet oxygen with a yield comparable to that of porphyrins.

3.4.1.3 Hypericin
Hypericin (Figure 16) has been known as a photosensitizing agent for about thirty years but its anti-cancer and anti-virus use was recently studied.

3.4.2 The xanthenes

The xanthene dyes can be divided into two main groups viz. the biphenylmethane derivatives (pyronins) and the triphenylmethane derivatives (mainly the phthaleins). Among the phthaleins are fluorescein (hydroxy group) and rhodamin (amino group) and the dual compounds (hydroxy and amino). Both fluorescein and rhodamin, currently used as biological dyes, are also photosensitizers.

3.4.2.1 Fluorescein
Spectral, photophysical and photochemical properties (particularly the singlet oxygen production) of fluorescein (Figure 17) and derivatives such as Rose Bengal, eosin and erythrosin, have been widely studied. Fluoresceins and particularly Rose Bengal were widely studied for micro-organism (bacteria or yeasts) photoin-activation.

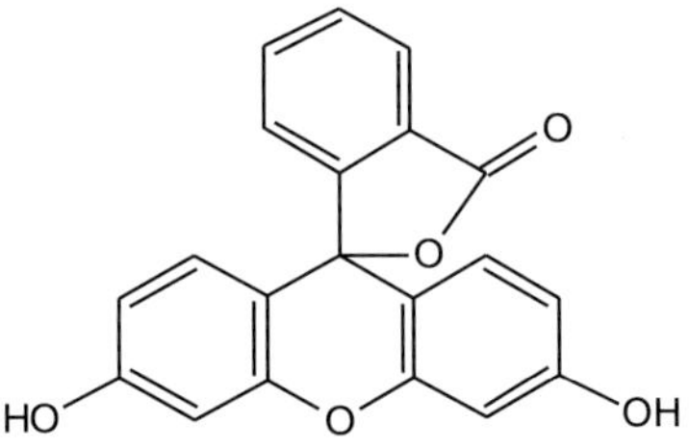

Figure 17. Fluorescein.

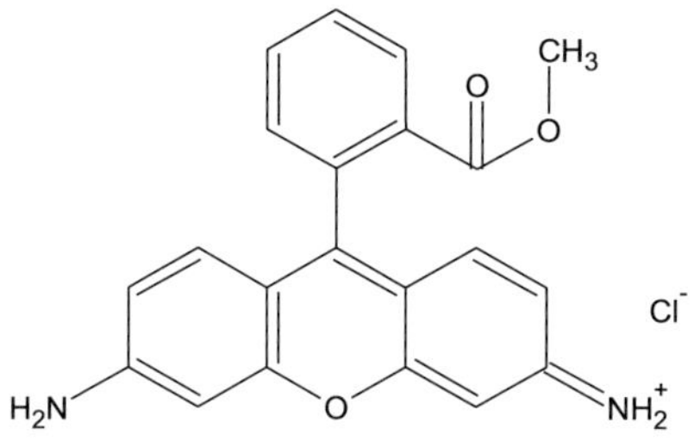

Figure 18. Rhodamin 123.

3.4.2.2 Rhodamins

Rhodamins are used as dyes for lasers and as potential photosensitizing agents. The most important is rhodamin 123 (Figure 18) which strongly absorbs at 500 nm with a shoulder at 475 nm. Its maximal fluorescence emission is localised at around 530 nm with a significant fluorescence quantum yield (equal to 0.9%) but its phototoxic effectiveness is controversial. Rhodamin 123 is a lipophilic molecule with a relocated positive charge which promotes its accumulation in mitochondria of live cells.

3.4.3 The cyanins

Originally developed for photographic emulsions, the cyanins have been considered as diagnostic and phototherapeutic agents since the middle of the 1970s. It is their high affinity for cancer cells that explains interest in their tumor PDT.

3.4.3.1 Merocyanin 540

The affinity and selective phototoxicity of merocyanin 540 (MC540) (Figure 19) for different types of normal cancer cells and viruses have been exploited for the selective purge cells of leukemia or lymphoma as part of autologus marrow graft. Its phototoxicity is relatively important compared with other agents, such as HPD, nevertheless its absorption at short wavelengths limits its applications to superficial tumors.

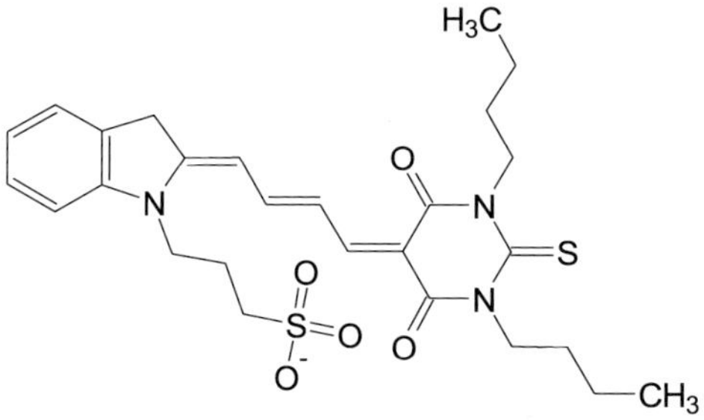

Figure 19. Merocyanin 540 (MC540).

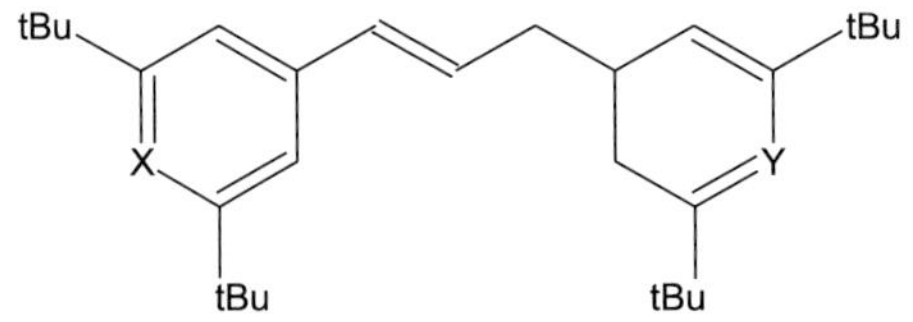

Figure 20. Chalcogenapyrylium dyes (X or Y= O, Se and Te).

3.4.3.2 Other cyanins

Unlike MC540, most of the cyanins used in PDT are compounds which easily penetrate the membrane. The kryptocyanins have a high tumor selectivity, a strong absorption in the red, a minimal toxicity in darkness, a low skin photosensitivity and a single action mode, properties which make them interesting photosensitizers.

Chalcogenapyrylium dye (Figure 20) induces mitochondrial damage both with and without light. In spite of their photochemical instability and their substantial toxicity in darkness, their selective effectiveness towards tumor cells might be an advantage in PDT.

3.4.4 Cationic dyes

Several dyes with photosensitizing properties have been reported including phenon-axine, phenothiazine (methylene blue) (Figure 21) and acridine orange. Thiazine dyes and acridine were the first agents used in antiviral PDT.

3.5 Conclusions

PDT has been actively studied for many therapeutic uses in oncology and also for many non-anticancer uses. Concerning its use in oncology, PDT is emerging as a promising therapy for a wide variety of malignant tumors [79,80–89]. Moreover, PDT is used as a post-operative adjuvant in the treatment of lung cancer or head and neck cancer [83,86]. The results of preclinical studies suggest that PDT could enhance the purge of bone marrow by eliminating malignant cells before the graft [90]. Clinical studies concerning this potential use are actually in progress. PDT is also considered in Barrett's oesophagus treatment where the tissue to be destroyed is often widely extended and multifocal [91].

Not only anticancer uses have been proposed but also, for example, recent studies have shown that PDT can reduce choroidal neovascularisation secondary to

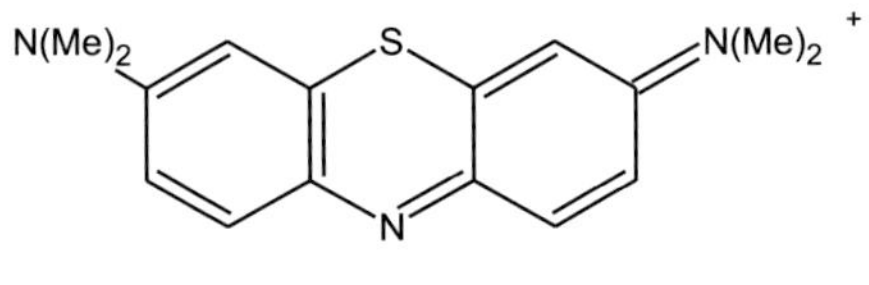

Figure 21. Methylene blue.

age-related macular degeneration [92]. BPD (Visudyne®) was approved by the Food and Drug Administration (FDA) in the USA in November 1999 and in France in September 2000 for this indication. PDT effectiveness has also been studied for the treatment of cardiovascular atherosclerotic lesions [93] or in the prevention of re-stenosis occurring after angioplasty [94] and very encouraging results have been obtained. For some conditions, PDT can modulate immunological processes by destroying a number of active immune cells. PDT could also have applications in virology; its potential use against the papilloma virus or the VIH agent has been actively studied [95,96]. Skin accessibility of light allows PDT great expediency in skin treatments. Its effectiveness against psoriasis, acne, alopecia, a number of angioma and in the area of hair removal is, on that account, being actively studied [97,98]. Actually, analysis of PDT potential indications shows a remarkable re-covering with those actual or past of ionising X-rays or implanted isotopes.

Another area of PDT application is photodiagnosis. These techniques are based on fluorescence measurement and the greatest part stems from PDT. Several studies have been performed in order to demonstrate the preferential uptake of photosensitizers in cancer tissues [96–102].

An actual PDT limitation is the tumor size that can be treated. This tumor volume is determined, for example, by the penetration depth of a laser beam through tissues (depending on the wavelength used). The evaluation of photosensitizing compounds, absorbing in the IR band, fullfils a real necessity as many authors agree that IR wavelengths penetrate more deeply into tissues [103–107] and could allow treatment of larger tumors.

PDT presents real opportunities to replace many invasive treatments of cancers but this relatively new therapy also appears well placed for use in different non-oncological indications such as AMD in ophthalmology or atherosclerosis, for example. Considering its efficacy, cost effectiveness and ease of use, it is easy to understand why we see a substantial expansion in the range of indications for this novel form of therapy.

References

1. K.R. Weishaupt, C.J. Gomer, T.J. Dougherty (1976). Identification of singlet oxygen as the cytotoxic agent in photoinactivation of murine tumors. *Cancer Res.*, **36**, 2326–2329.
2. J. Mitchell, S. McPherson, W. Degraff, J. Gamson, A. Zabell, A. Russo (1985). Oxygen dependence of hematoporphyrin derivative-induced photo-inactivation of Chinese hamster cells. *Cancer Res.*, **45**, 2008–2011.
3. N.L. Oleinick, H.H. Evans (1998). The photobiology of photodynamic therapy: cellular targets and mechanisms. *Radiation Res.*, **150**, S146–S156.
4. D. Kessel (1997). Symposium-in-print: subcellular localization of photosensitizing agents. *Photochem. Photobiol.*, **65**, 387–388.
5. Q. Peng, J. Moan, J.M. Nesland (1996). Correlation of subcellular and intratumoral photosensitizer localization with ultrastructural features after photodynamic therapy. *Ultrastruct. Pathol.*, **20**, 109–129.
6. R.W. Boyle, D. Dolphin (1996). Structure and biodistribution relationships of photodynamic sensitizers. *Photochem. Photobiol.*, **64**, 469–485.

7. D. Kessel (1984). Haematoporphyrin and HPD: photophysics, photochemistry, and phototherapy. *Photochem. Photobiol.*, **39**, 851–859.
8. C.J. Gomer, T.J. Dougherty (1979). Determination of [³H]- and [¹⁴C]-hematoporphyrin derivative distribution in malignant and normal tissue. *Cancer Res.*, **39**, 146–151.
9. R.L. Lipson, E.J. Baldes, M.J. Gray (1967). Haematoporphyrin derivative for detection and management of cancer. *Cancer*, **20**, 2255–2257.
10. H.B. Gregorie, E.O. Horger, J.L. Ward, J.F. Green, T. Richards, H.C. Robertson, T.B. Stevenson (1968). Haematoporphyrin-derivative fluorescence in malignant neoplasms. *Ann. Surg.*, **167**, 820–828.
11. Y. Hayata, H. Kato, C. Konaka, J. Ono, N. Takizawa (1982). Haematoporphyrin derivative and laser photoirradiation in the treatment of lung cancer. *Chest*, **81**, 269–277.
12. Q.L.T. sell Photofrin® to Axcan (2000). *Photodynam. News*, **3**, 9–10.
13. C.J. Byrne, L.V. Marshallsay, A.D. Ward (1990). The composition of Photofrin II. *J. Photochem. Photobiol. B*, **6**, 13–27.
14. T.J. Dougherty, W.R. Potter, K.R. Weishaupt (1987). Drugs comprising porphyrins. *US Patent* 4.649.151.
15. D.R. Doiron, L.O. Svaasand, A.E. Profio (1983). Light dosimetry in tissue: application to photoradiation therapy. In: D. Kessel, T.J. Dougherty (Eds), *Advances in Experimental Medicine and Biology* (Vol. 160, pp. 63–77). Plenum Press, New York.
16. A.E. Profio, D.R. Doiron (1981). Dosimetry considerations in phototherapy. *Med. Phys.*, **8**, 190–196.
17. J.A. Carruth (1998). Clinical applications of photodynamic therapy. *Int. J. Clin. Pract.*, **52**, 39–42.
18. J.R. Prous (1995). This year's new drugs (Overview). *Drug News Perspect.*, **8**, 24–37.
19. Ligand (1995). Ligand launches Proleukin in Canada. Press Release, 10 May.
20. QLT Phototherapeutics Inc (1995). Canadian health protection branch approves Photofrin for treatment of esophageal cancer. Press Release, 17 July.
21. QLT Phototherapeutics Inc (1995). 1st Quarterly Report, 31 March.
22. QLT Phototherapeutics Inc (1999). QLT Phototherapeutics Inc. announces Photofrin approved in Finland. Press Release, 25 February.
23. Sanofi launches Photofrin in the US (1996). *Script*, **22**, 2171.
24. QLT Phototherapeutics Inc (1997). QLT Phototherapeutics Inc. announces Photofrin approved in Germany for the treatment of early-stage lung cancer. Press Release, 07 August.
25. Photofrin approved for late-stage cancer (1999). *Bioworld Week*, **7**, 3.
26. QLT Phototherapeutics Inc (1997). QLT's Photofrin approved in UK for palliation of esophageal cancer and late-stage lung cancer. Press Release, 09 January.
27. QLT Phototherapeutics Inc (1996). QLT Photofrin approved in France for treatment of lung and esophageal cancer. Press Release, 16 April.
28. Photofrin (1996). *Clin. Trials Monitor*, **5**, 19413.
29. Z.P. Bernstein, B.D. Wilson, A.R. Oseroff, C.M. Jones, S.E. Dozier, J.S. Brooks, R. Cheney, L. Foulke, T.S. Mang, D.A. Bellnier, T.J. Dougherty (1999). Photofrin photodynamic therapy for treatment of AIDS-related cutaneous Kaposi's sarcoma. *A.I.D.S.*, **13**, 1697–1704.
30. Axcan Pharma Inc (2000). Photofrin Phase III trials – excellent results confirm efficiency in the treatment of Barrett's Esophagus and as a means of prevention of esophageal cancer. Press Release, 5 September.
31. J.D. Luketich, N.T. Nguyen, T.L. Weigel, R.J. Keenan, P.F. Ferson, C.P. Belani (1999). Photodynamic therapy for treatment of malignant dysphagia. *Surg. Laparosc. Endosc. Percutan. Tech.*, **9**, 171–175.

32. A. Maier, U. Anegg, B. Fell, P. Rehak, B. Ratzenhofer, F. Tomaselli, O. Sankin, H. Pinter, F.M. Smolle-Juttner, G.B. Friehs (2000). Hyperbaric oxygen and photodynamic therapy in the treatment of advanced carcinoma of the cardia and the esophagus. *Lasers Surg. Med.*, **26**, 308–315.

33. T. Amemiya, H. Nakajima, T. Katoh, H. Rakue, M. Miyagi, C. Ibukiyama (1999). Photodynamic therapy of atherosclerosis using YAG-OPO laser and Porfimer Sodium, and comparison with using argon-dye laser. *Jpn. Circ. J.*, **63**, 288–295.

34. A.M. del C. Batlle (1993). Porphyrins, porphyrias, cancer and photodynamic therapy – a model for carcinogenesis. *J. Photochem. Photobiol. B*, **20**, 5–22.

35. Q. Peng, T. Warloe, K. Berg, J. Moan, M. Kongshaug, K.E. Giercksky, J.M. Nesland (1997). 5-Aminolevulinic acid-based photodynamic therapy. *Cancer*, **79**, 2282–2308.

36. P. Uehlinger, M. Zellweger, G. Wagnières, L. Juillerat-Jeanneret, H. van den Bergh, N. Lange (2000). 5-Aminolevulinic acid and its derivatives: physical chemical properties and protoporphyrin IX formation in cells. *J. Photochem. Photobiol. B*, **54**, 72–80.

37. R.M. Szeimies, P. Calzavara-Pinton, S. Karrer, B. Ortel, M. Landthaler (1996). Topical photodynamic therapy in dermatology. *J. Photochem. Photobiol. B*, **36**, 213–219.

38. W. Stummer, S. Stocker, S. Wagner, H. Stepp, C. Fritsch, C. Goetz, A.E. Goetz, R. Kiefmann, D.H.J. Reulen (1998). Intraoperative detection of malignant gliomas by 5-aminolevulinic acid-induced porphyrin fluorescence. *Neurosurg.*, **42**, 518–526.

39. J.D. Spikes (1990). Chlorins as photosensitizers in biology and medicine. *J. Photochem. Photobiol. B*, **6**, 259–274.

40. B. Röder, H. Wabnitz (1987). Time-resolved fluorescence spectroscopy of hematoporphyrin, mesoporphyrin, pheophorbide a and chlorine e6 in ethanol and aqueous solution. *J. Photochem. Photobiol. B*, **1**, 103–113.

41. K.L. Molpus, M.R. Hamblin, I. Rivzi, T. Hasan (2000). Intraperitoneal photoimmunotherapy of ovarian carcinoma xenografts in nude mice using charged photoimmunoconjugates. *Gynecol. Oncol.*, **76**, 397–404.

42. M. Del Governatore, M.R. Hamblin, E.E. Piccinini, G. Ugolini, T. Hasan (2000). Targeted photodestruction of human colon cancer cells using charged 17.1A chlorin e6 immunoconjugates. *Br. J. Cancer*, **82**, 56–64.

43. PDT Inc (1997). PDT Inc. begins SnET2 trials in France and Australia; company's clinical program now spans three continents. Press Release, 15 January.

44. M. Kaplan, T. Wieman, J. Glaspy, T. Mang, R. Rifkin, T. Panella, D. Albrecht (1999). Phase II/III clinical study of tin ethyl etiopurpurin (SnET2) photodynamic therapy (PDT) in cutaneous metastatic breast cancer (CMBC). *Proc. Am. Soc. Clin. Oncol.*, **18**, 419.

45. R. Grekin, N. Razum, R. Trommer, D. Doiron, A. Synder (1996). Tin ethyl etiopurpurin (SnEt2) photodynamic therapy (PDT): results of a phase I-II clinical study conducted at UCSF for the treatment of AIDS-related cutaneous Kaposi's carcinomas. *Int. Conf. Aids*, **11**, 98.

46. Miravant Medical Technologies (2000). FDA designates Miravant's Purlytin as fast track product for macular degeneration. Press Release, 5 May.

47. Miravant Medical Technologies (1998). Miravant to begin Purlytin clinical trials for prostate cancer. Press Release, 2 September.

48. Miravant Medical Technologies (2000). Miravant present cardiovascular results at international meeting – Photopoint treatment shows promise in preclinical studies to prevent restenosis. Press Release, 5 July.

49. H.I. Pass (1993). Photodynamic therapy in oncology: mechanisms and clinical use. *J. Natl. Cancer Inst.*, **85**, 443–456.

50. D. Kessel (1989). In vitro photosensitization with a benzoporphyrin derivative. *Photochem. Photobiol.*, **49**, 579–582.

51. A.M. Richter, A.K. Jain, A.J. Canaan, E. Waterfield, E.D. Sternberg, J.G. Levy (1992). Photosensitizing efficiency of two regioisomers of the benzoporphyrin derivative monoacid ring A (BPD-MA). *Biochem. Pharmacol.*, **43**, 2349–2358.
52. L.J. Scott, K.L. Goa (2000). Verteporfin. *Drugs Aging*, **16**, 139–146.
53. IBC's 3rd Annual International Conference on Apoptosis Boston, USA (1999). *IDdb Meeting Report*, 20–21 November.
54. CIBA Vision Corp (2000). Visudyne supplemental filing granted priority review by FDA. Press Release, 29 August.
55. R. Bissonnette, D.I. McLean, G. Reid, J. Kelsall, H. Lui (1998). Photodynamic Therapy of psoriasis and psoriatic arthrisis with BPD verteporfin. *IPA Congress*, Abs RC87.
56. R. Bonnett, R.D. White, U.J. Winfield, M.C. Berenbaum (1989). Hydroporphyrins of the meso-tetra(hydroxyphenyl)porphyrin series as tumor photosensitizers. *Biochem. J.*, **261**, 277–280.
57. Biolitec pharma ltd (2002). Biolitec expects approx. EUR 100 million sales potential of approved cancer drug. Press Release, 30 September.
58. A.M. Ronn, M. Nbouri, A.L. Abramson, F. Pecci (1999). Evaluation of the third generation photosensitiser SC102 in two animal models. *Lasers Med. Sci.*, **14**, 307–318.
59. R. Hornung, M.K. Fehr, J. Monti-Frayne, B.J. Tromberg, M.W. Berns, Y. Tadir (1999). Minimally-invasive debulking of ovarian cancer in the rat pelvis by means of photodynamic therapy using the pegylated photosensitizer PEG-m-THPC. *Br. J. Cancer*, **81**, 631–637.
60. J.P. Rovers, A.E. Saarnack, M. de Jode, H.J. Sterenborg, O.T. Terpstra, M.F. Grahn (2000). Biodistribution and bioactivity of terapegylated meta-tetra(hydroxyphenyl) chlorin in a rat liver tumor model. *Photochem. Photobiol.*, **71**, 211–217.
61. I. Rosenthal (1991). Phthalocyanines as photodynamic sensitizers. *Photochem. Photobiol.*, **53**, 859–870.
62. M. Carcenac, C. Larroque, R. Langlois, J.E. van Lier, J.C. Artus, A. Pelegrin (1999). Preparation, phototoxicity and biodistribution studies of anti-carcinoembryonic antigen monoclonal antibody-phthalocyanine conjugates. *Photochem. Photobiol.*, **70**, 930–936.
63. N. Brasseur, R. Ouellet, C. La Madeleine, J.E. van Lier (1999). Water-soluble aluminium phthalocyanine-polymer conjugates for PDT: photodynamic activities and pharmacokinetics in tumor-bearing mice. *Br. J. Cancer*, **80**, 1533–1541.
64. G. Roncucci, D. Dei, M.P. Filippis, L. Fantelli, V. Masini, B. Cosimelli, G. Jori (1999). Zinc-Phthalocyanines and corresponding conjugates, their preparation and use in photodynamic therapy and as diagnostic agents. *L. Molteni C. Frantelli Alitti Soc. Esercizo SPA*, EP-00906758.
65. S.M. Gorun (2000). Novel substituted perhalogenated phthalocyanines useful as PDT agents for the treatment of cancer. *Brown Univ. Res. Found.*, WO-00021965.
66. M. Carcenac, C. Larroque, R. Langlois, J.E. van Lier, J.C. Artus, A. Pelegrin (1999). Preparation, phototoxicity and biodistribution studies of anti-carcinoembryonic antigen monoclonal antibody-phthalocyanines conjugates. *Photochem. Photobiol.*, **70**, 930–936.
67. G. Kostenich, A. Orenstein, L. Roitman, Z. Malik, B. Ehrenberg (1997). In vivo photodynamic therapy with the new near-IR absorbing water soluble photosensitizer lutetium texaphyrin and a high intensity pulsed light delivery system. *J. Photochem. Photobiol. B*, **39**, 36–42.
68. A.R. Yuen, T.J. Panella, T.K. Wieman, C. Julius, M. Panjehpour, S. Taber, V. Fingar, S.J. Horning, R.A. Miller, S.W. Young, M.F. Renischier (1997). Phase I trial of photodynamic therapy with lutetium texaphyrin (LU-TEX). *Proc. Am. Soc. Clin. Oncol.*, **16**, Abs. 768.5.

69. S.W. Young, K.W. Woodburn, M. Wright, T.D. Mody, Q. Fan, J.L. Sessler, W.C. Dow, R.A. Miller (1996). Lutetium texaphyrin (PCI-0123): a near-infrared, water-soluble photosensitizer. *Photochem. Photobiol.*, **63**, 892–897.

70. Pharmacyclics Inc (2000). Pharmacyclics receives milestone payment from Alcon to continue development of Optrin for ophthalomology. Press Release, 10 January.

71. Pharmacyclics Inc (1999). Pharmacyclics announces Antrin photoangioplasty phase II clinical trial patients treated during live case demonstration at TCT. Press Release, 27 September.

72. T.J. Wieman, T. Panella, R. Lustig, J. Liebmann, R. Carlson, L. Esserman, S. Dougherty, V. Fingar, D. Hoth, M. Renschler, D. Adelman (1999). Photodynamic therapy (PDT) of locally recurrent breast cancer (LRBC) with lutetium texaphyrin (lutrin): a phase IB/IIA trial. *Proc. Am. Soc. Clin. Oncol.*, **18**, 418.

73. I. Amato (1993). Hope for a magic bullet that moves at the speed of light. *Science*, **262**, 32–33.

74. D.R. Doiron (1984). Photophysics and instrumentation for porphyrin detection and activation. In: D.R. Doiron, C.J. Gomer (Eds), *Porphyrin Localization and Treatment of Tumors* (pp. 41–73). Liss, New York.

75. J.J. Schuitmaker, J.A. van Best, J.L. van Delft, T.M. Dubbelman, J.A. Oosterhuis, D. de Wolff-Rouendaal (1990). Bacteriochlorin a, a new photosensitizer in photodynamic therapy. In vivo results. *Invest. Ophthalmol. Vis. Sci.*, **31**, 1444–1450.

76. B.W. Henderson, A.B. Sumlin, B.L. Owczarcrak, T.J. Dougherty (1991). Bacteriochlorophyll-a as photosensitizer for photodynamic therapy of transplantable murine tumors. *J. Photochem. Photobiol. B*, **10**, 303–313.

77. J. Zilberstein, V. Rosenbalch-Belkin, M. Neeman, P. Bendel, F. Kohen, B. Gayer, A. Scherz, Y. Salomon (1998). Bacteriochlorophyll based PDT of solid tumors relies on vascular destruction. *Nantes IPA Congress*, 7–9 July.

78. Z. Diwu, J.W. Lown (1994). Phototherapeutic potential of alternative photosensitizers to porphyrins. *Pharmacol. Ther.*, **63**, 1–35.

79. R.M. Szeimies, P. Calzavara-Pinton, S. Karrer, B. Ortel, M. Landthaler (1996). Topical photodynamic therapy in dermatology. *J. Photochem. Photobiol. B*, **36**, 213–219.

80. J.C. Kennedy, S.L. Marcus, R.H. Pottier (1996). Photodynamic therapy (PDT) and photo-diagnosis (PD) using endogenous photosensitization induced by 5-aminolevulinic acid (ALA): mechanisms and clinical results. *J. Clin. Laser Med. Surg.*, **14**, 289–304.

81. H. Koren, G. Alth (1996). Photodynamic therapy in gynaecologic cancer. *J. Photochem. Photobiol.*, **36**, 189–191.

82. L. Corti, R. Mazzarotto, S. Belfontali, C. De Lucas, C. Baiocchi, C. Boso, F. Calzavara (1996). Photodynamic therapy in gynaecological neoplastic diseases. *J. Photochem. Photobiol. B*, **36**, 193–197.

83. M.A. Biel (1996). Photodynamic therapy as an adjuvant intraoperative treatment of recurrent head and neck carcinomas. *Arch. Otolaryngol. Head Neck Surg.*, **122**, 1261–1265.

84. T.G. Sutedja (1996). Posthumus Photodynamic therapy in lung cancer, a review. *J. Photochem. Photobiol. B*, **36**, 199–204.

85. D.A. Cortese, E.S. Edell, J.H. Kinsey (1997). Photodynamic therapy for early stage squamous cell carcinoma of the lung. *Mayo Clin. Proc.*, **72**, 595–602.

86. P. Baas, L. Murrer, F.A. Zoetmulder, F.A. Stewart, H.B. Ris, N. Van Zandwijk, J.L. Peterse, E.J. Rutgers (1997). Photodynamic therapy as adjuvant therapy in surgically treated pleural malignancies. *Br. J. Cancer*, **76**, 819–826.

87. R. Rifkin, B. Reed, F. Hetzel, K. Chen (1997). Photodynamic therapy using SnET2 for basal cell nevus syndrome: a case report. *Clin. Ther.*, **19**, 639–641.

88. G. Koderhold, R. Jindra, H. Koren, G. Alth, G. Schenk (1996). Experience of photodynamic therapy in dermatology. *J. Photochem. Photobiol. B*, **36**, 221–223.

89. H. Kostron, A. Obwegeser, R. Jacober (1996). Photodynamic therapy in neurosurgery. A review. *J. Photochem. Photobiol. B*, **36**, 157–168.
90. F. Sieber, G.J. Krueger (1989). Photodynamic therapy and bone marrow transplantation. *Semin. Hematol.*, **26**, 35–39.
91. B.F. Overholt, M. Panjehpour (2000). Photodynamic therapy in the management of Barrett's esophagus with dysplasia. *J. Gastrointest. Surg.*, **4**, 129–130.
92. H. Miller, B. Miller (1993). Photodynamic therapy of subretinal neovascularization in the monkey eye. *Arch. Ophthalmol.*, **111**, 855–860.
93. G.M. Lamuraglia, M.L. Klyachkin, F. Adili, W.M. Abbot (1995). Photodynamic therapy of vein grafts: suppression of intimal hyperplasia of the vein graft but not the anastomosis. *J. Vasc. Surg.*, **21**, 882–890.
94. P. Ortu, G.M. Lamuraglia, W.G. Roberts, T.J. Flotte, T. Hasan (1992). Photodynamic therapy of arteries. A novel approach for treatment of experimental intimal hyperplasia. *Circulation*, **85**, 1189–1196.
95. J. Feyh, R. Gutmann, A. Leunig (1993). Photodynamic laser therapy in the field of otolaryngology. *Laryngorhinologie*, **72**, 273–278.
96. Z. Smetana, Z. Malik, A. Orenstein, E. Mendelson, E. Ben-Hur (1997). Treatment of viral infections with 5-aminolevulinic acid and light. *Lasers Surg. Med.*, **21**, 351–358.
97. G. Monfrecola, F. D'Anna, M. Delfino (1987) Topical hematoporphyrin plus UVA for treatment of alopecia areata. *Photodermatology*, **4**, 305–306.
98. R.J. Ort, R.R. Anderson (1999). Optical hair removal. *Semin. Cutan. Med. Surg.*, **18**, 149–158.
99. S. Andersson-Engels, J. Ankerst, J. Johansson, K. Svanberg, S. Svanberg (1993). Laser-induced fluorescence in malignant and normal tissue of rats injected with benzo-porphyrin derivative. *Photochem. Photobiol.*, **57**, 978–983.
100. D. Braichotte, J.F. Savary, T. Glanzmann, P. Westermann, S. Folli, G. Wagnieres, P. Monnier, H. Van Den Bergh (1995). Clinical pharmacokinetic studies of tetra (meta-hydroxyphenyl) chlorin in squamous cell carcinoma by fluorescence spectroscopy at 2 wavelengths. *Int. J. Cancer*, **63**, 198–204.
101. D.R. Braichotte, G.A. Wagnieres, R. Bays, P. Monnier, H.E. Van Den Bergh (1995). Clinical pharmacokinetic studies of Photofrin by fluorescence spectroscopy in the oral cavity, the esophagus, and the bronchi. *Cancer*, **75**, 2768–2778.
102. T. Filbeck, W. Roessler, R. Knuechel, M. Straub, H.J. Kiel, W.F. Wieland (1999). 5-Aminolevulinic acid-induced fluorescence endoscopy applied at secondary transurethral resection after conventional resection of primary superficial bladder tumors. *Urology*, **53**, 77–81.
103. B.W. Henderson, A.B. Sumlin, B.L. Owczarczak, T.J. Dougherty (1991). Bacteriochlorophyll-a as photosensitizer for photodynamic treatment of transplantable murine tumors. *J. Photochem. Photobiol. B*, **10**, 303–313.
104. R.K. Pandey, F.Y. Shiau, A.B. Sumlin, T.J. Dougherty, K.M. Smith (1992). Structure/activity relationships among photosensitizers related to pheophorbides and bacteriopheophorbides. *Bioorg. Med. Chem. Lett.*, **5**, 491–496.
105. D. Kessel, K. Smith (1989). Photosensibilisation with derivative of chlorophyll. *Photochem. Photobiol.*, **49**, 157–160.
106. E.M. Beems, T. Dubbelman, J. Lugtenburg, J.A. Van Best, M.F. Smeets, J.P. Boegheim (1987). Photosensitizing properties of bacteriochlorophyllin-a and bacteriochlorin-a, two derivative of bacteriochlorophyll-a. *Photochem. Photobiol.*, **46**, 639–643.
107. J.K. Hoober, T.W. Sery, N. Yamamoto (1988). Photodynamic sensitizers from chloro-phyll: purpurin-18 and chlorin-p6. *J. Photochem. Photobiol. B*, **48**, 579–582.

Chapter 4

Photodynamic therapy using 5-aminolevulinic acid-induced protoporphyrin IX

E.F. Gudgin Dickson, J.C. Kennedy and R.H. Pottier

Table of contents

4.1 Introduction

The use of a harmless drug and ordinary visible light to treat disease may sound like wishful thinking, but the last two decades have shown remarkable progress in the development of such a therapy. However, the concept of a light-activated drug is far from new. Egyptian, Chinese, and Indian physicians have been using photo-sensitizing herbal preparations in the treatment of disorders such as vitiligo, rickets, psoriasis, skin cancer, and psychosis for over 3000 years [1,2], and clinical studies involving the use of chemically-defined photosensitizers for the treatment of cancer and other lesions of the skin were published at the beginning of last century [3–6]. It was found subsequently that photosensitizers could be used to induce phototoxic damage in many different types of biological systems namely plants, animals, cells, viruses, and biomolecules such as enzymes, toxins and proteins [7–13].

Photodynamic therapy (PDT) is a form of treatment in which a normally inactive drug is activated by exposing it in situ to an otherwise harmless dose of light. A molecule of drug absorbs a photon of energy and then quickly transfers some of that energy to a molecule of oxygen, thus producing a very reactive form of oxygen known as singlet oxygen, $^1O_2$. Chemical combination of the $^1O_2$ with various struc-tures in the target cells or tissue triggers a series of chemical and physiological reactions which damage the target [14–25]. The damage may be direct, leading to either necrosis or apoptosis of the target cells, or may be secondary resulting from damage to blood vessels serving the target tissue or even from diffusing photo-induced toxic molecules. If the singlet oxygen molecules encounter enough essen-tial components of the target cells, and react with them, the oxidized components become non-functional and the cells die.

The specificity of the phototoxic effect in any given case may depend upon multiple factors. Certain cells and tissues may preferentially accumulate the photo-sensitizing drug, and, all other factors being equal, the higher the concentration of photosensitizer (within limits), the greater will be the phototoxic damage. Also, different types of cells and tissues may show substantial differences in their intrinsic susceptibility to phototoxic damage, a difference which may result from variations in the efficiency of cellular repair of phototoxic damage, in the concentration of non-essential materials which may react with singlet oxygen before it encounters more essential cellular components, and/or in the sensitivity of the trigger to apoptosis of cells that experience phototoxic damage. Moreover, the inflammation that can be induced by phototoxic damage may trigger a rather specific response of the host defenses against abnormal cells or foreign organisms. Finally, the wavelength of light used to activate the photosensitizing drug, the intensity of that light, the duration and periodicity of illumination, and the total dose of light may also influ-ence cell and tissue specificity.

In order to achieve a phototoxic reaction in living tissue it is necessary for light to penetrate that tissue. Since most human tissues transmit visible light (especially red light) much more readily than ultraviolet radiation, biological photosensitizers tend to have extended conjugated systems that give rise to absorption bands in the visible part of the electromagnetic spectrum. This means that most photosensitizing drugs are relatively large coloured molecules. Figure 1 illustrates the structure of one such photosensitizer, protoporphyrin IX (PpIX).

It has been known since the beginning of the last century that certain metal-free porphyrins are both strongly fluorescent and good photosensitizers [26,27], and, in addition to that, some porphyrins accumulate preferentially in tumors [28]. However, it is only relatively recent that this information has been exploited for the detection and treatment of cancer. In 1967 it was reported that hematoporphyrin derivative (a complex mixture of porphyrin monomers and polymers) could be used to detect and treat carcinoma of the cervix [29]. Serious interest by other researchers began to develop about 10 years later, thanks mainly to the leadership of T.J. Dougherty, and clinical research in the field suddenly expanded as others began to realize its potential [30]. During the past decade several thousand patients have received PDT on an experimental basis, with promising results for a wide variety of tumors. However, basic research has not been neglected. Studies have been made of the molecular mechanisms involved [31–33], and many new photosensitizing compounds have been screened as potential photochemotherapeutic agents [34–37].

The porphyrin-based photosensitizer that is furthest along in clinical development is porfimer sodium (Photofrin®), a standardized preparation of hematoporphyrin derivative which is administered intravenously to patients 24–48 h prior to exposure to 630 nm light [38–40]. It has turned out to be a very useful photosensitizer in certain clinical situations [41], but it has the disadvantage that it is cleared from the skin relatively slowly and therefore may cause prolonged cutaneous photosensitivity. In addition, its relatively slow rate of photodegradation, which may be an advantage when deep tumors are exposed to photoactivating light via implanted optical fibres, may necessitate very complex light dosimetry calculations and beam geometry when more superficial lesions are treated with light delivery from the outside.

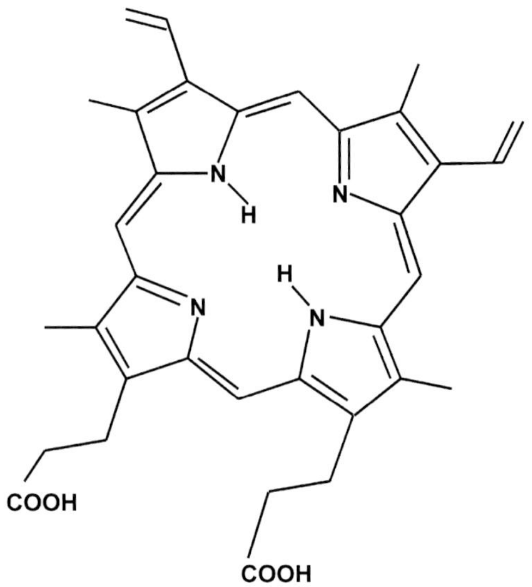

Figure 1. Protoporphyrin IX (PpIX), a biosynthesized porphyrin.

Another porphyrin-based photosensitizer, benzoporphyrin derivative monoacid A (Verteporphin, Visudyne®), has received approval for treatment of age-related macular degeneration, and a form of pathologic myopia.

There is now general agreement on the characteristics of an ideal photodynamic therapy agent. Such a drug must show relatively low toxicity in the dark toward normal cells, tissues, and organs, being either harmless or at worst producing only mild and reversible effects. It must accumulate selectively within the target cells or tissues, and reach potentially phototoxic concentrations within a few hours of administration. Its concentration in normal tissues should be so low in the days following treatment that there will be little danger of clinically significant damage to normal tissues from either room lighting or solar radiation. Of course the drug must be an efficient tissue photosensitizer, and preferably should be able to absorb energy from electromagnetic radiation in the red or near-IR spectral region where tissue transmittance is maximal. Finally, clinical use is facilitated if the drug is soluble in aqueous solutions.

A large number of modified porphyrins, chlorins, bacteriochlorins, purpurins and phthalocyanines have been studied with the aim of producing photosensitizers with reduced retention by skin and enhanced light absorption in the red or near-IR spectral region. Noteworthy examples of such second generation photosensitizers that produce less prolonged cutaneous photosensitivity are benzoporphyrin derivative monoacid ring A (BPD-MA), mono-2-aspartyl chlorin e_6 (NPe6), chloro-aluminium sulfonated phthalocyanine (CASPc) and tin ethyl etiopurpurin (SnET2) [42]. However, a porphyrin-based photosensitizer that was long overlooked is one that is normally synthesized by almost every cell, PpIX (see Figure 1). Originally there appeared to be good reasons for this neglect. It was difficult to give large enough doses of PpIX by intravenous injection because its solubility in most aqueous solutions at physiological pH is very limited, and in vitro data indicated that it was likely to persist in the body for a relatively long time and lead to persistent photosensitization. Consequently, PpIX was considered initially to have little promise as a tissue photosensitizer. However, PpIX differs from all other porphyrins, whether synthesized or biosynthesized, in that it is an intermediate in the biosynthetic pathway for heme. This fact turned out to be the key to its value for PDT.

4.2 ALA-PDT: the basics (exploiting nature's laboratories)

The basic concept upon which ALA-PDT is based was conceived as we studied the biochemical basis for the group of metabolic diseases known as the porphyrias [43]. Some of the porphyrias are associated with a generalized photosensitization which is caused by the accumulation of specific types of porphyrin in the blood and/or tissues, where each type of porphyrin accumulates as the result of a specific abnormality in the biosynthetic pathway for heme. Although at that time it was generally believed that the expression of such abnormalities was restricted to the liver and the hemopoietic system (those tissues which synthesize large amounts of heme), it was obvious that all nucleated cells must have at least some capacity to synthesize heme because they all use heme-containing enzymes for the tricarboxylic acid cycle.

We wondered therefore whether, under appropriate conditions, it might be possible to induce a transient state of "porphyria" in cells and tissues other than liver and bone marrow. Experiments which were carried out in mice and subsequently on human volunteers confirmed that it was in fact possible.

ALA-PDT is thus based upon a deliberate perturbation by the physician of the biosynthetic pathway for heme. Figure 2 shows the biochemical pathways of both the normal processes involved in heme biosynthesis, where in the cell they occur, and various abnormalities in those processes which may lead to an increase in the concentration of specific porphyrins. One of the normal intermediates in that pathway is PpIX, a metal-free porphyrin that is strongly fluorescent as well as strongly photosensitizing. Under certain abnormal conditions, uroporphyrin I, uroporphyrin III, coproporphyrin I, and/or coproporphyrin III can be generated by the heme biosynthetic pathway, but only as useless byproducts rather than intermediates. Heme biosynthesis usually is regulated so closely that the rate at which heme is being produced within a cell matches the rate at which it is being utilized, destroyed, or excreted by the same cell. Consequently, under normal conditions the intracellular concentrations of free heme and all of its precursors (including PpIX) are very low. However, if the level of activity of a specific enzyme in the pathway is sufficiently abnormal, (as occurs with each of the porphyrias), the concentration of one or more of the porphyrins listed above may become elevated and cause clinically significant photosensitization [44].

The heme biosynthetic pathway is an essential part of the aerobic energy-producing system, and much of the synthesis of heme takes place inside the mitochondria. The portion of the pathway that is involved in ALA-PDT is indicated in Figure 2 by heavy red arrows. These arrows link the steps which convert eight molecules of 5-aminolevulinic acid (ALA), a specialized amino acid whose only known function in the body is to be a precursor of heme, into PpIX, the immediate precursor of heme [46–56]. Each step in the pathway requires the activity of one or more enzymes (these are indicated by italics in Figure 2). The direct pathway from ALA to heme produces the intermediates porphobilinogen, uroporphyrinogen III, coproporphyrinogen III, protoporphyrinogen IX, and protoporphyrin IX [57–66]. Uroporphyrins and coproporphyrins which may be produced under some conditions are merely by-products which have no known function in the body.

Each type of porphyria (indicated in Figure 2 by a name on a white background) is associated with abnormal levels of activity of a specific enzyme in the heme biosynthetic pathway [67–91]. For example, a reduction in ferrochelatase activity is associated with protoporphyria, a condition in which photosensitizing concentrations of PpIX may accumulate because it is being produced faster than it can be converted into heme [92].

In principle, it should be possible to induce a transient uroporphyria, coproporphyria, or protoporphyria in normally non-porphyric cells and tissues by administering a drug which reversibly suppresses the synthesis and/or the activity of the appropriate enzyme(s) in the biosynthetic pathway for heme. However, this approach seems in general unwise, since it would suppress heme production in every cell in the body, and heme-containing enzymes are essential for aerobic energy production. An alternate approach is to administer large doses of ALA, the first

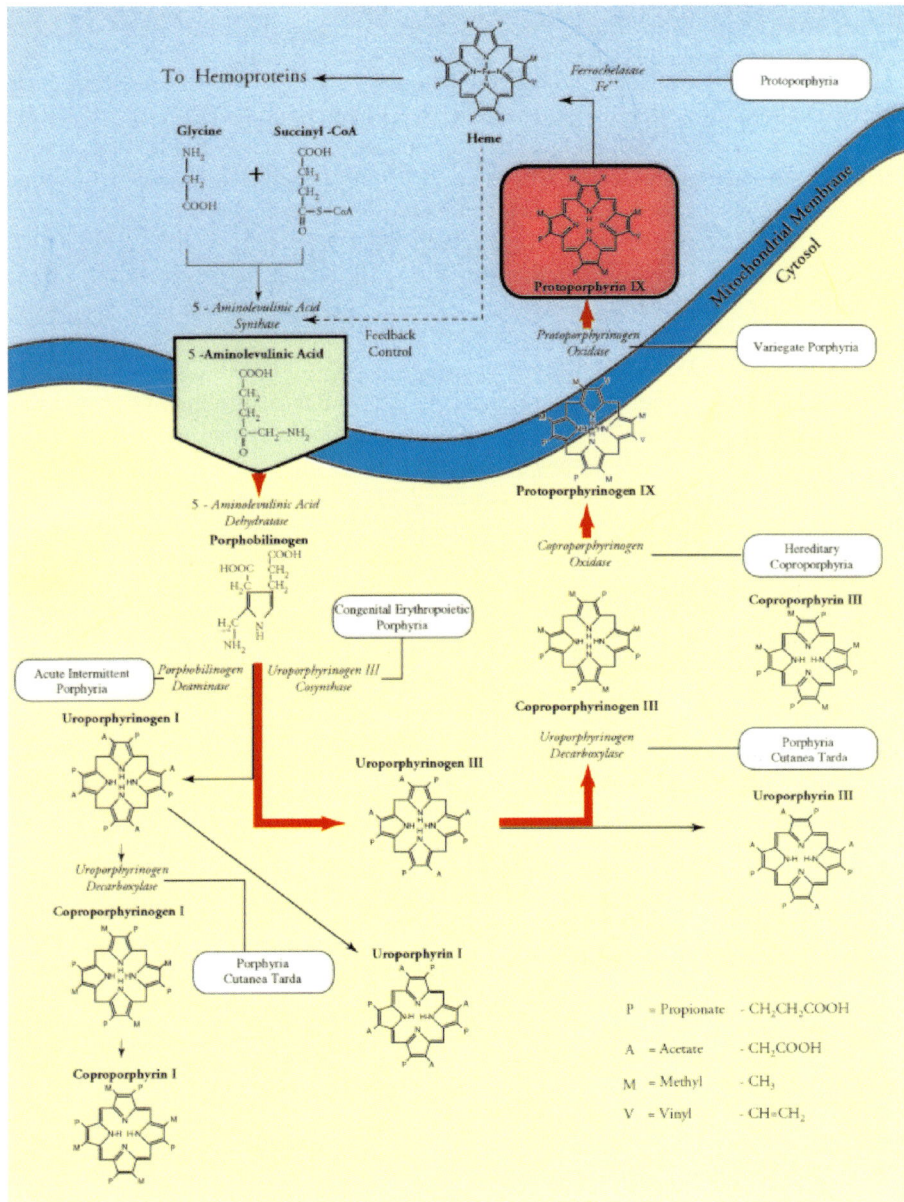

Figure 2. Heme biosynthetic pathway. Reproduced with permission from Ref. [45].

committed precursor of heme. Under these conditions, PpIX accumulates because it is being synthesized faster than it can be converted into heme, metabolized, or otherwise lost from the cell. The rate of synthesis of ALA normally is the rate-limiting step in the synthesis of heme [57,59]. Thus if a relatively high concentration of exogenous ALA is present it saturates the biosynthetic capacity of the pathway.

Under such conditions, ferrochelatase becomes the rate-limiting step in the pathway [93,94], and PpIX accumulates because the rate at which PpIX is being synthesized is greater than the rate at which it can be converted into heme. In effect, we have used exogenous ALA to induce a transient case of protoporphyria.

The duration of this ALA-induced protoprophyria is relatively brief. For example, the clearance time of PpIX that was injected intraperitoneally into mice was found to be 8 h (expressed as three times the $1/e$ lifetime), which is considerably shorter than that for either hematoporphyrin IX (HP) or hematoporphyrin derivative (HPD) [95]. ALA-induced overproduction of PpIX, likewise, is of relatively short duration in humans [96–98], limiting the likelihood of cutaneous photosensitivity post-treatment.

ALA-induced PpIX would be of little value in PDT if it failed to show a clinically useful degree of specificity for target tissues. Experiments in which mice and rats were given exogenous ALA by intraperitoneal injection revealed great variation in the extent to which normal tissues accumulated PpIX [43]. Relatively high concentrations of ALA-induced PpIX accumulated in most tissues of ectodermal and endodermal origin (primarily tissues that line body surfaces and their associated exocrine glands), but not in most tissues of mesodermal origin (primarily connective tissues). There were exceptions. Skin, mucosa of the respiratory, digestive, and urogenital tracts, and both major and minor exocrine glands accumulated significant amounts of PpIX, but most cells of the central nervous system did not although they are considered to be of ectodermal origin. Mesoderm-derived tissues that showed little increase in ALA-induced PpIX included muscle (striated, smooth, and cardiac), dermis, blood vessels, and nucleated blood cells, although thymocytes (which are considered to be of mesodermal origin) became strongly fluorescent. Subsequent clinical experience confirmed many of these findings, although there appears to be some degree of variation from species to species (men do not necessarily behave like mice).

The major mechanism responsible for tissue-to-tissue differences in the accumulation of ALA-induced PpIX almost certainly involves differences in the enzyme profile of the heme biosynthetic pathway. For example, a large enough alteration in the maximum biosynthetic capacity of any of the steps upstream from PpIX would alter the capacity of the pathway as a whole to synthesize PpIX, while a significant alteration downstream from PpIX would lead to an alteration in the rate of conversion of PpIX into heme.

Although differences in the accumulation of ALA-induced PpIX by different types of normal cells and tissues are important, the differences between abnormal cells and adjacent or overlying normal cells usually are of more importance clinically. Many different types of malignant cells and tissues have been studied: basal cell carcinomas, squamous cell carcinomas (of skin, oral mucosa, lung, and bladder), adenocarcinomas (of sebaceous gland, sweat gland, parotid gland, bowel, and breast), hepatocellular carcinoma, melanoma, rhabdomyosarcoma, and various types of leukemias and lymphomas (including mycosis fungoides and erythroleukemia). In all of these cancers there was more ALA-induced PpIX in the malignant tissues than in the corresponding normal tissues from which they were derived [43]. Certain non-malignant but abnormal tissues also show enhanced PpIX concentration (as measured by its characteristic fluorescence), including those of psoriasis, actinic keratosis, porokeratosis, and certain types of bacterial, fungal, and parasitic infections [96,99]. Thus the presence of ALA-induced PpIX fluorescence may additionally be used for the detection of certain types of cell or tissue abnormalities, for example bladder cancer [100].

It is apparent that normal and malignant cells differ in the manner in which they respond to exogenous ALA, and there is evidence that the activation of oncogenes is associated with alterations in the activity of certain enzymes in the heme biosynthetic pathway [101,102]. Defective iron metabolism is commonly associated with cancer [94], but it is not clear whether such a defect is responsible for the observed ALA-induced PpIX phenotypes of malignant and pre-malignant cells. In addition, in certain clinical situations there may be differences in the rate at which ALA reaches normal and malignant cells. The "leaky roof" theory proposes that the abnormal stratum corneum that is generated by superficial basal cell carcinomas and squamous cell carcinoma in situ facilitates rapid diffusion of topically-applied ALA into the deeper layers of the tumor [103,104]. Denuded skin lesions and ulcers in which there is no longer any stratum corneum offer even less of a barrier to entry of the ALA [43,104]. While this theory may be true, it does not explain the preferential accumulation of ALA-induced PpIX in malignant cells exposed to exogenous ALA either in vitro or following the systemic administration of ALA. Finally, it has been suggested that the concentration of PpIX in abnormal tissues may be increased by selective biodistribution from the circulation. When ALA is administered systemically, the liver synthesizes and then excretes a substantial amount of PpIX into the duodenum via the bile. Some of this may be reabsorbed and enter the blood via capillaries or lymphatics, where it might be expected to show the same biodistribution as PpIX that had been injected intravenously. However, this hypothesis does not explain the selectivity of ALA-induced PpIX for malignant cells and tissues when the ALA is administered either in vitro or topically. In summary, the observed difference between the response of normal and malignant cells to exogenous ALA is likely to be caused primarily by differences in the relative activities of the enzymes involved in heme biosynthesis.

4.3 The early history of ALA-PDT

In 1956, it was reported that of four volunteers who ingested ALA as part of a study of porphyrin metabolism, those who took the largest doses experienced transient photosensitization of the skin [105,106]. These experiments took place at a time when it was generally believed that the capacity for excessive synthesis of porphyrins was restricted to the liver and the bone marrow, because it was these organs which produced the porphyrins responsible for the skin photosensitization associated with certain types of porphyria [107]. Consequently, the skin photosensitization induced by ingestion of ALA was interpreted in similar terms. Despite this erroneous assumption about the mechanism involved, these researchers had established that systemic administration of ALA to humans could cause skin photosensitization. We verified these observations in mice given ALA systemically, but reported in 1981 that the localized administration of ALA by intradermal or subcutaneous injection led to development of protoporphyrin fluorescence that was restricted to the injection sites [108]. Such observations were incompatible with the concept that ALA-induced skin photosensitization was caused by porphyrins that had been synthesized by liver or bone marrow and then transported to the skin via the blood. This was the first clear evidence that cells of the skin could themselves synthesize and accumulate significant concentrations of PpIX when exposed in vivo to exogenous ALA. When we repeated these observations using human volunteers, it was soon discovered that the ALA did not have to be injected into the skin in order to induce localized PpIX fluorescence, but that it was effective when merely applied to the surface. In 1986, pharmacokinetic studies of the rate of conversion of ALA into PpIX in mice confirmed that, as with humans, PpIX rapidly appeared in the skin following systemic administration of ALA and was subsequently rapidly cleared [109]. In the same year we studied ALA-induced PpIX in various types of bacteria, and the following year used ALA-PDT in an attempt to cure mice bearing transplantable tumors and a cat with sunlight-induced squamous cell carcinoma involving the tips of the ears [unpublished, and 103]. In addition, groups in Israel and Norway further reported on the selective usage of ALA-induced PpIX for certain types of abnormal cells [110] and demonstrated that it could be used to destroy erythroleukemic cells in vitro [111]. Our first human patient was treated by ALA-PDT in 1987, using topical applications of ALA in aqueous solution to treat multiple superficial and ulcerated basal cell carcinomas [103]. Subsequent clinical studies using ALA applied topically in either aqueous or alcoholic solutions or in oil-in-water emulsions demonstrated the clinical usefulness of ALA-PDT for controlling many other types of malignant and premalignant tissues involving the skin and mucous membranes.

4.4 ALA-PDT in clinical practice

Since 1990 when we first described the technique of ALA-PDT and reported the preliminary results of our first clinical trial, there have been at least 130 reports in refereed journals of clinical studies by other groups who are using this particular form of PDT [112]. The use of ALA-PDT in clinical applications up to 1996 has been thoroughly reviewed by Kennedy et al. [45,43] and Peng et al. [113,114]; in the following paragraphs, only more recent publications are cited. At present, ALA-PDT is being used experimentally at major centres of research throughout the world, and is considered by many in the field to be the "gold standard" against which all other forms of PDT must be compared. The popularity of ALA-PDT, in comparison to PDT in general, can in part be judged by the rate of increase in the number of publications in these fields. Statistics reported by Brown [112] show that while there has been a doubling of the number of PDT papers published between 1980 and 2000, there has been a *fifty*-fold increase in the number of ALA-PDT publications occurring during the same time.

There are a number of very good reasons for this widespread acceptance of ALA-PDT. First, numerous researchers have demonstrated to their own satisfaction that it is a clinically effective technique for detecting and/or controlling many different types of malignant or premalignant cells and tissues. For example, malignant lesions of the skin are the most common form of cancer reported among people of European ancestry ("Europoids"), and the premalignant lesions of actinic keratosis are even more common in this population. Unfortunately, most agents that have been tested for PDT are rather large molecules that do not readily penetrate either normal or abnormal skin. They are therefore generally administered by intravenous injection, with consequent photosensitization of the whole body. Although generalized photosensitization can be tolerated if PDT is the only option available for the treatment of serious disease, such a side-effect becomes quite significant if the lesion in question involves only a small area of skin, is not life-threatening, and can be treated effectively (though possibly with worse cosmetic results) by some other well-established method. Consequently, prior to the discovery of ALA-PDT, photodynamic therapy was not often used for the treatment of superficial basal cell carcinomas, the premalignant lesions of actinic keratosis, or non-malignant lesions of the skin. Primary malignant lesions of the skin against which ALA-PDT has been since shown to be effective include basal cell carcinoma [115–119], squamous cell carcinoma [115,120], malignant melanoma, adenocarcinoma of the sebaceous gland, adenocarcinoma of the sweat gland, T-cell lymphoma of the skin (mycosis fungoides), and Kaposis sarcoma (non-epidemic type). In addition, other skin conditions studied for potential treatment include acne vulgaris [121,122], warts [123,124] and condylomata [125], scleroderma [126], nevus sebaceus [127], actinic keratosis [119,128,129], and psoriasis [130,131]. Primary malignant lesions of mucosal surfaces that have benefited from ALA-PDT include carcinomas of the digestive system (mouth to anus) [132–139], the respiratory system (nasal passages to bronchi), and the urogenital system (vagina, uterus, urinary bladder and urethra, but not the kidneys themselves) [140–145]. Leukoplakia, a premalignant lesion of the oral mucosa, sometimes responds very well but in other cases is quite resistant. Exocrine gland tumors that have been treated include carcinomas of the breast and of the salivary glands. Some of these tumors have been treated as primaries, and others as secondaries. Finally, ALA-induced PpIX fluorescence has been used to follow the response to chemotherapy of various types of leukemia, widespread mycosis fungoides, and intradermal secondaries of carcinoma of the breast. If the chemotherapy is effective, the malignant cells will no longer develop strong fluorescence when subsequently exposed to exogenous ALA either in vitro (leukemia) or in vivo (mycosis fungoides and breast carcinoma) [43,103,104].

A second reason for the widespread acceptance of ALA-PDT is its safety. Although topical ALA-PDT may cause quite significant discomfort to the patient, it is very rare for it to cause any medically serious complications. During the past 13 years we have had only one serious problem as a result of ALA-PDT that required hospitalization of the patient, a man with very extensive mycosis fungoides who fainted during ALA-PDT, apparently because of hypovolemia secondary to extensive edema involving the whole of his back. He was given some intravenous saline and kept in hospital overnight. A few patients have demonstrated blistering, or slow to heal lesions, and one elderly woman with diabetes developed an infection under a toenail a few days following treatment of that same toe by ALA-PDT, which was treated uneventfully. These complications of ALA-PDT are the worst we have seen while treating thousands of lesions over 13 years, demonstrating its safety.

The unusually rapid photobleaching of PpIX during treatment [146] is a major reason for the low incidence of serious complications with ALA-PDT. If a photosensitizer of a type that is relatively resistant to photobleaching is used, the phototoxic damage caused by a treatment is influenced by both the concentration of the photosensitizer in the tissues and the total dose of light that reaches those tissues. Under such conditions, any increase in the dose of light will result in an increase in phototoxic damage. However, if the photosensitizer is of a type that is rapidly destroyed by photoactivating light, then once its destruction is complete there can be no additional phototoxic damage no matter how much additional light is given. Thus the drug rather than the light is limiting, and light dosimetry is much less critical. For example, different light doses received within the same treatment field resulting from complicated surface geometry of the area being treated (for example the nose) should have little impact on the final result provided sufficient light is used to bleach all of the drug present.

Another aspect of patient safety is related to the relatively low risk of accidental phototoxic reactions in the days and weeks following ALA-PDT. When the ALA is applied topically, photosensitization is not generalized but is restricted to the site of treatment. Also, since the PpIX is photobleached during the treatment and any residual PpIX is rapidly cleared from the body, most patients can safely expose themselves to sunlight 24 h after treatment.

A third reason for the popularity of ALA-PDT has to do with its versatility in terms of delivery methods. ALA is effective when applied topically to any accessible surface (skin, oral cavity, urinary bladder, vagina, uterus), it can be injected intradermally or intralesionally in order to produce a higher local concentration of ALA than is possible by topical application, and it can be given systemically (orally, or by subcutaneous or intravenous injection). Under certain conditions it can even be given regionally, via the lymphatics. No other photosensitizer is known which can match ALA-induced PpIX for versatility.

4.5 ALA-PDT: the future

ALA-PDT in the form of Levulan® is now an approved therapy in the USA for actinic keratosis, and it is likely to be approved soon in several other jurisdictions for both oncological and non-oncological applications. Thus ALA-PDT seems to have a bright future. Of all the photosensitizers available, ALA-induced PpIX is without doubt the most versatile and easiest to use. It can be administered either topically, locally, or systemically, and neither the dose of drug nor the dose of light is very critical for most clinical applications. It appears to be a relatively safe drug, one that is rapidly cleared from the body and therefore does not cause lingering photosensitization. However, it certainly does have certain limitations, one being its poor absorption of light at the far-red end of the spectrum where the tissue trans-mittance of light generally is best.

Techniques that might be used to enhance the far-red/near-IR absorption of ALA-induced PpIX appear to be quite limited. PpIX is very photolabile and is easily converted into photoproducts that include at least one chlorin (known as photoprotoporphyrin) which has been identified as the hydroxyaldehyde chlorin derivative of PpIX [147]. All such photoproducts of PpIX absorb light at wavelengths longer than PpIX itself. For example, the hydroxyaldehyde absorbs strongly in the red in the vicinity of 667 nm, while PpIX absorbs around 635 nm. The relevance of these PpIX photoproducts to ALA-PDT as currently practiced is not yet clear. While it is relatively easy to produce PpIX photoproducts in vitro, their yields and photostability in vivo under conditions of clinical ALA-PDT are difficult to determine [148,149]. If treatment conditions that enhance the yield of ALA-induced PpIX photoproducts can be identified, then it might be possible to increase both the photosensitizing efficiency of ALA-PDT and the effective depth of the treatment field.

While topical ALA penetrates the outer layers of intact epidermis very much better than any pre-formed photosensitizers (porphyrins, chlorins, purpurins, or phthalocyanines), it still requires at least several hours for photosensitizing concen-trations of PpIX to accumulate within the underlying cells, and even longer if the cells are deeper. Two main approaches to enhancing the penetration of ALA through biological barriers such as the stratum corneum are current topics of research in several laboratories.

One approach is to use penetration enhancers such as dimethyl sulfoxide (DMSO) [150–153]. DMSO may play a dual role, since it also may facilitate the differentia-tion of certain types of malignant cells and thus alter their response to exogenous ALA [154].

A second approach involves modification of the chemical structure of ALA itself by turning it into an ester. This is somewhat reminiscent of the topical use of methyl salicylate (the methyl ester of salicylic acid) which releases salicylic acid in vivo following enzymatic cleavage of the ester bond. Certain esters of ALA can penetrate the skin more rapidly than ALA itself (presumably because the esters are more hydrophobic), and then subsequently release ALA in vivo following cleavage by tissue esterases. It certainly would be an advance if the use of such "pre-ALA" compounds increased the depth at which ALA-PDT was effective [155–159]. While the hexyl ester derivative of ALA seems to show better membrane penetration properties and improved PpIX production than other esters [159], a current drawback is the significantly greater toxicity of the hexyl ester compared with ALA itself. While there are still some problems to be worked out, derivatives of ALA may well prove useful in ALA-PDT.

Attempts have been made to increase the accumulation of ALA-induced PpIX by using drugs that reversibly inhibit the conversion of PpIX into heme by chelating the iron which is required for the process. These chelators include ethylenediamine tetraacetic acid (EDTA) [153,160–163], desferroxamine (DFO) [164,165] and 1,10-phenanthroline [166,167]. There has been some degree of success with this approach.

One very important area for future research is the relationship between ALA-PDT and the immune system. We have found that patients who were strongly immuno-suppressed for any of a variety of reasons did not respond well to ALA-PDT for squamous cell carcinoma of the skin, Kaposis sarcoma, fungal infections of the skin, or plantar warts. For example, when ALA-PDT was used to treat kidney transplant patients with squamous cell carcinoma, all those who were on strong immuno-suppressive therapy responded very poorly while all who were on minimal immuno-suppressive therapy showed what appeared to be a completely normal response. This difference was so definite that we have initiated a policy of refusing to even try ALA-PDT for squamous cell carcinoma if the patients are strongly immuno-suppressed. There is also evidence that PDT treatment of psoriasis may result from immune activation, again highlighting the possible role of the immune system in some PDT treatments [168].

The growth of interest in ALA-PDT since our initial clinical report in 1990 has been remarkable, with applications to oncological and non-oncological indications being developed the world over. For example, Sac-Morales [169 and unpublished results] has discovered that ALA-PDT may have applications in several areas of tropical medicine. Many different classes of single-celled and multi-celled parasites have been investigated, and all could be photosensitized if exposed to exogenous ALA under appropriate conditions. In addition, blood containing viable malarial parasites in a mouse malarial model can be rendered non-infective by incubating the blood with ALA and then exposing it to red light. This type of work opens up a new realm of potential applications for ALA-PDT in infectious diseases and blood treatment.

References

1. J.D. Spikes (1985). The historical development of ideas on applications of photo-sensitized reactions in the health sciences. In: R.V. Bensasson, G. Jori, E.J. Land, T.G. Truscott (Eds), *Primary Photoprocesses in Biology and Medicine* (pp. 209–227). Plenum Press, New York.
2. L.C. Harber, I.E. Kochevar, A.R. Shalita (1982). Mechanisms of photosensitization to drugs in humans. In: J.D. Regan, J.A. Parrish (Eds), *The Science of Photomedicine* (pp. 323–347). Plenum Press, New York.
3. H.v. Tappeiner (1900). Uber die wirkung fluorescierender stoffe auf infusorien nach versuchen von O. Raab. *Münch. Med. Vochenschr.*, **47**, 5.
4. O. Raab (1900). Uber die wirkung fluorescierender stoffe auf infusorien. *Z. Biol.*, **39**, 524.
5. H.v. Tappeiner, A. Josionek (1903). Therapeutische versuch mit fluoreszierenden stoffen. *Münich. Med. Wochenschr*, **50**, 2042.
6. H.v. Tappeiner, A. Jodlbaur (1904). Die sensibilizierende wirkung fluorescierender substanzer. *Dusch. Arch. Klin. Med.*, **80**, 524.
7. H.F. Blum (1941). *Photodynamic Action and Diseases Caused by Light*. Reinhold, New York.
8. G. Jori, J.D. Spikes (1984). Photochemistry of porphyrins. In: K. Smith (Ed.), *Topics in Photomedicine* (pp. 183–318). Plenum Press, New York.
9. J. Piette, M.-P. Merville, J. Decuyper (1986). Damages induced in nucleic acids by photo-sensitization. *Photochem. Photobiol.*, **44**, 793–803.

10. C. Salet, S. Passarella, E. Quagliarello (1987). Effects of selective irradiation on mammalian mitochondria. *Photochem. Photobiol.*, **45**, 433–438.

11. F. Dall'Acqua (1988). Psoralens: a review. In: G. Moreno, R.H. Pottier, T.G. Truscott (Eds), *Photosensitization: Molecular, Cellular and Medical Aspects* (pp. 269–278). Springer-Verlag, Berlin.

12. E.L. Oginsky, G.S. Green, D.G. Griffith (1959). Lethal photosensitization of bacteria with 8-methoxypsoralen to long wavelength ultraviolet radiation. *J. Bacteriol.*, **78**, 821–833.

13. M.M. Mathews (1963). Comparative study of lethal photosensitization of *Dacrina lutea* by 8-methoxypsoralen and by toluidine blue. *J. Bacteriol.*, **85**, 322–328.

14. A.W. Girotti (1983). Mechanisms of photosensitization. *Photochem. Photobiol.*, **38**, 745–751.

15. N.I. Krinsky (1983). Singlet oxygen in biological systems. *Trends Biochem. Sci.*, **2**, 35–38.

16. K. Suwa, T. Kimura, A.P. Schapp (1977). Reactivity of singlet molecular oxygen with cholesterol in a phospholipid membrane matrix. A model for oxidative damage to membranes. *Biochem. Biophys. Res. Commun.*, **75**, 785–792.

17. C.S. Foote (1984). Mechanisms of photooxidation. In: R. Doiron, C.J. Gomer (Eds), *Porphyrin Localization and Treatment of Tumors*, (pp. 3–18) Alan R. Liss, New York.

18. L.I. Grossweiner, A.S. Patel, J. Grossweiner (1982). Type I and Type II mechanisms in the photosensitized lysis of phosphatidylcholine liposomes by hematoporphyrin. *Photochem. Photobiol.*, **36**, 159–167.

19. J. Moan (1984). The photochemical yield of singlet oxygen from porphyrins in different states of aggregation. *Photochem. Photobiol.*, **39**, 445–449.

20. M. Krieg, D.G. Whitten (1984). Self-sensitizing photooxidation of protoporphyrin IX and related porphyrins in erythrocyte ghosts and microemulsions: a novel photooxidation pathway involving singlet oxygen. *J. Photochem.*, **25**, 235–252.

21. G. Jori (1985). Molecular and cellular mechanisms in photomedicine: porphyrins in micro-heterogeneous environments. In: R.V. Bensasson, G. Jori, E.J. Land, T.G. Truscott (Eds), *Primary Photoprocesses in Biology and Medicine* (pp. 349–355). Plenum Press, New York.

22. J.D. Spikes (1989). Photosensitization. In: K.C. Smith (Ed.), *The Science of Photobiology* (2nd edn., pp. 79–110). Plenum Press, New York.

23. G.R. Buettner (1989). Hematoporphyrin derivative and light produces the vitamin E radical. In: E.V. Bensasson, G. Jori, E.J. Land, T.G. Truscott (Eds), *Primary Photoprocesses in Biology and Medicine* (pp. 341–344). Plenum Press, New York.

24. G.R. Buettner, L.W. Oberly (1980). The apparent production of superoxide and hydroxyl radicals by hematoporphyrin and light as seen by spin-trapping. *FEBS Lett.*, **121**, 161–164.

25. B.W. Henderson, T.J. Dougherty, P.B. Malone (1984). Studies on the mechanism of tumor destruction by photoradiation therapy. *Prog. Clin. Biol. Res.*, **170**, 601–612.

26. W. Hausemann (1911). The sensitizing action of haematoporphyrin. *Biochem. Z.*, **30**, 276–316.

27. F. Meyer-Betz (1913). Investigation on the biological (photodynamic) action of hemato-porphyrin and other derivatives of blood and bile pigments. *Dtsch. Arch. Klin. Med.*, **112**, 476–503 (in German).

28. D.S. Rassmussen-Taxdal, G.E. Ward, G.H.E. Figge (1955). Fluorescence of human lymphatic and cancer tissue following high doses of intravenous hematoporphyrin. *Cancer*, **8**, 78–81.

29. M.J. Gray, R. Lipson, J.V. Marck, L. Parker, D. Romeyn (1967). Use of hematoporphyrin derivative in detection and management of cervical cancer. *Am. J. Obstet. & Gynecol.*, **99**, 766–771.

30. Photodynamic therapy (special issue) (1987). *Photochem. Photobiol.*, **46**, 563–952.
31. D. Kessel, T.J. Dougherty (Eds) (1983). *Porphyrin Photosensitization*. Plenum Press, New York.
32. D. Kessel (1983). *Methods in Porphyrin Photosensitization*. Plenum Press, New York.
33. S. Spinelli, M. Dal Fante, R. Marchesini (Eds) (1992). *Photodynamic Therapy and Biomedical Lasers*. Elsevier Science, Amsterdam.
34. *Photosensitizing Compounds: Their Chemistry, Biology and Clinical Use* (1989). Ciba Foundation Symposium 146 (pp. 1–241). Wiley, Chichester.
35. T.J. Dougherty (1987). Photosensitizers: therapy and detection of malignant tumors. *Photochem. Photobiol.*, **45**, 879–889.
36. C.J. Gomer, N. Rucker, A. Ferrario, S. Wong (1989). Properties and application of photodynamic therapy. *Radiat. Res.*, **120**, 1–18.
37. G. Jori, J.D. Spikes (1984). Photochemistry of porphyrins. In: K. Smith (Ed.), *Topics in Photomedicine* (pp. 183–318). Plenum Press, New York.
38. R.L. Lipson, E.J. Baldes, M.J. Gray (1967). Hematoporphyrin derivative for detection and management of cancer. *Cancer*, **20**, 2255–2257.
39. H.I. Pass (1993). Photodynamic therapy in oncology: mechanisms and clinical use. *J. Natl. Cancer Inst.*, **85**, 443–456.
40. D. Eton, M.D. Colburn, V. Shim, W. Panek, D. Lee, W.S. Moore, S.S. Ahn (1992). Inhibition of intimal hyperplasia by photodynamic therapy using photofrin. *J. Surg. Res.*, **53**, 558–562.
41. S. Brown (1999). Photodynamic therapy: a bright future. *Photodyn. News* (*Special Issue*) 1–2.
42. C.C. Dierickx, R.R. Anderson (1996). Why use PDT in dermatology? *Int. Photodyn.*, **1**, 2–6.
43. R. Pottier, J.C. Kennedy (1996). Photodynamic therapy with 5-aminolevulinic acid: basic principles and applications. In: B. Ehrenberg, G. Jori, J. Moan (Eds), *Photochemotherapy: Photodynamic Therapy and Other Modalities* (Proc. SPIE 2625, pp. 2–10).
44. K.E. Anderson (1990). Erythrocyte disorders: diseases related to abnormal heme or porphyrin metabolism. In: W.J. Williams, E. Beutler, A.J. Erslev, M.A. Lichtman (Eds), *Hematology*, 4th Edn., (pp. 722–741). McGraw-Hill, New York.
45. J.C. Kennedy, S.L. Marcus, R.H. Pottier (1996). Photodynamic therapy (PDT) and photodiagnosis (PD) using endogenous photosensitization induced by 5-aminolevulinic acid (ALA): mechanisms and clinical results. *J. Clin. Laser Med. Surg.*, **14**, 289–304.
46. D. Shemin, C.S. Russell (1953). δ-Aminolevulinic acid, its role in the biosynthesis of porphyrins and purines. *J. Am. Chem. Soc.*, **75**, 4873–4874.
47. S.I. Beale (1990). Biosynthesis of the tetrapyrrole pigment precursor, δ-Aminolevulinic acid, from glutamate. *Plant Physiol.*, **93**, 1273–1279.
48. N.I. Berlin, A. Neuberger, J.J. Scott (1956). The metabolism of delta-aminolaevulinic acid. 1. Normal pathways, studied with the aid of ^{15}N, and 2. Normal pathways studied with the aid of ^{14}C. *Biochem. J.*, **64**, 80–100.
49. H.L. Bonkovsky, J.F. Healey, P.R. Sinclair, J.F. Sinclair (1985). Conversion of 5-aminolevulinate into heme by homogenates of human liver. Comparison with rat and chick-embryo liver homogenates. *Biochem. J.*, **227**, 893–901.
50. L. Eriksen, N. Eriksen (1976). Possible pathways in protoporphyrin biosynthesis. In: M. Doss (Ed.), *Porphyrins in human diseases* (pp.105–110). Karger, Basel.
51. D. He, E. Karas, S. Sassa, H.W. Lim (1993). Porphyrin synthesis by murine epidermal cells. *Skin Pharmacol.*, **6**, 20–25.
52. J.F. Healey, J.L. Bonkowsky, P.R. Sinclair, J.F. Sinclair (1981). Conversion of 5-aminolaevulinate into heme by liver homogenates. Comparison of rat and chick embryo. *Biochem. J.*, **198**, 595–604.

53. E. Ivanov, M. Pisanets (1982). Studies on the biosynthesis of porphyrins in erythrocytes after incubation with δ-aminolevulinic acid: An attempt to investigate the pathogenesis of nephrogenic anemia. *Acta Biol. Med. Germ.*, **41**, 307–313.

54. V.M. Sardesai, J. Waldman, J.M. Orten (1964). A comparative study of porphyrin biosynthesis in different tissues. *Blood*, **24**, 178–186.

55. D.L. Stout, F.F. Becker (1990). Heme synthesis in normal mouse liver and mouse liver tumors. *Cancer Res.*, **50**, 2337–2340.

56. W.O. Whetsell Jr., S. Sassa, D. Bickers, A. Kappas (1978). Studies on porphyrin-heme biosynthesis in organotypic cultures of chick dorsal root ganglion. I. Observations on neuronal and non-neuronal elements. *J. Neuropath. Exp. Neurol.*, **37**, 497–507.

57. H. Fujita, R. Yamamoto, M. Ikeda (1985). In vivo regulation of δ-aminolevulinate dehydratase activity. *Toxic Appl. Pharm.*, **77**, 66–75.

58. L.C. Gardner, S.J. Smith, T.M. Cox (1991). Biosynthesis of δ-aminolevulinic acid and the regulation of heme formation by immature erythroid cells in man. *J. Biol. Chem.*, **266**, 22010–22018.

59. A. Goldbert, W.D.M. Paton, J.W. Thompson (1954). Pharmacology of the porphyrins and porphobilinogen, *Br. J. Pharmacol. Chemother.*, **9**, 90–94.

60. S. Granik, S. Sassa (1971). δ-Aminolevulinic acid synthetase and control of heme and chlorophyll synthesis. In: H.J. Vogel (Ed.), *Metabolic Regulation* (pp. 77–141) McGraw-Hill, New York.

61. J. Ho, R. Gutherie, H. Tieckelmann (1986). Detection of δ-aminolevulinic acid, porphobilinogen and porphyrins related to heme biosynthesis by high performance liquid chromatography. *J. Chromatogr.*, **375**, 56–63.

62. A. Hradilek, J. Neuwirt (1990). The relationship between heme synthesis and iron uptake in erythroid cells. *Biomed. Biochim. Acta*, **49**, S94-99.

63. S.M. Mayer, S.I. Beale (1992). Succinyl-Coenzyme A Synthetase and its role in δ-aminolevulinic acid biosynthesis in euglena gracilis. *Plant Physiol.*, **99**, 482–487.

64. G.P. O'Neill, D. Soll (1990). Transfer RNA and the formation of the heme and chlorophyll precursor, 5-aminolevulinic acid. *Biofactors*, **2**, 227–235.

65. R.R. Sinclair, S. Granick (1975). Heme control on the synthesis of delta-aminolevulinic acid synthetase in cultured chick embryo liver cells. *Ann. NY Acad. Sci.*, **244**, 509–520.

66. W.O. Whetsell Jr., S. Sassa, D. Bickers, A. Kappas (1978). Studies on porphyrin-heme biosynthesis in organitypic cultures of chick dorsal root ganglion. I. Observations on neuronal and non-neuronal elements. *J. Neuropath. Exp. Neurol.*, **37**, 497–507.

67. H. Baart de la Faille (1975). Biochemical features of EPP; pathogenesis of the photodermatosis. In: H. Gaart de la Faille (Ed.), *Erythropoietic protoporphyria* (pp. 14–20). Oosthrek, Scheltema, Holkema, Utrecht.

68. M.M. Berenson, R. Kimura, W. Samowitz, D. Bjorkman (1992). Protoporphyrin overload in unrestrained rats: biochemical and histopathologic characterization of a new model of protoporphyric hepatopathy. *Int. J. Path.*, **73**, 665–673.

69. D.R. Bickers, L. Keogh, A.B. Rifkind, L.C. Harber, A. Kappas (1977). Biosynthesis of porphyrins in mammalian skin and in the skin of porphyric patients. *J. Inv. Derm.*, **68**, 5–9.

70. H.L. Bonokwosky, D.P. Tschudy, J. Collins, I. Doherty, R. Bossenmaier, A. Cardina, C.J. Watson (1971). Repression of the overproduction of porphyrin precursors in acute intermittent porphyria by intravenous infusions of hematin. *Proc. Acad. Sci.*, **68**, 2725–2729.

71. A. Brun, S. Sandberg (1991). Mechanisms of photosensitivity in porphyric patients with special emphasis on erythropoietic protoporphyria. *J. Photochem. Photobiol. B Biol.*, **10**, 285–302.

72. A. Brun, A. Western, Z. Malik, S. Sandberg (1990). Erythropoietic protoporphyrin: photodynamic transfer of protoporphyrin from intact erythrocytes to other cells. *Photochem. Photobiol.*, **51**, 573–577.

73. A. Gorchein, R. Weber (1987). δ-Aminolaevulinic acid in plasma, cerebrospinal fluid, saliva and erythrocytes: studies in normal, uraemic and prophyric subjects. *Clin. Sci.*, **72**, 103–112.

74. S. Granick, H.G. van den Schrieck (1955). Porphobilinogen and delta-aminolevulinic acid in acute porphyria. *Proc. Soc. Exp. Biol.*, **88**, 270–273.

75. A. Jarrett, C. Rimington, D.A. Willoughby (1956). Delta-aminolaevulinic acid and porphyria, *Lancet*, **i**, 125–127.

76. A. Kappas, S. Sassa, R.A. Galbraith, Y. Norman (1989). The porphyrias: In: C.R. Scriver (Ed.). *Metabolic Basis of Inherited Disease*, 6th Edn. (pp. 1305–1365). McGraw-Hill, NY.

77. K. Konrad, H. Honigsmann, F. Gschnait, K. Wolff (1975). Mouse model for protoporphyria. Cellular and subcellular events in the photosensitivity flare of the skin. *J. Invest. Dermatol.*, **65**, 300–310.

78. J.D. Laskey, P. Ponka, H.M. Schulman (1986). Control of heme synthesis during friend cell differentiation: role of iron and transferrin. *J. Cell Physiol.*, **129**, 185–192.

79. H.W. Lim, D. Sooper, S. Sassa, H. Doski, M.R. Buchness, N.A. Soter (1992). Photosensitivity, abnormal porphyrin profile, and sideroblastic anemia. *J. Am. Acad. Dermatol.*, **27**, 287–292.

80. R.L.P. Lindberg, C. Procher, B. Grandchamp, B. Lederman, K. Burki, S. Aguzzi, U.A. Meyer (1996). Porphobilinogen deaminase deficiency in mice causes a neuropathy resembling that of human hepatic porphyria. *Nat. Gen.*, **12**, 195–199.

81. I.A. Magnus (1968). The cutaneous porphyrias. *Semin. Hematol.*, **4**, 380–408.

82. M.R. Moore, F.B. McGillion, A. Goldberg (1976). Some pharmacological and behavioral effects of delta-aminolaevulinic acid. In: M. Doss (Ed.). *Porphyrins in Human Diseases* (pp. 148–154). Karger, Basel.

83. D.A. Paslin (1992). The porphyrias. *Inter. J. Dermatol*, **31**, 527–539.

84. M.G. Perlroth (1988). The porphyrias. In: E. Rubenstein, D.D. Federman (Eds), *Scientific American Medicine* (pp. 1–9). Section 9V, New York.

85. M.B. Poh-Fitzpatrick (1989). Erythropoietic porphyrias: current mechanistic, diagnostic, and therapeutic considerations. *Semin. Hematol.*, **14**, 211–219.

86. S. Sassa, S. Schwartz, G. Ruth (1981). Accumulation of protoporphyrin IX from δ-aminolevulinic acid in bovine skin fibroblasts with hereditary erythropoietic protoporphyria. A gene-dosage effect. *J. Exp. Med.*, **153**, 1094–1101.

87. B.C. Shanley, V.A. Percy, A.C. Neethling (1976). Neurochemistry of acute porphyria. Experimental studies on delta-aminolaevulinic acid and porphobilinogen. In: M. Doss (Ed.), *Porphyrins in Human Diseases* (pp. 156–162). Karger, Basel.

88. B.C. Shanley, J.J.F. Taljaard, W.M. Deppe, S.M. Joubert (1972). Delta-aminolaevulinic acid in acute porphyria. *S. Afr. Med. J.*, **46**, 84.

89. V.P. Sweeney, M.A. Pathak, A.K. Asbury (1970). Acute intermittent porphyria. Increased ALA-synthetase activity during an acute attack. *Brain*, **3**, 369–380.

90. O. Visser, J.W.O. van den Berg, H. Koole-Lesuuis, G. Voortman, J.H.P. Wilson (1991). Porphyrin synthesis by human hepatocytes and HepG2 cells - effect of enzyme inducers and delta-aminolevulinic acid. *Toxicology*, **67**, 75–83.

91. A.C. Yeung Laiwah, M.R. Moore, A. Goldberg (1987). Pathogenesis of acute porphyria *Q. J. Med.*, **63**, 377–392.

92. H.W. Lim, H.D. Perez, M. Poh-Fitzpatrick, I.M. Goldstein, I. Gigli (1981). Generation of chemotactic activity in serum from patients with erythropoietic protoporphyria and cutanea tarda. *New Engl. J. Med.*, **304**, 212–216.

93. S.V. Torti, F.M. Torti (1994). Iron and ferritin in inflammation and cancer. *Adv. Inorg. Biochem.*, **10**, 119–137.

94. E.L. Elliot, M.C. Elliot, F. Wang, J.F. Head (1993). Breast carcinoma and the role of iron metabolism. A cytochemical, tissue culture, and ultrastructural study. *Ann. N. Y. Acad. Sci.*, **698**, 159–166.

95. J.C. Kennedy, P. Nadeau, Z.J. Petryka, R.H. Pottier, G. Weagle (1992). Clearance times of porphyrin derivatives from mice as measured by in vivo fluorescence spectroscopy. *J. Photochem. Photobiol. B Biol.*, **55**, 729–734.

96. A.L. Golub, E.F. Gudgin Dickson, J.C. Kennedy, S.L. Marcus, Y. Park, R.H. Pottier (1999). The monitoring of ALA-induced protoporphyrin IX accumulation and clearance in patients with skin lesions by in vivo surface-detected fluorescence spectroscopy. *Lasers Med. Sci.*, **14**, 112–122.

97. J. Webber, D. Kessel, D. Fromm (1997). On-line fluorescence of human tissues after oral administration of 5-aminolevulinic acid, *J. Photochem. Photobiol. B - Biol.*, **38**, 209–214.

98. J. Webber, D. Kessel, D. Fromm (1997). Side effects and photosensitization of human tissues after aminolevulinic acid. *J. Surgical Res.*, **68**, 31–37.

99. A. Sac-Morales, J.C. Kennedy, unpublished results.

100. N. Lange, P. Jichlinski, M. Zellweger, M. Forrer, A. Marti, L. Guillou, P. Kucera, G. Wagnieres, H. van den Bergh (1999). Photodetection of early human bladder cancer based on the fluorescence of 5-aminolaevulinic acid hexylester-induced protoporphyrin IX: a pilot study. *Br. J. Cancer*, **80**, 185–193.

101. G. Li, R. Pottier, M.R. Szewczuk, J.C. Kennedy (1999). Effect of mammalian cell differentiation on response of malignant cells to exogenous 5-aminolevulinic acid. *Photochem. Photobiol.*, **69**, 231–235.

102. G. Li, M.R. Szewczuk, L. Raptis, J.C. Johnson, W.E. Weagle, R. Pottier, J.C. Kennedy (1999). Rodent fibroblast model for studies of response of malignant cells to exogenous 5-aminolevulinic acid. *Br. J. Cancer*, **80**, 676–684.

103. J.C. Kennedy, R.H. Pottier (1992). Endogenous protoporphyrin IX, a clinically useful photosensitizer for photodynamic therapy. *J. Photochem. Photobiol. B Biol.*, **6**, 275–292.

104. J.C. Kennedy, R.H. Pottier, D.C. Pross (1990). Photodynamic therapy with endogenous protoporphyrin IX: Basic principles and present clinical experience. *J. Photochem. Photobiol. B Biol.*, **6**, 143–148.

105. N.I. Berlin, A. Neuberger, J.J. Scott (1956). The metabolism of δ-aminolevulinic acid. 1. Normal pathways studied with the aid of ^{15}N. *Biochem. J.*, **64**, 80–90.

106. N.I. Berlin, A. Neuberger, J.J. Scott (1956). The metabolism of δ-aminolevulinic acid. 2. Normal pathways studied with the aid of ^{14}C. *Biochem. J.*, **64**, 90–100.

107. D.P. Tschudy (1983). The Porphyrias. In: W.J. Williams, E. Beutler, A.J. Erslev, M.A. Lichtman (Eds), *Hematology* 3rd Edn., (pp. 691–703). McGraw-Hill, New York.

108. A. Sima, J.C. Kennedy, D. Blakeslee, D.M. Robertson (1981). Experimental porphyric neuropathy: a preliminary report. *Can. J. Neurol. Sci.*, **8**, 105–114.

109. R.H. Pottier, Y.F.A. Chow, J.P. LaPlante, T.G. Truscott, J.C. Kennedy, L.A. Beiner (1986). Non-invasive technique for obtaining fluorescence excitation and emission spectra in vivo. *Photochem. Photobiol.*, **44**, 679–687.

110. Q. Peng, J.F. Svensen, C. Rimington, J. Moan (1987). A comparison of different photo-sensitizing dyes with respect to uptake of C3H-tumors and tissues of mice. *Cancer Lett.*, **36**, 1–10.

111. Z. Malik, H. Lugaci (1987). Destruction of erythroleukaemic cells by photoactivation of endogenous porphyrins. *Br. J. Cancer*, **56**, 589–595.

112. J.E. Brown (2000). Awareness of PDT among clinicians. *Photodyn. News*, **3**, 2–5.

113. Q. Peng, K. Berg, J. Moan, M. Kongshaug, J.M. Nesland (1997). 5-Aminolevulinic acid-based photodynamic therapy: principles and experimental research. *Photochem. Photobiol.*, **65**, 235–251.

114. Q. Peng, T. Warloe, K. Berg, J. Moan, M. Kongshaug, K.-E. Giercksky, J.M. Nesland (1997). 5-Aminolevulinic acid-based photodynamic therapy. *Clinical research and future challenges. Cancer*, **78**, 2282–2308.

115. Y. Harth, B. Hirshovitz (1998). [Topical photodynamic therapy in basal and squamous cell carcinoma and penile Bowen's disease with 20% aminolevulinic acid, and exposure to red light and infrared light]. [Hebrew]. *Harefuah*, **134**, 602–605, 672, 671.

116. Y. Itoh, T. Henta, Y. Ninomiya, S. Tajima, A. Ishibashi (2000). Repeated 5-aminolevulinic acid-based photodynamic therapy following electro-curettage for pigmented basal cell carcinoma. *J. Dermatol.*, **27**, 10–15

117. A.F. Hurlimann, G. Hanggi, R.G. Panizzon (1998). Photodynamic therapy of superficial basal cell carcinomas using topical 5-aminolevulinic acid in a nanocolloid lotion. *Dermatol.*, **197**, 248–254.

118. R. Fink-Puches, P. Wolf, H. Kerl (1997). Photodynamic therapy of superficial basal cell carcinoma by instillation of aminolevulinic acid and irradiation with visible light [see comments]. *Arch. Dermatol.*, **133**, 1494–1495.

119. M. Stefanidou, A. Tosca, G. Themelis, E. Vazgiouraki, C. Balas (2000). In vivo fluorescence kinetics and photodynamic therapy efficacy of delta-aminolevulinic acid-induced porphyrins in basal cell carcinomas and actinic keratoses; implications for optimization of photodynamic therapy. *Eur. J. Dermatol.*, **10**, 351–356.

120. G.I. Stables, M.R. Stringer, D.J. Robinson, D.V. Ash (1997). Large patches of Bowen's disease treated by topical aminolaevulinic acid photodynamic therapy. *Br. J. Dermatol.*, **136**, 957–960.

121. Y. Itoh, Y. Ninomiya, S. Tajima, A. Ishibashi (2000). Photodynamic therapy for acne vulgaris with topical 5-aminolevulinic acid. *Arch. Dermatol.*, **136**, 1093–1095.

122. W. Hongcharu, C.R. Taylor, Y. Chang, D. Aghassi, K. Suthamjariya, R.R. Anderson (2000). Topical ALA-photodynamic therapy for the treatment of acne vulgaris. *J. Invest. Dermatol.*, **115**, 183–192.

123. I.M. Stender, J. Lock-Andersen, H.C. Wulf (1999). Recalcitrant hand and foot warts successfully treated with photodynamic therapy with topical 5-aminolaevulinic acid: a pilot study. *Clin. Exp. Dermatol.*, **24**, 154–159.

124. I.M. Stender, R. Na, H. Fogh, C. Gluud, H.C. Wulf (2000). Photodynamic therapy with 5-aminolaevulinic acid or placebo for recalcitrant foot and hand warts: randomised double-blind trial. *Lancet*, **355**, 963–966.

125. E.V. Ross, R. Romero, N. Kollias, C. Crum, R.R. Anderson (1997). Selectivity of protoporphyrin IX fluorescence for condylomata after topical application of 5-aminolaevulinic acid: implications for photodynamic treatment. *Br. J. Dermatol.*, **137**, 736–42.

126. S. Karrer, C. Abels, M. Landthaler, R.M. Szeimies (2000). Topical photodynamic therapy for localized scleroderma. *Acta Dermato-Venereol.*, **80**, 26–27.

127. C.C. Dierickx, M. Goldenhersh, P. Dwyer, A. Stratigos, M. Mihm, R.R. Anderson (1999). Photodynamic therapy for nevus sebaceus with topical delta-aminolevulinic acid. *Arch. Dermatol.*, **135**, 637–640.

128. Y. Itoh, Y. Ninomiya, T. Henta, S. Tajima, A. Ishibashi (2000). Topical delta-aminolevulinic acid-based photodynamic therapy for Japanese actinic keratoses. *J. Dermatol.*, **27**, 513–518.

129. E.W. Jeffes, J.L. McCullough, G.D. Weinstein, P.E. Fergin, J.S. Nelson, T.F. Shull, K.R. Simpson, L.M. Bukaty, W.L. Hoffman, N.L. Fong (1997). Photodynamic therapy

of actinic keratosis with topical 5-aminolevulinic acid. A pilot dose-ranging study. *Arch. Dermatol.*, **133**, 727–732.

130. P. Collins, D.J. Robinson, M.R. Stringer, G.I. Stables, R.A. Sheehan-Dare (1997). The variable response of plaque psoriasis after a single treatment with topical 5-amino-laevulinic acid photodynamic therapy. *Br. J. Dermatol.*, **137**, 743–749.

131. D.J. Robinson, P. Collins, M.R. Stringer, D.I. Vernon, G.I. Stables, S.B. Brown, R.A. Sheehan-Dare (1999). Improved response of plaque psoriasis after multiple treatments with topical 5-aminolaevulinic acid photodynamic therapy. *Acta Dermato-Venereol.*, **79**, 451–455.

132. H. Barr (2000). Barrett's esophagus: treatment with 5-aminolevulinic acid photodynamic therapy. *Gastroint. Endoscopy Clinics N. Am.*, **10**, 421–437.

133. W.C. Tan, C. Fulljames, N. Stone, A.J. Dix, N. Shepherd, D.J. Roberts, S.B. Brown, N. Krasner, H. Barr (1999). Photodynamic therapy using 5-aminolaevulinic acid for oesophageal adenocarcinoma associated with Barrett's metaplasia. *J. Photochem. Photobiol. B - Biol.*, **53**, 75–80.

134. S.G. Bown, A.Z. Rogowska (1999). New photosensitizers for photodynamic therapy in gastroenterology. *Can. J. Gastroenterol.*, **13**, 389–392.

135. H. Kashtan, F. Konikoff, R. Haddad, Y. Skornick (1999). Photodynamic therapy of cancer of the esophagus using systemic aminolevulinic acid and a non laser light source: a phase I/II study. *Gastroint. Endoscopy*, **49**, 760–764.

136. A. Kubler, T. Haase, M. Rheinwald, T. Barth, J. Muhling (1998). Treatment of oral leukoplakia by topical application of 5-aminolevulinic acid. *Int. J. Oral & Maxillofacial Surg.*, **27**, 466–469.

137. L. Gossner, M. Stolte, R. Sroka, A. May, E.G. Hahn, C. Ell (1998). [Photodynamic therapy of early squamous epithelial carcinomas and severe squamous epithelial dysplasias of the esophagus with 5-aminolevulinic acid]. [German]. *Z. Gastroenterol.*, **36**, 19–26.

138. M. Ortner, H. Lochs (1998). [Eradication of high-grade dysplasia in Barrett esophagus by photodynamic therapy with endogenously generated protoporphyrin IX]. [German]. *Z. Gastroenterol.*, **36**, 111–113.

139. L. Gossner, M. Stolte, R. Sroka, K. Rick, A. May, E.G. Hahn, C. Ell (1998). Photodynamic ablation of high-grade dysplasia and early cancer in Barrett's esophagus by means of 5-aminolevulinic acid [see comments]. *Gastroenterology*, **114**, 448–455.

140. P. Hillemanns, M. Untch, C. Dannecker, R. Baumgartner, H. Stepp, J. Diebold, H. Weingandt, F. Prove, M. Korell (2000). Photodynamic therapy of vulvar intraepithelial neoplasia using 5-aminolevulinic acid. *Int. J. Cancer*, **85**, 649–653.

141. T. Henta, Y. Itoh, M. Kobayashi, Y. Ninomiya, A. Ishibashi (1999). Photodynamic therapy for inoperable vulval Paget's disease using delta-aminolaevulinic acid: successful management of a large skin lesion. *Br. J. Dermatol.*, **141**, 347–349.

142. F. Wierrani, A. Kubin, R. Jindra, M. Henry, K. Gharehbaghi, W. Grin, J. Soltz-Szotz, G. Alth, W. Grunberger (1999). 5-aminolevulinic acid-mediated photodynamic therapy of intraepithelial neoplasia and human papillomavirus of the uterine cervix – a new experimental approach. *Cancer Detect. & Prevent.*, **23**, 351–355.

143. P. Hillemanns, M. Korell, M. Schmitt-Sody, R. Baumgartner, W. Beyer, R. Kimmig, M. Untch, H. Hepp (1999). Photodynamic therapy in women with cervical intraepithelial neoplasia using topically applied 5-aminolevulinic acid. *Int. J. Cancer*, **81**, 34–38.

144. P. Hillemanns, M. Untch, F. Prove, R. Baumgartner, M. Hillemanns, M. Korell (1999). Photodynamic therapy of vulvar lichen sclerosus with 5-aminolevulinic acid. *Obstetrics & Gynecol.*, **93**, 71–74.

145. R. Waidelich, A. Hofstetter, H. Stepp, R. Baumgartner, E. Weninger, M. Kriegmair (1998). Early clinical experience with 5-aminolevulinic acid for the photodynamic therapy of upper tract urothelial tumors. *J. Urol.*, **159**, 401–404.

146. D. Robinson (2000). Photobleaching – Vice or virtue. *Photodyn. News*, **3**, 4–7.

147. H.H. Inhoffen, H. Brockmann, K.-M. Blieserner (1969). Further knowledge of chlorophyll and of hemin, XXX, Protoporphyrins and their transformation into spirographis and isospirographis-porphyrins (in German). *Leibigs Ann. Chem.*, **730**, 173–185.

148. K. König, H. Schneckenbruger, A. Ruck, R. Steiner (1993). In vivo photoproduct formation during PDT with ALA-induced endogenous porphyrin. *J. Photochem. Photobiol. B Biol.*, **18**, 287–290.

149. E.F. Gudgin Dickson, R.H. Pottier (1995). On the role of protoporphyrin IX photo-products in photodynamic therapy. *J. Photochem. Photobiol. B Biol.*, **29**, 91–93.

150. F. Iwasa, R.A. Galbraith, S. Sassa (1988). Effects of dimethyl sulfoxide on the synthesis of plasma proteins in human hepatoma HepG2. Induction of an acute-phase-like reaction. *Biochem. J.*, **253**, 927–930.

151. H. Fujita, M. Yamamoto, T. Yamagami, N. Hayashi, T.R. Bishop, H. De Verneuil, T. Yoshinaga, S. Shibahara, R. Morimoto, S. Sassa (1991). Sequential activation of genes for heme pathway enzymes during erythroid differentiation of mouse Friend virus-transformed erythroleukemia cells, *Biochem. Biophys. Acta*, **1090**, 311–316.

152. Y. Jukuda, H. Fujita, S. Taketani, S. Sassa (1993). Dimethyl sulphoxide and haemin induce ferrochelatase mRMA by different mechanisms in murine erythroleukemia cells. *Br. J. Haematol.*, **83**, 480–484.

153. A.M. Soler, E. Angell-Petersen, T. Warloe, J. Tausjo, H.B. Steen, J. Moan, K.E. Giercksky (2000). Photodynamic therapy of superficial basal cell carcinoma with 5-aminolevulinic acid with dimethyl sulfoxide and ethylendiaminetetraacetic acid: a comparison of two light sources. *Photochem. Photobiol.*, **71**, 724–729.

154. P.S. Ebert, I. Wars, D.N. Buell (1976). Erythroid differentiation in cultured Friend leukemia cells treated with metabolic inhibitors. *Cancer Res.*, **36**, 1809–1813.

155. Q. Peng, J. Moan, T. Warloe, V. Iani, H.B. Steen, A. Bjorseth, J.M. Nesland (1996). Build-up of esterified aminolevulinic-acid-derivative-induced porphyrin fluorescence in normal mouse skin. *J. Photochem. Photobiol. B Biol.*, **34**, 95–96.

156. J.M. Gaulier, K. Berg, Q. Peng, H. Anholt, P.K. Selbo, L.W. Ma, J. Moan (1997). Use of 5-aminolevulinic acid esters to improve photodynamic therapy on cells in culture. *Cancer Res.*, **57**, 1481–1486.

157. J. Kloek, W. Akkermans, G.M.J. Beirjersbergen van Henegouwen (1998). Derivatives of 5-aminolevulinic acid for photodynamic therapy: enzymatic conversion into proto-porphyrin. *Photochem. Photobiol.*, **76**, 150–154.

158. C. Fritsch, B. Homey, W. Stahl, P. Lehmann, T. Ruzicka, H. Sies (1998). Preferential relative porphyrin enrichment in solar keratoses upon topical application of δ-aminole-vulinic acid methylester. *Photochem. Photobiol.*, **68**, 218–221.

159. H. van den Bergh, N. Lange, P. Jichlinski (1999). ALA hexyl ester: A second-genera-tion precursor for protoporphyrin IX in photodynamic therapy and photodetection of early bladder cancer. *Photodyn. News*, **2**, 4–8.

160. K. Berg, H. Anholt, Ø. Bech, J. Moan (1996). The influence of iron chelators on the accumulation of protoporphyrin IX in 5-aminolevulinic acid-treated cells. *Br. J. Cancer*, **74**, 688–697.

161. Z. Malik, M. Djaldetti (1979). 5-Aminolevulinic acid stimulation of derivative, and hemoglobin synthesis by uninduced friend erythroleukemic cells. *Cell Differ.*, **8**, 223–233.

162. L.H. Conder, S. I. Woodard, H. A. Dailey (1991). Multiple mechanisms for the regulation of heme synthesis during erythroid cell differentiation. Possible role for coproporphyrinogen oxidase. *Biochem. J.*, **275**, 321–326.

163. Z. Malik, G. Kostenich, L. Roitman, B. Ehrenberg, A. Orenstein (1995). Topical application of 5-aminolevulinic acid, DMSO and EDTA: protoporphyrin IX accumulation in skin and tumors of mice. *J. Photochem. Photobiol. B Biol.*, **28**, 213–218.

164. B. Ortel, A. Tenew, H. Honigsmann (1993). Lethal photosensitization by endogenous porphyrins of PAM cells – modification by desferrioxamine. *J. Photochem. Photobiol. B Biol.*, **17**, 273–278.

165. Q. Peng, J. Moan, V. Iani, J.M. Nesland (1996). Effect of desferrioxamine on production of ALA-induced protoporphyrin IX in normal mouse skin. *Proc. SPIE*, **2625**, 51–57.

166. N. Rebeiz, C.C. Rebeiz, S. Arkins, K.W. Kelley, C.A. Rebeiz (1992). Photodestruction of tumor cells by induction of endogenous accumulation of protoporphyrin IX: enhancement by 1,10-phenanthroline. *Photochem. Photobiol.*, **55**, 431–435.

167. N. Rebeiz, S. Arkins, C.A. Rebeiz, J. Simon, J.F. Zachary, K.W. Kelley (1996). Induction of tumor necrosis by δ-aminolevulinic acid and 1,10-phenanthroline photodynamic therapy. *Cancer Res.*, **56**, 339–344.

168. D.W.C. Hunt, A.H. Chan, J. Levy (1998). Immunological aspects of photodynamic therapy. *Photodyn. News*, **1**, 2–4.

169. A. Sac Morales (1995). Antimalarial activity of 5-aminolevulinic acid-induced protoporphyrin IX photodynamic therapy in rodent model. M.Sc. Thesis, Queen's University, Kingston, Canada.

Chapter 5

Sensitizers for PDT: phthalocyanines

Nicole Brasseur

Table of contents

5.1 Introduction

Phthalocyanines (Pc) are synthetic porphyrins that have an additional benzo ring on each of the four pyrrole subunits, which are linked by a nitrogen atom. These modifications result in a redshift in the absorption spectrum and in high molar absorption coefficients (10^5 M^{-1} cm^{-1} ~ 670 nm). This allows for better tissue penetrating properties as compared to clinically used porphyrins. Pc form stable complexes with a variety of metals and their photophysical properties vary according to the presence and nature of the central metal ion. Closed shell diamagnetic Pc, such as Zn^{2+}, Al^{3+} and Ga^{3+} result in higher yields and longer lifetimes of the triplet state as compared to paramagnetic Pc. The photodynamic effect could result from energy and/or electron transfer of the triplet state of the photosensitizer (PS) to the substrate to yield radicals (Type I). However, Type II photooxidation involving interaction between the triplet state of the PS and molecular oxygen to give singlet oxygen seems to be the main intermediate in the photodynamic activity of Pc. The relative importance of Type I and Type II mechanisms has been reviewed [1–4]. The production of singlet oxygen by various metallo-Pc is consistent with their triplet state properties with a quantum yield of around 0.4 for Zn, Al and Ga sulfonated metallo-Pc [5]. Pc usually show good photochemical stability [6].

Unsubstituted Pc are not water soluble, and therefore various chemical substitutions have been applied to the macrocycle or to the central metal ion to allow their direct biological administration.

Water-soluble Pc usually aggregate in physiological solvents even at low concentrations (dimerization constant 10^5–10^8 M^{-1}) but they monomerize in the presence of detergents or organic solvents and in a lipophilic environment. This is of importance since only the monomeric species are photoactive [7].

The photobiological properties of phthalocyanines will be reviewed in relation to their degree of hydrophobicity and two classes of photosensitizers will be defined, i.e. the hydrophilic Pc which are freely soluble in physiological solvents and the hydrophobic Pc, which require appropriate drug formulations, such as oil emulsion, liposomes or other colloidal systems. This review will focus on the in vivo evaluation of Pc. A recent review deals with more chemical aspects of this class of photosensitizers [8].

5.2 Hydrophilic phthalocyanines

5.2.1 Peripheral substitutions

5.2.1.1 Sulfonated Pc
Water-soluble Pc are obtained by direct sulfonation of the corresponding unsubstituted Pc with fuming sulfuric acid [9] or alternatively by the condensation of various ratios of 4-sulfophthalic acid and phthalic acid (or substituted phthalic acid, phthalonitrile, phthalamide, phthalic anhydride) [10]. In both cases, this yields a mixture of many isomeric differently sulfonated Pc, which require time

consuming and tedious purification procedures. Determination of the degree of sulfonation of the various fractions can be carried out by an oxidative degradation process [11].

The effect of the degree of sulfonation on the photobiological properties of various metallo-Pc has been extensively studied. Decreasing the degree of sulfonation increases the degree of hydrophobicity and leads to higher photodynamic activities of Ga, Al and ZnPc both in vitro and in vivo [12–16]. In particular, the amphiphilic disulfonated Pc bearing two sulfo groups on adjacent benzo rings is the most active, due to better cell penetrating properties [17]. The subcellular localization of the differently sulfonated AlPc, studied by means of confocal laser scanning microscopy, showed that the mono- and di-sulfonated AlClPc localize diffusely in the cytoplasm while the tri and tetra derivatives are found in association with lysosomes. At high concentrations of the dyes, low laser light exposure resulted in a translocation from the lysosomes to the cytoplasm and an increase in fluorescence intensity due to disaggregation [18]. An attempt to correlate the ultrastructural changes seen in treated cells and tumors with the subcellular/intratumoral localization of differently sulfonated AlPc has been reviewed [19]. In vivo, $AlClPcS_2$, and to a lesser extent $AlClPcS_4$, accumulated in higher levels in tumor associated macrophages (TAM) rather than in malignant or other tumor associated cells [20]. Also, it has been shown to induce tumor regression mainly via direct tumor cell kill rather than damage to the tumor microvasculature, in contrast to Photofrin and $AlClPcS_1$ [21–27]. The potential of $AlClPcS_2$ to treat localized prostate cancer has been demonstrated in normal canine prostate [28].

More recently, an alternative pathway to obtain sulfonated Pc containing less possible isomers was developed via the ring expansion of a boron subphthalocyanine [29]. Differently substituted trisulfonated ZnPc obtained in this manner [30] were evaluated for their photobiological potential in the EMT-6 tumor model. The t-butylbenzo and t-butylnaphtho derivatives were equally photoactive in vitro while the bulky diphenylpyrazino substituent decreases the degree of cell photoinactivation. In vivo, the t-butylnaphtho group induced a photoactivity superior to that of the tetrasulfonated ZnPc at 2 μmol kg^{-1} and showed good promise for clinical applications owing to its broad absorption spectrum (λ_{max} 706 and 678 nm) and its water solubility [31].

5.2.1.2 Tetrasubstituted Pc
The water-soluble tetraphosphonate ZnPc was synthesized and tested for the PDT-induced EMT-6 tumor response in mice. Even at 10 μmol kg^{-1}, this dye was ineffective [32].

Four tetrasubstituted ZnPc varying in charge and hydrophobicity were evaluated in vivo using two rat tumors of different vascularity, the mammary carcinoma, LMC_1 and the fibrosarcoma ($LSBD_1$). The cationic bis-methylene-pyridinium derivative (ZnPPc) was more potent, as determined by the tumor growth delay, than the anionic derivatives, tetrasulfonate (ZnTSPc) and dicarboxylic acid dicarboxyamide (ZnTCPc) and the neutral dye, sulfonamide (ZnTSAPc). The more vascular tumor ($LSBD_1$) was more sensitive to PDT, suggesting an action mechanism acting mainly via vascular injury [33].

5.2.1.3 Octasubstituted Pc

A water-soluble octacarboxy ZnPc has been produced [34]. Despite interesting local tumor damage, this dye was not fully photoactive in inducing EMT-6 tumor response in Balb/c mice even at high concentration (10 μmol kg^{-1}) (Brasseur et al., unpublished results).

5.2.1.4 Binding to BSA and mal-BSA

Di- and tri-sulfonated Pc were covalently coupled to native BSA, via an amide bond using the carbodiimide method [35]. Maleylation of AlClPcS-BSA was performed in order to take advantage of the high affinity of the scavenger receptor for mal-BSA to selectively deliver water-soluble Pc to macrophages. Competition studies of the conjugates with ^{125}I-mal-BSA showed recognition of the BSA- and mal-BSA-Pc conjugates by the scavenger receptor present in J774 cells of monocyte-macrophage origin and enhanced binding affinity of mal-BSA-Pc conjugates over BSA-Pc conjugates. Cellular photoinactivation induced by the conjugates paralleled their relative affinity, with mal-BSA-AlPcS$_4$ coupled via two sulfon-amide-hexanoic-amide spacer chains showing the best activity. Such water-soluble conjugates may find applications in various macrophage-associated disorders, i.e. restenosis after transluminal angioplasty and rheumatoid arthritis.

5.2.2 Axial ligation on the central metal ion

Pc featuring metals of Group IV allow for one (Al) or two (Si, Ge) axial coordination on the central metal ion. This substitution results in single isomeric products and a reduced tendency to aggregate. Also, varying the type of axial group on the central metal allows modulation of the drug solubility.

5.2.2.1 Binding to polymers

Two water-soluble AlClPc derivatives were prepared by axial ligation of one poly(ethylene glycol) (PEG, MW 2000) or one poly(vinyl alcohol) (PVA, MW 13000–23000) with the Al(III) ion [36] and their photobiological potential was compared to that of AlClPc formulated in Cremophor. All three dyes injected iv at 0.25 μmol kg^{-1} 24 h before PDT induced complete EMT-6 tumor regression in 75–100% of mice. In the Colo-26 tumor model, 2 μmol kg^{-1} were required to achieve complete tumor regression in 30% of mice, while in the non-cured animals, AlPcPVA induced the most significant tumor regrowth delay. In addition, AlPcPVA showed a prolonged plasma half-life ($t_{1/2}$: 6.8 h) as compared to AlClPc ($t_{1/2}$: 2.6 h) and AlPcPEG ($t_{1/2}$: 23 min), lower retention by liver and spleen and higher tumor–skin and tumor–muscle ratios. These advantageous pharmacokinetics render AlPcPVA a valuable water-soluble candidate for clinical PDT of cancer.

The water-soluble silicon Pc bearing two PEG at the axial position on the central Si ion was also prepared by displacement of the chloro groups of the bis(chloro-propyldimethylsiloxy) SiPc (SiPcClPr) by the methoxypoly(ethylene glycol) amine (Brasseur, et al., unpublished results). In vitro, SiPcPEG$_2$ and SiPcClPr formulated, respectively, in PBS and in Cremophor emulsion were much more active than

AlPcPEG in photoinactivating EMT-6 cells after 1 h incubation while after 24 h the differences were less pronounced. Plasmatic concentration of $SiPcPEG_2$ injected iv at 2 μmol kg^{-1} in Balb/c mice showed a biexponential decay with a relatively fast absorption phase ($t_{1/2\alpha}$: 1 h) and a slower elimination phase ($t_{1/2\beta}$: 18.6 h). The extended circulation time of $SiPcPEG_2$ as compared to AlPcPEG could be explained by the presence of two PEG on the Si ion. In fact, a study by Yamaoka and colleagues [37] related that the half-life of PEG extended from 18 min to one day as the molecular weight increased from 6000 to 190000. Uptake by the reticulo-endothelial system was much lower while tumor retention was higher for $SiPcPEG_2$ as compared to AlPcPEG. However, muscle, normal skin and skin overlying the tumor accumulated as much $SiPcPEG_2$ as the tumor at least at 24 h post injection (pi). Phototreatment of EMT-6 tumor with $SiPcPEG_2$ at 0.75 μmol kg^{-1} 24 h pi induced mortality within 48 h after PDT. At non-lethal doses (0.3 μmol kg^{-1}), this dye induced poor tumor control and important edema. Similar systemic toxicity was observed with the hydrophobic SiPcClPr and was therefore attributed to the silicon metal rather to the PEG substitution. This toxicity limits the PDT potential of the water-soluble $SiPcPEG_2$ derivative.

5.3 Hydrophobic phthalocyanines

Hydrophobic Pc are usually prepared in Cremophor oil emulsions. However, because Cremophor is known to induce anaphylactic reactions in clinic [38], other drug formulations have been proposed.

5.3.1 Cremophor emulsions

5.3.1.1 Unsubstituted Pc
The unsubstituted AlClPc formulated in Cremophor oil emulsion (10% in saline) was shown to be preferentially retained by a gliosarcoma and was able to induce tumor necrosis in this model [39]. The same formulation induced complete EMT-6 tumor control at 0.25 μmol kg^{-1} in Balb/c mice, which renders it more photoactive than the mono- and disulfonated AlClPc [25].

Unsubstituted ZnPc, which is poorly soluble in the usual organic solvents, was initially dissolved in 1-methyl-2-pyrrolidinone or pyridine and delivered to mice in Cremophor. While the pyridine formulation induced complete EMT-6 tumor regression at 2 μmol kg^{-1}, the 1-methyl-2-pyrrolidinone preparation was inactive at the maximal concentration available of 1 μmol kg^{-1} [40].

5.3.1.2 Fluorinated Pc
Fluorinated ZnPc were synthesized in order to improve the solubility of the molecule so as to facilitate drug formulation and also as potential agents for fluorine magnetic resonance imaging. The hexadecafluorinated analog, $ZnPcF_{16}$, prepared in Cremophor from acetone solution was fully active in EMT-6 tumor-bearing mice at 5 μmol kg^{-1} while significant tumor growth delay was obtained at

2 μmol kg^{-1} [40]. The monosulfododecafluoro ZnPc (ZnPcF$_{12}$S$_1$) was obtained via substitution of a monoamino ring using the Meerwein procedure [41] and formulated in Cremophor emulsion. In spite of improved pharmacokinetics, this compound showed a very narrow therapeutic window in mice with a high mortality rate at low drug dose (0.5 μmol kg^{-1}). This was related to a very efficient cell photoinactivation [42].

5.3.1.3 Hydroxylated Pc

Structure–activity relationships between differently hydroxylated ZnPc prepared in Cremophor emulsion showed that ZnPc substituted with four hydroxy groups fused directly to the Pc skeleton was inactive in contrast to the tetrahydroxy derivatives bearing an aliphatic spacer chain. The tetrapropylhydroxy and tetrahexylhydroxy ZnPc induced 100% EMT-6 tumor cure at 0.5 and 1 μmol kg^{-1} respectively [44]. Unsymmetrical mono- and dihydroxy ZnPc were also evaluated. The 2,9-OH ZnPc induced complete EMT-6 tumor regression in 75% of mice at 2 μmol kg^{-1} while the 2-OH and 2,3-OH ZnPc resulted in poor tumor control [44]. These data confirm the higher photoefficiency of unsymmetrically substituted Pc bearing substituents on two adjacent benzene rings.

5.3.1.4 Octasubstituted Pc

The new potential photosensitizer octa-alpha-butyloxy ZnPc (8-alpha-bo-ZnPc) is characterised by a high absorption coefficient in the far-red (735 nm). However, this compound showed no accumulation in the Lewis lung carcinoma in mice over a one-week investigation period [45].

Three tetradibenzobarreleno-octabutoxy SiPc (TDiBOPc) bearing different axial ligands on the metal ion were evaluated as Cremophor emulsions (0.35 μmol kg^{-1}) for their localizing and photosensiting properties in Balb/c mice bearing an MS-2 fibrosarcoma. The tumor response was correlated with tumor uptake values, the bis(trihexylsiloxy) substituted Si-TDiBOPc derivative showing the best tumor–muscle ratio and the highest tumor growth delay [46].

Among a set of isomerically pure octa-alkyl ZnPc [47], the octapentyl (ZnOPPc) and octadecyl (ZnODPc) derivatives were prepared in Cremophor emulsions and evaluated in Balb/c mice bearing the MS-2 fibrosarcoma in comparison with ZnPc formulated in dipalmitoylphosphaditylcholine (DPPC) liposomes. The efficiency and selectivity of tumor targeting slightly increased upon increasing the length of the alkyl substituents. Phototreatment of the tumor 24 h after drug administration at 1.46 μmol kg^{-1} induced a significant delay of tumor growth which was largest for ZnPc and smallest for ZnODPC [48–50].

5.3.1.5 Axially ligated Pc

A series of aluminium and silicon Pc substituted axially with one or two mono- or diamino bearing ligands were evaluated for in vitro structure–activity relationships [51]. The silicon Pc4 (HOSiPc-OSi(CH$_3$)$_2$[CH$_2$]$_3$N(CH$_3$)$_2$) bearing one mono-aminosiloxy ligand offered the most desirable structural features [52]. The silicon Pc dyes, Pc4, Pc10 (one diaminosiloxy ligand), Pc12 (two aminosiloxy ligands) and Pc18 (two diaminosiloxy ligands), were administered in C3H mice in Cremophor

5%/ethanol 5% saline solutions. All four dyes were found to be highly effective in causing the regression of the RIF-1 tumor at low drug dosage (0.83 μmol kg^{-1}) while inducing little cutaneous phototoxicity [53,54]. The potential of Pc4 was recently confirmed in a human tumor model, the ovarian epithelial carcinoma OVCAR-3, transplanted SC in athymic nude mice [55].

5.3.2 Liposomes

The pharmacokinetics and photodynamic activity of ZnPc formulated in small unilamellar DPPC liposomes have been extensively studied [56,57]. A preparation of ZnPc in liposomes consisting of a mixture of 1-palmitoyl-2-oleoylphosphatidyl-choline (POPC) and 1,2-dioleoylphosphatidylserine (OOPS) (9:1) showed advantageous pharmacokinetics and has been proposed for clinical use as a proprietary formulation [60,61].

Four octabutoxygermanium (IV) phthalocyanine (GePc-OBu$_8$) bearing two triethyl-, dimethyl(3-acetoxypropyl)-, trihexyl- or tridecyl-siloxy ligands were synthesized. All dyes showed lower tumor uptake and tumor-to-muscle ratios when formulated in DPPC liposomes as compared to Cremophor emulsions [60,61]. The bis(trihexyl)siloxy-GePc-OBu$_8$ in Cremophor showed the highest phototherapeutic activity towards MS-2 fibrosarcoma [61]. GePc bearing two axially ligated cholesterol moieties was also encapsulated in POPC/OOPS (9:1) liposomes and showed good tumor localization in the MS-2 fibrosarcoma implanted IM in mice. PDT performed 24 h pi of 1.52 mg kg^{-1} dye resulted in fast and massive tumor necrosis [62]. The pharmacokinetics of various photosensitizers including ZnPc and GePc derivatives in these different formulations have been reviewed [63].

Dichloro SiPc (SiPcCl$_2$) and two SiPc bearing two axial alkylsiloxy ligands were incorporated into (7/3) DPPC/cholesterol liposomes and tested in vitro [64]. The bis(tri-n-hexylsiloxy)SiPc (PcHEX) was less active than AlClPc whereas the bis(triphenylsiloxy)SiPc (PcPHE) was inactive in photoinactivating TF-1 and Daudi leukemic cells. The phototoxicity of SiPcCl$_2$ and PcHEX was also tested against the human amelanotic melanoma cell line M6 and healthy human keratinocytes and melanocytes and was related to the vehicle of administration, solvent-PBS mixture, egg-yolk lecithin liposomes or Cremophor micelles [65]. We have synthesized various bis(alkyl-siloxy)SiPc and tested their PDT potential in the EMT-6 tumor model when formulated in Cremophor emulsion. Complete tumor regression was obtained at 2 μmol kg^{-1} respectively in 75, 66 and 33% of mice with the bis(tri-n-hexylsiloxy) SiPc (PcHEX), the bis(dimethyloctylsiloxy) SiPc and the bis (dimethyl-thexylsiloxy) SiPc. However, mortality was reported in a few cases with all three dyes (Brasseur et al., unpublished).

5.3.3 Nanoparticles

The hexadecafluoro ZnPc ZnPcF$_{16}$ has also been formulated in biodegradable poly(D,L-lactic acid) (PLA) nanoparticles and in PEG-coated PLA nanoparticles. As expected, PEG-coated nanoparticles displayed a reduced uptake by the reticuloendothelial system leading to an enhanced tumor uptake as compared to plain nanoparticles. Plasma clearance of $ZnPcF_{16}$ in PEG-coated nanoparticles was faster than in CRM formulation, while the tumor uptakes were quite similar with both preparations [66]. Formulation in PEG-coated nanoparticles improved the photodynamic response of the EMT-6 tumor (complete tumor regression in 63% of mice at 1 μmol kg^{-1}) while providing prolonged tumor sensitivity towards PDT as compared to the Cremophor emulsion [67].

5.3.4 Nanomicelles

Recently, polymeric nanomicelles of AlClPc have been prepared showing photo-dynamic in vitro activity similar to that of AlClPc Cremophor [68]. Preliminary results of its in vivo evaluation confirmed the potential of this new vehicle in PDT while its inocuity remains to be tested.

5.3.5 Binding to BSA

An alternative to the formulation of the unsubstituted ZnPc is through its non-covalent binding to bovine serum albumin (BSA). BSA-delivered ZnPc readily redistributes from BSA to high density lipoproteins after intravenous administration, as has been shown for liposomal preparations of ZnPc [56]. Tumor control was obtained at 0.5 μmol kg^{-1} both in EMT-6 tumor-bearing Balb/c mice and in T 380 human colon carcinoma in nude mice [69].

5.3.6 Binding to LDL

ZnPc has also been complexed with human low-density lipoprotein (LDL). This results in higher tumor uptake and tumor–muscle ratio as compared with the DPPC liposome preparation [56,70].

5.4 Clinical trials

While several phthalocyanine sensitizers show interesting properties for clinical PDT, to date only a few are undergoing clinical evaluation. A mixture of AlClPc bearing 2 to 4 sulfonato groups, called Photosense, is presently used successfully in clinical PDT in Russia [71,72]. Also, a liposomal formulation of ZnPc is in Phase I/II clinical trials in Switzerland [58,73].

References

1. I. Rosenthal, E. Ben-Hur (1995). Role of oxygen in the phototoxicity of phthalocyanines. *Int. J. Radiat. Biol.*, **67**, 85–91.
2. M. Ochsner (1997). Photophysical and photobiological processes in the photodynamic therapy of tumors. *J. Photochem. Photobiol. B Biol.*, **39**, 1–18.

3. D. Phillips (1997). Chemical mechanisms in photodynamic therapy with phthalocyanines (review). *Progr. React. Kinet.*, **22**, 175–300.

4. W.M. Sharman, C.M. Allen, J.E. van Lier (1999). The role of activated oxygen species in photodynamic therapy. In: H. Sies, L. Packer (Eds), *Singlet Oxygen UV-A and Ozone*, Methods in Enzymology, Academic Press: San diego, CA.

5. J.E. van Lier and J.D. Spikes (1989), in *Photosensitizing compounds: their chemistry, biology and clinical use*, T.J. Dougherty, G. Bock and S. Harnett (eds). (CIBA foundation Symposium 146), Wiley, Chichester, p 17.

6. I. Rosenthal (1991). Phthalocyanines as photodynamic sensitizers. Yearly review. *Photochem. Photobiol.*, **53**, 859–870.

7. J.E. Spikes (1986). Yearly review: phthalocyanines as photosensitizers in biological systems and for the photodynamic therapy of tumors. *Photochem. Photobiol.*, **43**, 691–699.

8. H. Ali, J.E. van Lier (1999). Metal complexes as radio- and photosensitizers.

9. R.P. Linstead and F.T. Weiss (1950). Phthalocyanines and related compounds – Part XX. Further investigations on tetrabenzoporphin and allied substances. *J. Chem. Soc.*, 2975.

10. J.H. Weber and D.H. Busch (1965). Complexes derived from strong field ligands – XIX. Magnetic properties of transition metal derivatives of 4,4',4",4'''-tetrasulfophthalocyanines. *Inorg. chem*, **4**, 469–471.

11. H. Ali, R. Langlois, J.R. Wagner, N. Brasseur, B. Paquette, J.E. van Lier (1988). Biological activities of phthalocyanines – X. Synthesis and analyses of sulfonated phthalocyanines. *Photochem. Photobiol.*, **47**, 713–717.

12. N. Brasseur, H. Ali, R. Langlois, J.E. van Lier (1987). Biological activities of phthalocyanines – VII. Photoinactivation of V-79 Chinese hamster cells by selectively sulfonated gallium phthalocyanines. *Photochem. Photobiol.*, **46**, 739–744.

13. N. Brasseur, H. Ali, R. Langlois, J.E. van Lier (1988). Biological activities of phthalocyanines -IX. Photosensitization of V-79 Chinese hamster cells and EMT-6 mouse mammary tumor by selectively sulfonated zinc phthalocyanines. *Photochem. Photobiol.*, **47**, 705–711.

14. K. Berg, J.C. Bommer, J. Moan (1989). Evaluation of sulfonated aluminum phthalocyanines for use in photochemotherapy. A study on the relative efficiencies of photoinactivation. *Photochem. Photobiol.*, **49**, 587–594.

15. W.-S. Chan, J.F. Marshall, R. Svensen, J. Bedwell, I.R. Hart (1990). Effect of sulfonation on the cell and tissue distribution of the photosensitizer aluminum phthalocyanine. *Cancer Res.*, **50**, 4533–4538.

16. P. Margaron, M.-J. Grégoire, V. Ščasnár, H. Ali, J.E. van Lier (1996). Structure-photodynamic activity relationships of a series of 4-substituted zinc phthalocyanines. *Photochem. Photobiol.*, **63**, 217–223.

17. B. Paquette, H. Ali, R. Langlois, J.E. van Lier (1988). Biological activities of phthalocyanines – VIII. Cellular distribution in V-79 Chinese hamster cells and phototoxicity of selectively sulfonated aluminum phthalocyanines. *Photochem. Photobiol.*, **47**, 215–220.

18. Q. Peng, G.W. Farrants, K. Madslien, J.C. Bommer, J. Moan, H.E. Danielsen, J.M. Nesland (1991). Subcellular localization, redistribution and photobleaching of sulfonated aluminum phthalocyanines in a human melanoma cell line. *Int. J. Cancer*, **49**, 290–295.

19. Q. Peng, J. Moan, J.M. Nesland (1996). Correlation of subcellular and intratumoral photosensitizer localization with ultrastructural features after photodynamic therapy (review). *Ultrastruct. pathol.*, **20**, 109–129.

20. M. Korbelik (1993) Distribution of disulfonated aluminium pthalocyanine between malignant and host cell populations of a murine fibrosarcoma. *J. Photochem. Photobiol. B*, **32**, 173–181.

21. Q. Peng, J. Moan, J.M. Nesland, C. Rimington (1990). Aluminium phthalocyanines with asymmetrical lower sulfonation and with symmetrical higher sulfonation: a comparison of localizing and photosensitizing mechanism in human tumor LOX xenografts. *Int. J. Cancer*, **46**, 719–726.

22. V.H. Fingar, T.J. Wieman, P.S. Karavolos, K.W. Doak, R. Ouellet, J.E. van Lier (1993). The effects of photodynamic therapy using differently substituted zinc phthalocyanines on vessel constriction, vessel leakage and tumor response. *Photochem. Photobiol.*, **58**, 251–258.

23. H.L.L.M. Van Leengoed, N. van der Veen, A.A.C. Versteeg, R. Ouellet, J.E. van Lier, V.M. Star (1993). In vivo photodynamic effects of phthalocyanines in a skin fold observation chamber model: role of central metal ion and degree of sulfonation. *Photochem. Photobiol.*, **58**, 575–580.

24. W.-S. Chan, N. Brasseur, C. La Madeleine, J.E. van Lier (1996). Evidence for different mechanisms of EMT-6 tumor necrosis by photodynamic therapy with disulfonated aluminum phthalocyanine or photofrin: tumor cell survival and blood flow. *Anticancer Res.*, **16**, 1887–1892.

25. W.-S. Chan, N. Brasseur, C. La Madeleine, R. Ouellet, J.E. van Lier (1997). Efficacy and mechanism of aluminum phthalocyanine and its sulfonated derivatives mediated photodynamic therapy on murine tumors. *Eur. J. Cancer*, **33** (11), 1855–1859.

26. P. Margaron, P. Madarnas, R. Ouellet, J.E. van Lier (1996). Biological activities of phthalocyanines XVII. Histopathologic evidence for different mechanisms of EMT-6 tumor necrosis induced by photodynamic therapy with disulfonated aluminum phthalocyanine or Photofrin. *Anticancer Res.*, **16**, 613–620.

27. D. Lapointe, N. Brasseur, J. Cadorette, C. La Madeleine, S. Rodrigue, J.E. van Lier, R. Lecomte (1999). High-resolution PET imaging for in vivo monitoring of tumor response after photodynamic therapy in mice. *J. Nucl. Med.*, **40**, 876–882.

28. S.C. Chang, G.A. Buonaccorsi, A.J. MacRobert, S.G. Bown (1997). Interstitial photodynamic therapy in the canine prostate with disulfonated aluminum phthalocyanine and 5-aminolevulinic acid-induced protoporphyrin IX. *Prostate*, **32**, 89–98.

29. N. Kobayashi, R. Kondo, S. Nakajima, T. Osa (1990). New routes to unsymmetrical phthalocyanine analogues by the use of structurally distorted subphthalocyanines. *J. Am. Chem. Soc.*, **112**, 9640–9641.

30. S.V. Kudrevich, S. Gilbert, J.E. van Lier (1996). Syntheses of trisulfonated phthalocyanines and their derivatives using boron(III) subphthalocyanines as intermediates. *J. Org. Chem.*, **61**, 5706–5707.

31. S.V. Kudrevich, N. Brasseur, C. La Madeleine, S. Gilbert, J.E. van Lier (1997). Syntheses and photodynamic activities of novel trisulfonated zinc phthalocyanine derivatives. *J. Med. Chem.*, **40**, 3897–3904.

32. W.M. Sharman, S.V. Kudrevich, J.E. van Lier (1996). Novel water-soluble phthalocyanines substituted with phosphonate moieties on the benzo rings. *Tetrahedron Lett.*, **37**, 5831–5834.

33. J.E. Cruse-Sawyer, J. Griffiths, B. Dixon, S.B. Brown (1998). The photodynamic response of two rodent tumor models to four zinc(II) substituted phthalocyanines. *Br. J. Cancer*, **77**, 965–972.

34. S.V. Kudrevich, M.G. Galpern and J.E. van Lier (1994). Synthesis of Octacarboxytetra (2,3-pyrazino)porphyrazine: novel water soluble photosensitizers for photodynamic therapy. *Synthesis*, **8**, 779–781.

35. N. Brasseur, R. Langlois, C. La Madeleine, R. Ouellet, J.E. van Lier (1999). Receptor mediated targeting of phthalocyanines to macrophages via covalent coupling to native or maleylated bovine serum albumin. *Photochem. Photobiol.*, **69**, 345–352.

36. N. Brasseur, R. Ouellet, C. La Madeleine, J.E. van Lier (1999). Water-soluble aluminium phthalocyanine-polymer conjugates for PDT: photodynamic activities and pharmacokinetics in tumor-bearing mice. *Br. J. Cancer*, **80**, 1533–1541.

37. T. Yamaoka, Y. Tabata, Y. Ikada (1994). Distribution and tissue uptake of poly(ethylene glycol) with different molecular weights after intravenous administration to mice, *J. Pharm. Sci.*, **83**, 601–606.

38. D. Dye, J. Watkins (1980). Suspected anaphylactic reaction to Cremophor EL. *Br. Med. J.*, **280**, 1353.

39. M.O. Dereski, L. Madigan, M. Chopp (1994). Brain response to photodynamic therapy with Photofrin, monsulfonated aluminum phthalocyanine and tin purpurin. In: D.A. Cortese (Ed.), PROC SPIE 2371, 579–581.

40. R.W. Boyle, J. Rousseau, S.V. Kudrevich, M.O.K. Obochi, J.E. van Lier (1996). Hexadecafluorinated zinc phthalocyanine: photodynamic properties against the EMT-6 tumor in mice and pharmacokinetics using ^{65}Zn as a radiotracer. *Br. J. Cancer*, **73**, 49–53.

41. S.V. Kudrevich, H. Ali, J.E. van Lier (1994). Syntheses of monosulfonated phthalocyanines, benzonaphthoporphyrazines and porphyrins via the Meerwein reaction. *J. Chem. Soc., Perkin Trans.*, **1**, 2767–2774.

42. E. Allémann, N. Brasseur, S.V. Kudrevich, C. La Madeleine, J.E. van Lier (1997). Photodynamic activities and biodistribution of fluorinated zinc phthalocyanine derivatives in the murine EMT-6 tumor model. *Int. J. Cancer*, **72**, 289–294.

43. R.W. Boyle, C.C. Leznoff, J.E. van Lier (1993). Biological activities of phthalocyanines – XVI. Tetrahydroxy- and tetraalkylhydroxy zinc phthalocyanines. Effect of alkyl chain length on in vitro and in vivo photodynamic activities. *Br. J. Cancer*, **67**, 1177–1181.

44. M. Hu, N. Brasseur, S.Z. Yildiz, J.E. van Lier, C.C. Leznoff (1998). Hydroxyphthalocyanines as potential photodynamic agents for cancer therapy. *J. Med. Chem.*, **41**, 1789–1802.

45. M.S. Ismail, C. Dressler, P. Koeppe, R.G. Senz, B. Roder, H. Weitzel, H.P. Berlien (1997). Pharmacokinetic analysis of octa-alpha-butyloxy-zinc phthalocyanine in mice bearing Lewis lung carcinoma. *J. Clin. Laser Med. Surgery*, **15**, 157–161.

46. M. Soncin, A. Busetti, E. Reddi, G. Jori, B.D. Rihter, M.E. Kenney, M.A. Rodgers (1997). Pharmacokinetic and phototherapeutic properties of axially substituted Si(IV)-tetradibenzobarreleno-octabutoxyphthalocyanines. *J. Photochem. Photobiol. B Biol.*, **40**, 163–167.

47. M.J. Cook, I. Chambrier, S.J. Cracknell, D.A. Mayes, D.A. Russell (1995). Octa-alkyl zinc phthalocyanines: potential photosensitizers for use in the photodynamic therapy of cancer. *Photochem. Photobiol.*, **62**, 542–545.

48. C. Ometto, C. Fabris, C. Milanesi, G. Jori, M.J. Cook, D.A. Russell (1996). Tumor-localising and -photosensitising properties of a novel zinc(II) octadecylphthalocyanine. *Br. J. Cancer*, **74**, 1891–1899.

49. C. Fabris, C. Ometto, C. Milanesi, G. Jori, M.J. Cook, D.A. Russell (1997). Tumor-localizing and tumor-photosensitizing properties of zinc(II)-octapentyl-phthalocyanine. *J. Photochem. Photobiol. B Biol.*, **39**, 279–284.

50. G. Jori, C. Fabris (1998). Relative contributions of apoptosis and random necrosis in tumor response to photodynamic therapy: effect of the chemical structure of Zn(II)-phthalocyanines. *J. Photochem. Photobiol. B Biol.*, **43**, 181–185.

51. N.L. Oleinick, A.R. Antunez, M.E. Clay, B.D. Rihter, M.E. Kenney (1993). New phthalocyanine photosensitizers for photodynamic therapy. *Photochem. Photobiol.*, **57**, 242–247.

52. J. He, H.E. Larkin, Y.-S. Li, B.D. Rihter, S.I.A. Zaidi, M.A.J. Rodgers, H. Mukhtar, M.E. Kenney, N.L. Oleinick (1997). The synthesis, photophysical and photobiological properties and in vitro structure–activity relationships of a set of silicon phthalocyanine PDT photosensitizers. *Photochem. Photobiol.*, **65**, 581–586.

53. C. Anderson, S. Hrabovsky, Y. McKinley, K. Tubesing, H.P. Tang, R. Dunbar, H. Mukhtar, C.A. Elmets (1997). Phthalocyanine photodynamic therapy: disparate effects of pharmacologic inhibitors on cutaneous photosensitivity and on tumor regression. *Photochem. Photobiol.*, **65**, 895–901.

54. C.Y. Anderson, K. Freye, K.A. Tubesing, Y.S. Li, M.E. Kenney, H. Mukhtar, C.A. Elmets (1998). A comparative analysis of silicon phthalocyanine photosensitizers for in vivo photodynamic therapy of RIF-1 tumors in C3H mice. *Photochem. Photobiol.*, **67**, 332–336.

55. V.C. Colussi, D.K. Feyes, J.W. Mulvihill, Y.S. Li, M.E. Kenney, C.A. Elmets, N.L. Oleinick, H. Muktar (1999). Phthalocyanine 4 (Pc4) photodynamic therapy of human OVCAR-3 tumor xenografts. *Photochem. Photobiol.*, **69**, 236–241.

56. E. Reddi, C. Zhou, R. Biolo, E. Menegaldo, G. Jori (1990). Liposome- or LDL-administered Zn(II)-phthalocyanine as a photodynamic agent for tumors. I. Pharmacokinetics properties and phototherapeutic efficiency. *Br. J. Cancer*, **61**, 407–411.

57. M. Shopova, V. Mantareva, K. Krastev, H. Hadjiolov, A. Milev, K. Spirov, G. Jori, F. Ricchelli (1992). Comparative pharmacokinetic and photodynamic studies with zinc(II) phthalocyanine in hamsters bearing an induced or transplanted rhabdomyosarcoma. *J. Photochem. Photobiol. B Biol.*, **16**, 83–89.

58. K. Schieweck, H.-G. Capraro, U. Isele, P. Van Hoogevest, M. Ochsner, T. Maurer, E. Batt (1994). CGP 55847, liposome-delivered zinc(II)-phthalocyanine as a phototherapeutic agent for tumors. *Proc. Int. Soc. Opt. Eng.*, **2078**, 107–118.

59. U. Isele, K. Schieweck, R. Kessler, P. van Hoogevest, H.-G. Capraro (1995). Pharmacokinetics and body distribution of liposomal zinc phthalocyanine in tumor-bearing mice: influence of aggregation state, particle size and composition. *J. Pharm. Sci.*, **84**, 166–173.

60. V. Cuomo, G. Jori, B. Rihter, M.E. Kenney, M.A.J. Rodgers (1991). Tumor-localizing and photosensitizing properties of liposomes-delivered Ge(IV)-octabutoxyphthalocyanine. *Br. J. Cancer*, **64**, 93–95.

61. M. Soncin, L. Polo, E. Reddi, G. Jori, M.E. Kenney, G. Cheng, M.A.J. Rodgers (1995). Effect of axial ligation and delivery system on the tumor-localizing and photosensitizing properties of Ge(IV)-octabutoxy-phthalocyanines. *Br. J. Cancer*, **71**, 727–732.

62. A. Segalla, C. Milanesi, G. Jori, H.-G. Capraro, U. Isele, K. Schieweck (1994). CGP 55394, a liposomal Ge(IV) phthalocyanine bearing two axially ligated cholesterol moieties: a new potential agent for photodynamic therapy of tumors. *Br. J. Cancer*, **69**, 817–825.

63. R.W. Boyle and D. Dolphin (1996). Structure and biodistribution relationships of photodynamic sensitizers. *Photochem. Photobiol.*, **64**, 469–485.

64. J.P. Daziano, S. Steenken, C. Chabannon, P. Mannoni, M. Chanon, M. Julliard (1996). Photophysical and redox properties of a series of phthalocyanines. Relation with their photodynamic activities on TF-1 and Daudi leukemic cells. *Photochem. Photobiol.*, **64**, 712–719.

65. R. Decreau, M.J. Richard, P. Verrando, M. Chanon, M. Julliard (1999). Photodynamic activities of silicon phthalocyanines against achromic M6 melanoma cells and healthy human melanocytes and keratinocytes. *J. Photochem. Photobiol. B. Biol.*, **48**, 48–56.

66. E. Allémann, N. Brasseur, O. Benrezzak, J. Rousseau, S.V. Kudrevich, R.W. Boyle, J.C. Leroux, R. Gurny, J.E. van Lier (1995). PEG-coated poly(lactic acid) nanoparticles for the delivery of hexadecafluoro zinc phthalocyanine to EMT-6 mouse mammary tumors. *J. Pharm. Pharmacol.*, **47**, 382–387.

67. E. Allémann, J. Rousseau, N. Brasseur, S.V. Kudrevich, K. Lewis, J.E. van Lier (1996). Photodynamic therapy of tumors with hexadecafluoro zinc phthalocyanine formulated in PEG-coated poly(lactic acid) nanoparticles. *Int. J. Cancer*, **66**, 1–4.

68. J. Taillefer, M.-C. Jones, N. Brasseur, J.E. van Lier, J.-C. Leroux (2000). Preparation and characterisation of pH-responsive polymeric micelles for the delivery of anticancer photosensitizing drugs. *J. Pharm. Sci.*, **89**, 52–62.

69. C. Larroque, A. Pelegrin and J.E. van Lier (1996). Serum albumin as a vehicle for zinc pthalocyanine: photodynamic activities in solid tumor models. *Br. J. Cancer*, **74**, 1886–1890.

70. E. Reddi, S. Cernuschi, R. Biolo, G. Jori (1990). Liposome- or LDL-administered Zn(II)-phthalocyanine as a photodynamic agent for tumors. III. Effect of cholesterol on pharmacokinetics and phototherapeutic properties. *Lasers. Med. Sci.*, **5**, 339–343.

71. N.N. Zharkova, D.N. Kozlov, V.V. Smirnov, V.V. Sokolov, V.I. Chissov, E.V. Filonenko, D.G. Sukhin, M.G. Galpern, G.N. Vorozhtsov (1994). *SPIE - Photodynamic Therapy of Cancer II*, **2325**: 400.

72. A.S. Sobolev, E.F. Stranadko. Photodynamic therapy in Russia. Clinical and fundamental aspects. In: S. Brown (Ed.), International photodynamics. A PDT forum. Eurocommunication Publications (Vol. 1, No 6. April 1997). West Sussex, UK.

73. C. Hadjur, G. Wagnières, F. Ihringer, P. Monnier, H. van den Bergh (1997). Production of the free radicals $O2^-$ and $\cdot OH$ by irradiation of the photosensitizer zinc(II) phthalocyanine. *J. Photochem. Photobiol. B Biol.*, **38**, 196–202.

Chapter 6

Combining photodynamic therapy with antiangiogenic therapy

Charles J. Gomer, Angela Ferrario, Karl von Tiehl, Margaret A. Schwartz, Prakash S. Gill and Natalie Rucker

Table of contents

6.1 Photodynamic therapy

Photodynamic therapy (PDT) is an effective localized procedure for the clinical treatment of solid malignancies [1–3]. Properties of photosensitizer localization in tumor tissue and photochemical generation of reactive oxygen species are combined with a precise delivery of laser generated light to produce a procedure offering effective local tumoricidal activity [4,5]. A porphyrin, Photofrin (PH) is the photosensitizer used in the majority of clinical trials. PH-mediated PDT has received FDA approval for treatment of esophageal and endobronchial carcinomas and PH has also received regulatory approval in Canada, Japan and several European countries [3]. PDT is also used for treating tumors of the brain, bladder, skin, head and neck, and cervix, as well as for non-malignant disorders such as psoriasis and age-related macular degeneration [1,6].

The development and clinical evaluation of new photosensitizers exhibiting improved pharmacological, photochemical and/or photophysical properties is an active research area [7]. These compounds exhibit a number of properties thought to be comparable or superior to Photofrin, including chemical purity, increased photon absorption at longer wavelengths, improved tumor tissue retention, rapid clearance from normal tissues, high quantum yields of reactive oxygen species and minimal dark toxicity [8–10].

As with cancer treatments, PDT does not uniformly lead to a total eradication of tumors in all patients. It is, therefore, important to examine the mechanisms associated with tumor recurrence following PDT and to evaluate methods to improve the efficacy of PDT.

6.2 Angiogenesis and tumor treatment strategies

Tumor growth depends on the generation of new blood vessels [11]. The discovery of positive and negative regulators of angiogenesis has led to heightened understanding of angiogenesis and to the development of new strategies for inhibiting pathological angiogenesis, particularly in cancer [12]. The establishment of new blood vessels occurs in a series of sequential steps. An endothelial cell forming the wall of an existing blood vessel becomes activated, produces matrix metalloproteinases to degrade the extracellular matrix, invades into the matrix and begins to divide, and creates hollow vascular tubes [13]. Numerous proteins can activate endothelial cell growth and movement [14]. VEGF and bFGF are among the most important angiogenic activators for sustaining tumor growth [15,16]. A variety of MMPs as well as integrins such as $\alpha v\beta 3$ and $\alpha v\beta 5$ are also intimately involved in endothelial cell viability and movement [17–20]. Novel strategies for inhibiting angiogenesis and subsequent tumor growth are being developed and evaluated in many laboratories [11,12,21]. Targets include molecules involved with the various biochemical pathways of angiogenesis. Generalized antiangiogenic approaches include: (a) blocking the ability of endothelial cells to degrade the extracellular matrix; (b) inhibiting the endothelial cells directly; (c) blocking the action of factors such as VEGF which stimulate angiogenesis; and (d) blocking the action of integrins. There are approximately 20 antiangiogenic agents in various clinical trials [22].

Some of the most promising treatment strategies involve the use of angiogenesis inhibitors in combination with surgery, chemotherapy or radiotherapy [23–27].

6.3 PDT induction of pro-angiogenic molecules

As mentioned before, numerous laboratories have demonstrated that PDT induces tumor tissue hypoxia [28–31]. Many studies have also demonstrated that hypoxia is strongly associated with gene activation [32,33]. A primary step in hypoxia-mediated gene activation is the formation of the HIF-1 transcription factor complex [34]. HIF-1 is a heterodimeric complex of two helix-loop-helix proteins, ARNT and HIF-1α [35]. ARNT is constitutively expressed while HIF-1α is rapidly degraded under normoxic conditions. Hypoxia stabilizes the HIF-1α subunit allowing the formation of the transcriptionally active protein complex [35,36]. Our recent data show that PDT treatment of mouse mammary tumors induces expression of HIF-1α. A number of HIF-1 responsive genes have been identified including VEGF and erythropoietin [35]. We also have recent data demonstrating that PDT can increase protein levels of the HIF-1 target gene, VEGF, within treated tumors. Hypoxia associated with PDT can, therefore, act as a potent stimulator of VEGF expression within tumor tissue [37]. VEGF in turn can activate additional pro-angiogenic molecules and pathways [15,20,38,39]. It is currently not known whether PDT-mediated oxidative stress, as compared to PDT-induced hypoxia, is also involved in this process. Interestingly, PDT induces expression of early response genes (c-fos, c-jun, c-myc, egr-1) and other transcription factors such as NF-κB [40–43]. These factors (AP-1 and NFκB) have also been reported to play a role in the transcriptional activation of pro-angiogenic genes [44–50].

VEGF is a potent endothelial cell mitogen involved with the induction and maintenance of tumor neovasculature [51,52]. VEGF increases the endothelial cell permeability, stimulates endothelial cell migration, induces differentiation, and acts as an endothelial cell survival factor [51,52]. The VEGF gene is naturally expressed through alternative splicing as four mRNA transcripts, coding for secreted proteins of 206, 189, 165, and 121 residues [53]. VEGF expression increases in tumor tissue under hypoxia due to both transcriptional activation and increased stabilization [54,55]. VEGF proteins exert their functions on endothelial cells through two specific receptors, flt-1 and KDR [56].

Our recent experiments indicate that PDT can elicit the up-regulation of a variety of pro-angiogenic genes. This could have significant clinical implications. It is important to characterize both the mechanism of activation (oxidative stress and/or hypoxia related) as well as the PDT parameters producing activation and the biological relevance of this activation.

6.4 Antiangiogenic treatment and PDT

There are a growing number of studies evaluating the combination of antiangiogenic therapy with cytotoxic procedures such as chemotherapy and radiation therapy

[21,23–27]. Antiangiogenic agents are more effective against small tumors and this factor supports the adjuvant therapy approach [11]. Combining radiation and angiostatin resulted in enhanced tumoricidal activity without increased toxicity towards normal tissue. Our recent work indicates a significant involvement of PDT with angiogenic molecules [37]. The Roswell Park group showed that antiangiogenic activity of receptor tyrosine kinase inhibitors (targeting growth factor receptors) enhanced the effectiveness of pyropheophorbide-mediated PDT in a mouse tumor model [57]. We hypothesize that combining PDT with antiangiogenic therapy can enhance long-term local tumor control without increasing normal tissue toxicity. This combination may also allow for a reduction in the PDT dose required to achieve complete tumor responses. We have shown that tumor bearing mice treated with PH–PDT and antiangiogenic therapy (IM862 or EMAP-II) have improved tumoricidal responses when compared to individual treatments. EMAP-II is a single chain polypeptide with potent antiangiogenic activity [58]. EMAP-II is in preclinical development and induces apoptosis in growing capillary endothelial cells and prevents vessel ingrowth. IM862 is a dipeptide of L-glutamyl-L-tryptophan and is currently being evaluated as an antiangiogenic agent in several clinical trials [59]. IM862 blocks angiogenesis by inhibiting the production of VEGF and by activating natural killer cells. Our study strongly supports our research goal of evaluating PDT plus antiangiogenic therapy.

Acknowledgements

This work was performed in conjunction with the Clayton Foundation for Research, and was supported by USPHS grant CA 31230 and DOD Army Medical research Grant BC981102.

References

1. T. Reynolds (1997). Photodynamic therapy expands its horizons. *J. Natl. Cancer Inst.*, **89**, 112–114.
2. A.M.R. Fisher, A.L. Murphree, C.J. Gomer (1995). Clinical and preclinical photodynamic therapy. *Laser Surgery Med.*, **17**, 2–31.
3. T.J. Dougherty, C.J. Gomer, B.W. Henderson, G. Jori, D. Kessel, M. Korbelik, J. Moan, Q. Peng (1998). Photodynamic Therapy. *J. Natl. Cancer Inst.*, **90**, 889–905.
4. C. Gomer (1989). Photodynamic therapy in the treatment of malignancies. *Sem. Hematol.*, **26**, 27–34.
5. B.W. Henderson, T.J. Dougherty (1992). How does photodynamic therapy work? *Photochem. Photobiol.*, **55**, 931–948.
6. U. Schmidt Erfurth, J.W. Miller, M. Sickenberg, *et al.* (1999). Photodynamic therapy with verteporfin for choroidal neovascularization caused by age-related macular degeneration: results of retreatments in a phase 1 and 2 study. *Arch. Ophthalmol.*, **117**, 1177–1178.
7. C.J. Gomer (1991). Preclinical examination of first and second generation photosensitizers used in photodynamic therapy. Yearly Review. *Photochem. Photobiol.*, **54**, 1093–1107.

8. T.J. Dougherty (1993). Photodynamic therapy. *Photochem. Photobiol.*, **58**, 895–900.
9. J. Moan, K. Berg (1992). Photochemotherapy of cancer, experimental research. *Photochem. Photobiol.*, **55**, 931–948.
10. Kessel (1989). Determinants of photosensitization by purpurins. *Photochem. Photobiol.*, **50**, 169–174.
11. A.L. Harris (1997). Antiangiogenesis for cancer therapy. *Lancet*, **349**, 13–15.
12. M. Klagsburn (1999). Angiogenesis and cancer, AACR special conference in cancer research. *Cancer Res.*, **59**, 487–490.
13. R. Bicknell, C.E. Lewis, N. Ferrara (1997). *Tumor angiogenesis*. Oxford University Press, Oxford.
14. R.K. Jain, K. Schlenger, M. Hockel, F. Yuan (1997). Quantitative angiogenesis assays: progress and problems. *Nat. Med.*, **3**, 1203–1208.
15. H.F. Dvorak, L.F. Brown, M. Detmar, A.M. Dvorak (1995). Vascular permeability factor/vascular endothelial growth factor, microvascular hyperpermeability, and angiogenesis. *Am. J. Pathol.*, **146**, 1029–1039.
16. G.T. Stavri, I.C. Zachary, P.A. Baskerville, J.F. Martin, J.D. Erusalimsky (1995). Basic fibroblast growth factor upregulates the expression of vascular endothelial growth factor in vascular smooth muscle cells. Synergistic interaction with hypoxia. *Circulation*, **92**, 11–14.
17. W.-H. Zhu, X. Guo, S. Villaschi, R.F. Nicosia (2000). Regulation of vascular growth and regression by matrix metalloproteinases in the rat aorta model of angiogenesis. *Lab. Invest.*, **4**, 545–555.
18. J.R. MacDougall, L.M. Matrisian (1995). Contributions of tumor and stromal matrix metalloproteinases to tumor progression, invasion and metastasis. *Cancer Metastasis Rev.*, **14**, 351–362.
19. B.P. Eliceiri, D.A. Cheresh (1999). The role of alpha integrins during angiogenesis: insights into potential mechanisms of action and clinical development. *J. Clin. Invest.*, **103**, 1227–1230.
20. D.R. Senger, K.P. Claffey, J.E. Benes, C.A. Perruzzi, A.P. Sergiou, M. Detmar (1997). Angiogenesis promoted by vascular endothelial growth factor; regulation through integrins. *Proc. Natl. Acad. Sci. U.S.A.*, **94**, 13612–13617.
21. J. Folkman (1998). Antiangiogenic gene therapy. *Proc. Natl. Acad. Sci. U.S.A.*, **95**, 9064–9066.
22. NIH Clinical Trials, http://cancertrials.nci.nih.gov/news
23. R.S. Herbst, H. Takeuchi, B.A. Teicher (1998). Paclitaxel/carboplatin administration along with antiangiogenic therapy in non-small cell lung and breast carcinoma models. *Cancer Chemo. Pharm.*, **41**, 497–504.
24. B.A. Teicher, E.A. Sotomayor, Z.D. Huang (1992). Antiangiogenic agents potentiate cytotoxic cancer therapies against primary and metastatic disease. *Cancer Res.*, **52**, 6702–6704.
25. H.J. Mauceri, N.N. Hanna, M.A. Beckett, D.H. Gorski, M.J. Staba, K.A. Stellato, K. Bigelow, R. Heimann, S. Gately, M. Dhanabal, G.A. Soff, V.P. Sukhatme, D.W. Kufe, R.R. Weichselbaum (1998). Combined effects of angiostatin and ionizing radiation in antitumor therapy. *Nature*, **394**, 287–291.
26. D.H. Gorski, M.A. Beckett, N.T. Jaskowiak, D.P. Calvin, H.J. Mauceri, R.M. Salloum, S. Seetharam, A. Koons, D.M. Hari, D.W. Kufe, R.R. Weichselbaum (1999). Blockage of the vascular endothelial growth factor stress response increases the antitumor effects of ionizing radiation. *Cancer Res.*, **59**, 3374–3378.
27. D.R. Shalinsky, J. Brekken, H. Zou, L.A. Bloom, C.D. McDermott, S. Zook, N.M. Varki, K. Appelt (1999). Marked antiangiogenic and antitumor efficacy of AG3340 in

chemoresistant human non-small cell lung cancer tumors: single agent and combination chemotherapy studies. *Clin. Cancer Res.*, **5**, 1905–1917.

28. I.P.J. van Geel, H. Oppelaar, P.F.J.W. Rijken, H.J.J.A. Bernsen, N.E.M. Hagemeier, A.L. van der Kogel, R.J. Hodgkiss, F.A. Stewart (1996). Vascular perfusion and hypoxic areas in RIF-1 tumors after photodynamic therapy. *Br. J. Cancer*, **73**, 288–293.

29. T.M. Sitnik, J.A. Hampton, A. B.W. Henderson (1998). Reduction of tumor oxygenation during and after photodynamic therapy in-vivo: effects of fluence rate. *Br. J. Cancer*, **77**, 1386–1394.

30. T.H. Foster, R.S. Murant, R.G. Byrant, R.S. Knox, S.L. Gibson, R. Hilf (1991). Oxygen consumption and diffusion effects in photodynamic therapy. *Radiat. Res.*, **126**, 296–303.

31. T.M. Busch, S.M. Hahn, S.M. Evans, C.J. Koch (2000). Depletion of tumor oxygenation during photodynamic therapy: detection by the hypoxia marker EF3. *Cancer Res.*, **60**, 2636–2642.

32. A.C. Koong, N.C. Denko, K.M. Hudson, C. Schindler, L. Swiersz, C. Koch, S. Evans, H. Ibrahim, Q.T. Le, D.J. Terris, A.J. Giaccia (2000). Candidate genes for the hypoxic tumor phenotype. *Cancer Res.*, **60**, 883–887.

33. P.J. Ratcliffe, J.F. O'Rourke, P.H. Maxwell, C.W. Pugh (1998). Oxygen sensing, hypoxia-inducible factor-1 and the regulation of mammalian gene expression. *J. Exp. Biol.*, **201**, 1153–1162.

34. G.L. Wang, G.L. Semenza (1993). General involvement of hypoxia-inducible factor in transcriptional response to hypoxia. *Proc. Natl. Acad. Sci. U.S.A.*, **90**, 4303–4308.

35. J.A. Forsythe, B.-H. Jiang, N.V. Iyer, F. Agani, S.W. Leung, R.D. Koos, G.L. Semenza (1996). Activation of vascular endothelial growth factor gene transcription by hypoxia-inducible factor 1. *Mol. Cell Biol.*, **16**, 4604–4613.

36. L.E. Huang, J. Gu, M. Schau, H.F. Bunn (1998). Regulation of hypoxia-inducible factor 1a is mediated by an oxygen dependent degradation domain via the ubiquitin proteasome pathway. *Proc. Natl. Acad. Sci. U.S.A.*, **95**, 7987–7992.

37. A. Ferrario, K. von Tiehl, N. Rucker, M. Schwarz, P. Gill, C.J. Gomer (2000). Anti-angiogenic treatment enhances photodynamic therapy responsiveness in a mouse mammary carcinoma. *Cancer Res.*, **60**, August 1st issue.

38. J. Folkman (1995). Angiogenesis in cancer, vascular, rheumatoid and other disease. *Nat. Med.*, **1**, 27–31.

39. K. Suzuma, H. Takagi, A. Otani, Y. Honda (1998). Hypoxia and vascular endothelial growth factor stimulate angiogenic integrin expression in bovine retinal microvascular endothelial cells. *Invest. Ophthal. Vis. Sci.*, **39**(6),1028–1035.

40. S.W. Ryter, C.J. Gomer (1993). Nuclear factor κB binding activity in mouse L1210 following Photofrin II-mediated photosensitization. *Photochem. Photobiol.*, **58**, 753–756.

41. M.C. Luna, S. Wong, C.J. Gomer (1994). Photodynamic therapy mediated induction of early response genes. *Cancer Res.*, **54**, 1374–1380.

42. G. Kick, G. Messer, A. Goetz, G. Plewig, P. Kind (1995). Photodynamic therapy induces expression of interleukin 6 by activation of AP-1 but not NF-kB DNA binding. *Cancer Res.*, **55**, 237–239.

43. D.J. Granville, C.M. Carthy, H. Jiang, J.G. Levy, B.M. McManus, J.Y. Matroule, J. Piette, D.W. Hunt (2000). Nuclear factor-kappa B activation by the photochemotherapeutic agent verteporfin. *Blood*, **95**, 256–262.

44. C.W. Lin, H.I. Georgescu, C.H. Evans (1993). The role of AP-1 in matrix metalloproteinase gene expression. *Agents & Actions*, **39**, 215–218.

45. K. Asakuno, M. Isono, Y. Wakabayashi, T. Mori, S. Hori, K. Kohno, M. Kuwano (1995). The exogenous control of transfected c-fos gene expression and angiogenesis in cells implanted into the rat brain. *Brain Res.*, **702**, 23–31.

46. P. Chiao, W. Wang, L. Ellis, D. Shen (1996). NF-κB regulates VEGF gene expression in human pancreatic cancers. *Am. Assoc. Cancer Res. Proc.*, **37**, 525.

47. T. Shono, M. Ono, H. Izumi, S. Jimi, K. Matsushima, T. Okamoto, K. Kohno, M. Kuwano (1996). Involvement of the transcription factor NF-κB in tubular morphogenesis of human microvascular endothelial cells by oxidative stress. *Mol. Cell. Biol.*, **16**, 4231–4239.

48. J. Moffett, E. Kratz, R. Florkiewicz, M.K. Stachowiak (1996). Promoter regions involved in density-dependent regulations of basic fibroblast growth factor gene expression human astrocytic cells. *Proc. Natl. Acad. Sci. U.S.A.*, **93**(6), 2470–2475.

49. Y.J. Le, P.M. Corry (1999). Hypoxia-induced bFGF gene expression is mediated through the JNK signal transduction pathway. *Mol. Cell Biochem.*, **202**, 1–8.

50. T.L. Haas, D. Stitelman, S.J. Davis, S.S. Apte, J.A. Madri (1999). Egr-1 mediates extracellular matrix-driven transcription of membrane type 1 matrix metalloproteinase in endothelium. *J. Biol. Chem.*, **274**(32), 22679–22685.

51. J. Folkman, Y. Shing (1992). Angiogenesis. *J. Biol. Chem.*, **267**, 10931–10934.

52. J. Grunstein, W.G. Roberts, O. Mathieu-Costello, D. Hanahan, R.S. Johnson (1999). Tumor-derived expression of vascular endothelial growth factor is a critical factor in tumor expansion and vascular function. *Cancer Res.*, **59**, 1592–1598.

53. N. Ferrara (1999). Vascular endothelial growth factor, molecular and biological aspects. *Curr. Top. Microbiol. Immunol.*, **237**, 1–30.

54. E. Ikeda, M.C. Achen, G. Breier, W. Risau (1995). Hypoxia induced transcriptional activation and increased mRNA stability of vascular endothelial growth factor in C6 glioma cells. *J. Biol. Chem.*, **270**, 19761–19766.

55. A. Damert, M. Machein, G. Breier, M.Q. Fujita, D. Hanahan, W. Risau, K.H. Plate (1997). Up-regulation of vascular endothelial growth factor expression in a rat glioma is conferred by two distinct hypoxia-driven mechanisms. *Cancer Res.*, **57**, 3860–3864.

56. W. Kolch, M.-G. Baron, A. Kieser, D. Marme (1995). Regulation of the expression of the VEGF/VPS and its receptors: role in tumor angiogenesis. *Breast Cancer Res. Treat.*, **36**, 139–155.

57. C.J. Dimitroff, W. Kloh, A. Sharma, P. Pera, D. Driscoll, J. Veith, R. Steinkampf, M. Schroeder, S. Klutchko, A. Sumlin, B. Henderson, T.J. Dougherty, R.J. Bernacki (1999). Anti-angiogenic activity of selected receptor tyrosine kinase inhibitors, PD166285 and PD173074, implications for combination treatment with photodynamic therapy. *Invest New Drugs*, **17**, 121–135.

58. M.A. Schwarz, J. Kandel, J. Brett, J. Li, J. Hayward, R.E. Schwarz, O. Chappey, J.-L. Wautier, J. Chabot, P.L. Gerfo, D. Stern (1999). Endothelial-monocyte activating polypeptide II, a novel antitumor cytokine that suppresses primary and metastatic tumor growth and induces apoptosis in growing endothelial cells. *J. Exp. Med.*, **190**, 341–353.

59. A. Tulpule, D.T. Scadden, B.M. Espina, S. Cabriales, W. Howard, K. Shea, P.S. Gill (2000). Results of a randomized study of IM862 nasal solution in the treatment of AIDS-related Kaposi's sarcoma. *J. Clin. Oncol.*, **18**, 716–723.

Chapter 7

Technologies and biophysical techniques for PDT

Brian C. Wilson

Table of contents

In this chapter the biophysical aspects and technologies for photodynamic therapy are considered. The former includes the tissue optics and factors governing the light distributions in vivo, and dynamic changes in photosensitizer and tissue oxygen during treatment. The "hardware" includes light sources, light delivery systems, and techniques and devices for measuring the various biophysical factors affecting the photodynamic response ("dosimetry"). In addition, methods used for monitoring the tissue response itself will be presented briefly viz. both spectroscopic and imaging.

As with many other aspects of PDT, these various techniques and technologies continue to evolve rapidly, especially in the past five years. Partly, this is in response to the needs in PDT itself, but mainly it is due to technology transfer from other fields, particularly from photonics (the science and technology of light generation, manipulation and measurement, that is emerging as a major branch of science with many applications, from fiber-optic telecommunications to materials processing to remote sensing and imaging). PDT has benefited also from developments in other areas of biophotonics, such as specialty medical optical fibers for light delivery, and advanced optical microscopy that has enabled the study of the microdistribution of photosensitizers in cells and tissues. PDT can also take credit for stimulating developments that have had much wider applications, as in the case of mathematical/ computational models and experimental methods to determine light propaga-tion in tissues that have become the standard tools for optimizing many other therapeutic and diagnostic clinical applications of lasers, and in vivo fluorescence imaging of photosensitizers that has spun off new endoscopic and surgical guidance techniques.

7.1 Light sources

There are currently three main classes of light source used in PDT: lasers, light emitting diodes (LEDs) and wavelength-filtered lamps. The basic requirements for PDT light sources, and how these are met by the different source types, are as follows.

7.1.1 Wavelength

The source must generate light at a wavelength, or over a wavelength range, that matches the in vivo activation spectrum of the photosensitizer. Figure 1 shows the absorption spectra of several photosensitizers used clinically at present. While most photosensitizers can be activated over a wide range of wavelengths, it is clearly best to activate the drug at its maximum absorption, since this will generate the greatest concentration of toxic photoproducts such as singlet oxygen ($^1O_2$). However, as illustrated in Figure 2 the penetration of light in tissues is strongly dependent on wavelength, due to the combined effects of light scattering by cells and other microstructures and absorption by specific tissue molecules, particularly hemoglobin, melanin and water. Above about 600 nm the absorption of hemoglobin falls off rapidly, so that the light penetration in most tissues correspondingly increases. Beyond about 800 nm, there is little gain in penetration and, in addition,

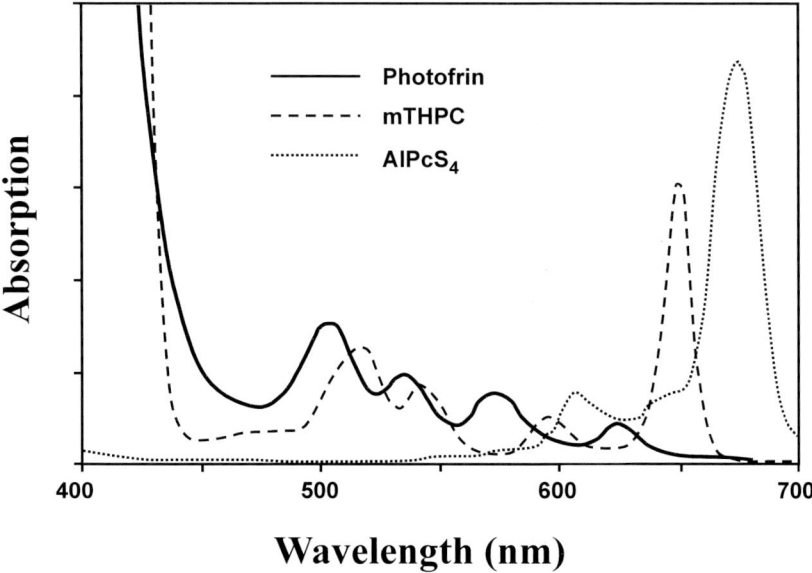

Figure 1. Absorption spectra of several photosensitizers used in PDT. Note the strong absorption peak in the far-red region (above 650 nm) for second-generation photosensitizers compared to the weak absorption at the maximum wavelength around 630 nm for the first-generation drug Photofrin.

the energy of the light photons is inversely proportional to wavelength so that there is not enough energy to generate singlet oxygen. In order to treat larger depths or, in the case of interstitial techniques, greater diameters of tissue, it may be better to use a longer wavelength even if the absorption of the photosensitizer is not at a maximum. The classic example is that of hematoporphyrin derivative (Photofrin), where most treatments have used red light (around 630 nm), activating the smallest of the absorption peaks, since this gives greater effective treatment depth than using blue light to activate the large (Soret) peak.

It should also be noted that the effective activation spectrum of photosensitizers is not necessarily exactly the same as their absorption spectrum in solution. The peak wavelengths may be shifted by several nanometers, due to binding of the photosensitizer to biomolecules (e.g. proteins). If a narrow wavelength source (e.g. laser) is used, it is important to match the wavelength to the true in vivo activation peak, so this needs to be determined, typically by studies in animal models. With broad-band sources (LEDs, lamps), this is less critical.

As illustrated in Figure 3, with broad-band sources there will be some overlap (not always 100%) between the source and the photosensitizer. Once a light photon is absorbed by the photosensitizer, the probability of generating ($^1O_2$) is independent of wavelength. Hence, the effective power of the source is given by:

$$P_{eff} = \int P(\lambda)A(\lambda)d\lambda \tag{1}$$

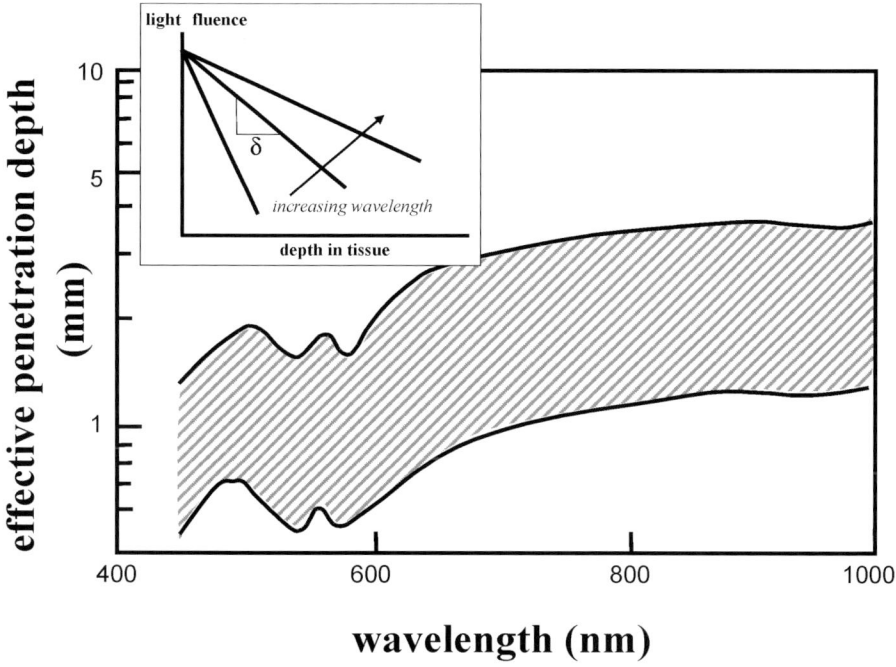

Figure 2. Penetration depth of light as a function of wavelength in tissue. The curves indicate approximate upper and lower bounds for the penetration depth for typical soft tissues. For example, liver is highly absorbing and has a penetration depth spectrum similar to the lower curve, whereas tissues such as brain or muscle have much lower pigmentation and so have larger penetration depths typically like the top curve. The insert shows the approximately exponential decrease in light fluence or fluence rate with depth in tissue and the definition of the effective penetration depth, δ.

i.e. by the overlap of the output power spectrum of the source, $P(\lambda)$, and the absorption (more correctly, activation) spectrum of the photosensitizer, $A(\lambda)$. For lasers, the power spectrum is a spike at a single wavelength, so that the effective power is the same as the actual power. However, for LEDs and filtered lamps, if the power spectrum is the same width as the photosensitizer spectrum, then the effective power will be only about 50% of the actual power delivered. That is, a 1 W laser at the absorption peak would be equivalent, in terms of singlet oxygen generation and hence PDT effect, to an LED or lamp source of about 2 W light output.

7.1.2 Power

For most photosensitizers used clinically at present for cancer treatment, and with surface irradiation, the typical energy densities required are around 100 J cm^{-2}. The energy densities for some other treatments (e.g. macular degeneration or actinic keratosis) are much less, and some photosensitizers (e.g. mTHPC) require much less

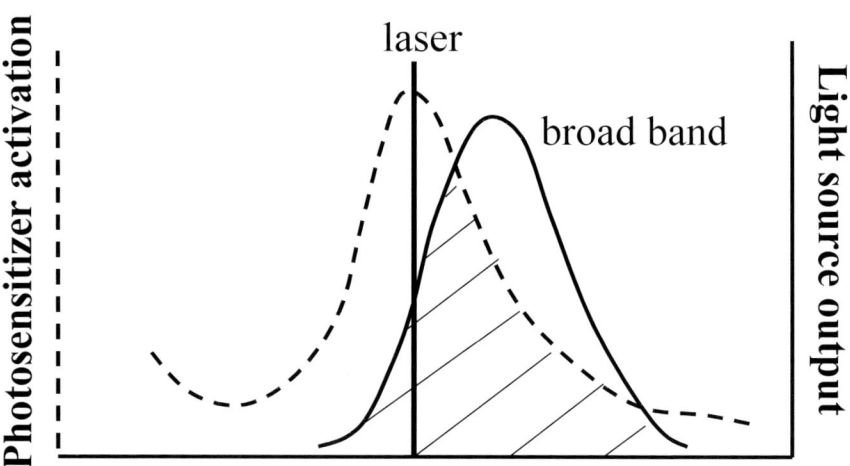

Figure 3. Illustration of the overlap between a photosensitizer absorption (activation) peak and a broad-band light source. The product of power and absorption is calculated at each wavelength and then summed over all wavelengths (equation 1) to obtain the effective power of the source for this photosensitizer. Also indicated is a laser source at a wavelength matching the photosensitizer peak.

light for the same PDT effect. However, we will use 100 J cm^{-2} as a representative value. For a tissue surface area of, say, 5 cm × 5 cm, the total energy required is then 2500 J. For a treatment time of 15 min (~900 s), the power required is then 2500/900 = 2.8 W. Note that this is the effective energy delivered. With a broad-band source, the actual power delivered would need to be about 5 W. Also note that the delivered power will be less than the source output, due to coupling and transmission losses. With laser-optical fiber systems the loss is only about 10–15%. However, it may be much larger with LEDs or lamps. For example, lamps generate light in all directions, but only a fraction can be collected and delivered to a local tissue area.

7.1.3 Laser sources

The greatest advantages of lasers are the high efficiency of coupling into single optical fibers and their monochromaticity (specific wavelength). Historically, the first generation of laser used for PDT was the argon-pumped dye laser. Although having the advantages of adequate power (several watts of red light using a 10–20 W argon ion laser pump) and wide wavelength tunability from the dye laser, these systems were generally expensive, cumbersome and unreliable in a clinical setting. They were superceded by dye lasers pumped by frequency-doubled Nd:YAG surgical lasers (KTP) that were reliable and mobile, and many of these systems are still in clinical use. Recently, solid–state diode lasers have become available with adequate power (several watts) at relevant wavelengths to match current clinical photosensitizers. These are reliable, operate from standard electrical power supplies, do not require

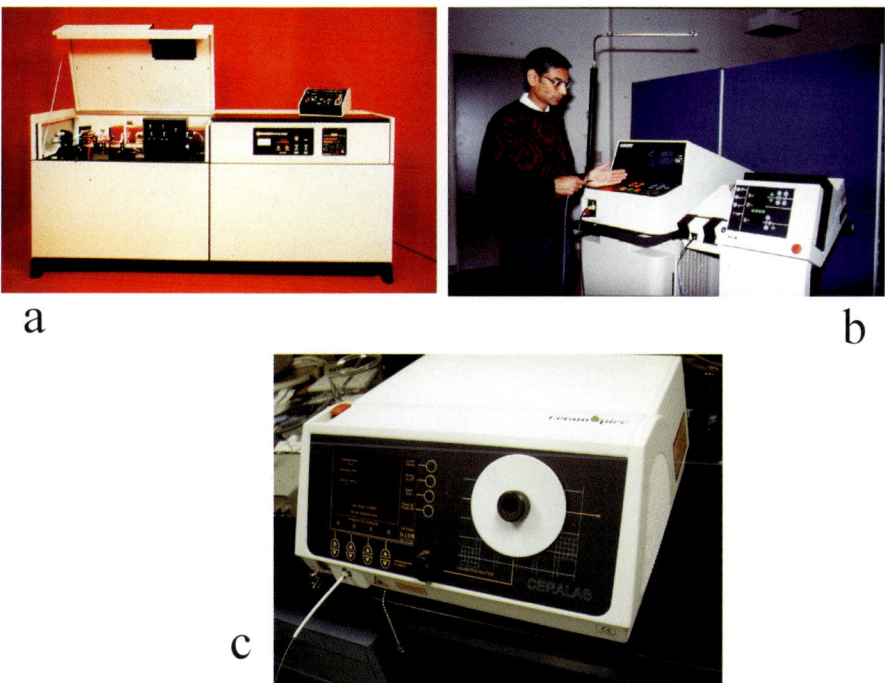

Figure 4. Clinical PDT lasers. (a) argon-pumped dye laser, (b) KTP-pumped dye laser, (c) diode laser (courtesy Ceram Optics, Germany).

water cooling and are relatively inexpensive (typically ~$50000 for clinical units). The main drawback is that they are not tunable, so that a different diode is required for each wavelength used, thus increasing the cost. Figure 4 shows these 3 generations of clinical laser sources. A number of other types of lasers (copper- or gold-vapor, pulsed diode, alexandrite) have been/are used for PDT to a limited extent.

7.1.4 LEDs

Light emitting diodes are small (~mm), incoherent sources that can be ubiquitously found in consumer devices (e.g. displays). Single LEDs have a power of typically only a few mW. However, by forming arrays of individual LEDs, substantial area power density can be achieved (tens to hundreds of mW per cm^2), so that these devices are particularly useful for surface applications. Two such devices are illustrated in Figure 5. LED sources are reliable, portable, require only standard electrical power and are less expensive per watt than diode lasers. However, coupling into optical fibers is not easy. They have been used as intra-operative devices (see below), and linear arrays of LEDs for interstitial application have been developed (Figure 5(c)). The main limitation until recently has been the restricted power and wavelengths (particularly for short wavelengths), but this is improving rapidly.

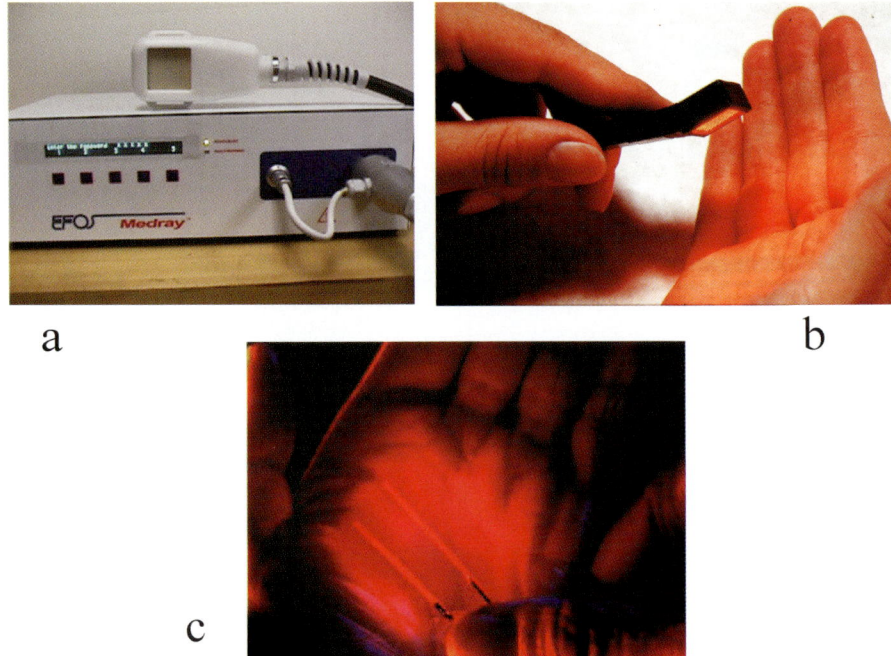

Figure 5. Light Emitting Diode (LED) sources for clinical PDT. (a) array for skin lesion treatments (courtesy EXFO Inc, Canada), (b) small array for intra-oral application (courtesy PRP Optoelectronics, England), (c) interstitial linear arrays (courtesy Light Sciences Inc., USA).

7.1.5 Filtered lamps

There are a number of lamp sources now available for PDT, using optical filters to select the appropriate region of their broad output spectrum to match the photosensitizer and desired tissue penetration. High-brightness arc lamps, as illustrated in Figure 6(a), may be coupled into optical fiber bundles or liquid light guides (5–10 mm diameter) for surface or intraoperative application. These are very flexible in terms of choosing the wavelength and effective powers of tens of mW per cm^2 may be obtained. Figure 6(b) shows a source designed specifically for PDT of actinic keratosis of the face and scalp, based on an array of filtered fluorescent tubes in a U-shape. This illustrates a major advantage of lamps, which is to deliver adequate power over large surface areas, as long as the geometry of the lamp(s) can be formed to suit the surface topography. Hence, lamps have also been used for irradiation of the peritoneal cavity and bowel for ovarian metastasis, where square metres of surface are involved.

7.2 Light delivery devices

The purpose of light delivery devices is to transport the light from the source efficiently to the target tissue, whether this be surface, intraluminal or interstitial,

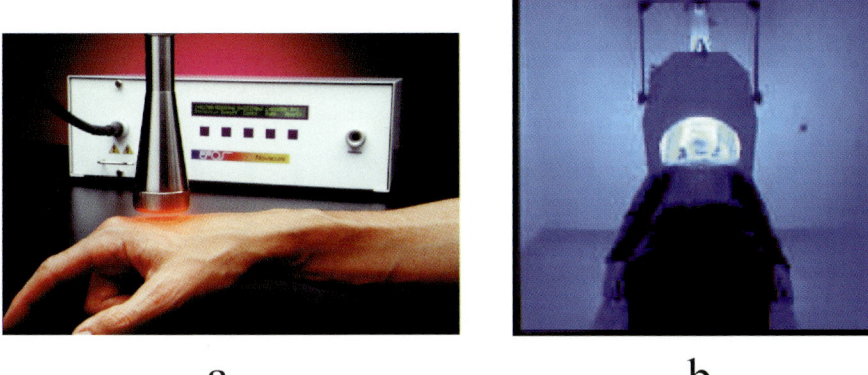

a   b

Figure 6. Filtered lamp sources for clinical PDT. (a) arc lamp coupled into a liquid light-guide (courtesy EXFO Inc., Canada), (b) bank of fluorescent tubes (courtesy DUSA Inc, USA). Note the use of the blue part of the spectrum for treating superficial actinic keratosis lesions with topical aminolevulinic acid.

and to have the light emitted spatially to match the shape and size (geometry) of the target volume. Many different target geometries are used, depending on the organ and disease being treated, as illustrated in Figure 7. These include illumination of external/accessible tissue surfaces such as the skin, oral cavity and certain intra-operative sites; interstitial illumination, in which diffusing optical fibers are inserted into tissue to produce roughly spherical or cylindrical treatment zones; intraluminal treatments as in the case of the bronchial and gastrointestinal tracts; intracavitary irradiation as in the urinary bladder and thoracic or peritoneal cavities; and transcorneal delivery in the case of retinal treatment for age-related macular degeneration. Note that intraluminal and intracavitary treatments can also involve surface or interstitial approaches.

In many of these irradiation techniques there is reliance on optical fibers with specially designed diffusing tips [1]. The cylindrical type, which is simply an extension of the simple spherical point diffusing tip, are typically small enough to fit through the instrument channel of a standard endoscope ($< \sim 2$ mm diameter), and are available up to about 10 cm in length with an output that is uniform to within ± 10 or 20% along the length and within $\pm 10\%$ radially. These are presently made by coating the optical fiber core over the required length with a material that is highly light scattering. The power output of these fibers is stated in terms of watts or joules for point diffusers and W cm^{-1} or J cm^{-1} for cylindrical diffusers, where this refers to the power or energy emitted radially in all directions along each cm of the diffusing tip. It is not simple to relate this directly to the usual measure of delivered light "dose" for surface irradiation (W cm^{-2} or J cm^{-2}) (see below). Figure 8 shows specific applications that illustrate the interstitial, intraluminal, intracavitary and retinal irradiation techniques. Surface treatments have been illustrated in Figures 5 and 6.

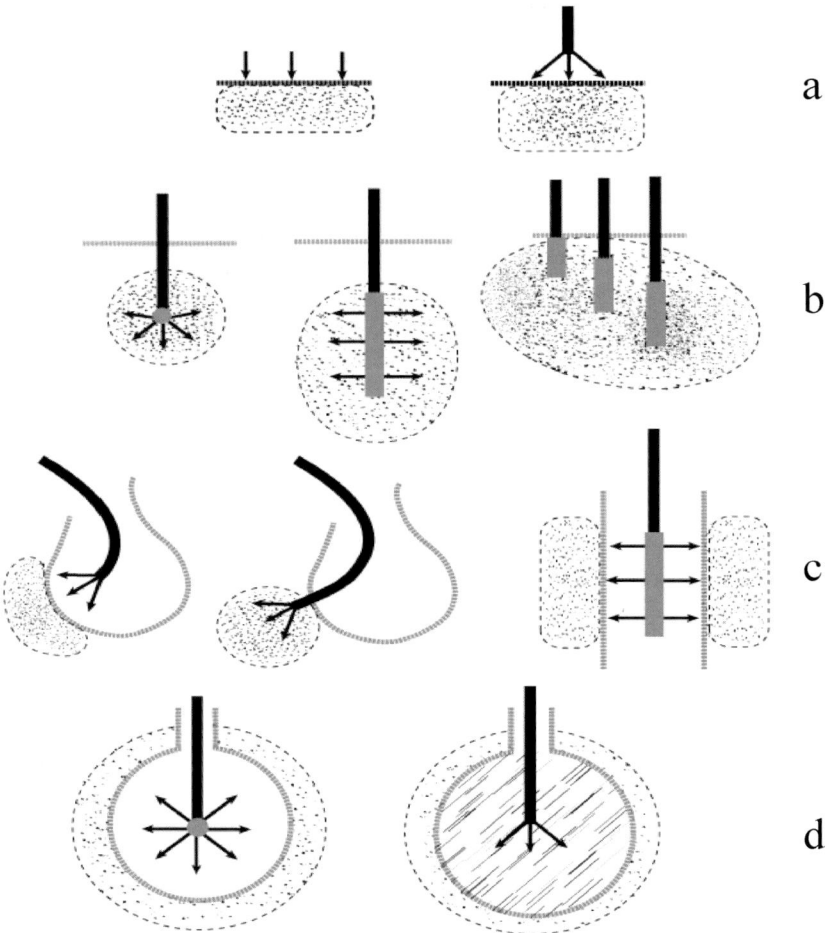

Figure 7. Target light geometries used commonly for different clinical applications. (a) surface irradiation, (b) interstitial irradiation, (c) intraluminal (endoscopic), (d) intracavity or intraoperative.

These optical fiber diffusers have been extended in a number of ways, for example the use of multiple diffusing fibers simultaneously to produce either a large treatment volume or to match a specific organ shape (Figures 8a,c), and various "balloon" applicators (Figure 9) in which the light is delivered into an extendable (elastic or inelastic) balloon whose shape matches the (hollow) organ surface. In the case of balloons used for esophageal treatments (Barrett's, dysplasia or superficial cancer), the balloon also serves to extend the esophagus, thereby reducing light "shadowing" by the tissue folds, and to center the diffusing fiber. The latter has been shown to be critical [2] to obtain uniform light dose to the whole circumference. When using a balloon applicator for treating a spherical cavity, an alternative to using a point diffuser at the center with the cavity filled with saline is to use

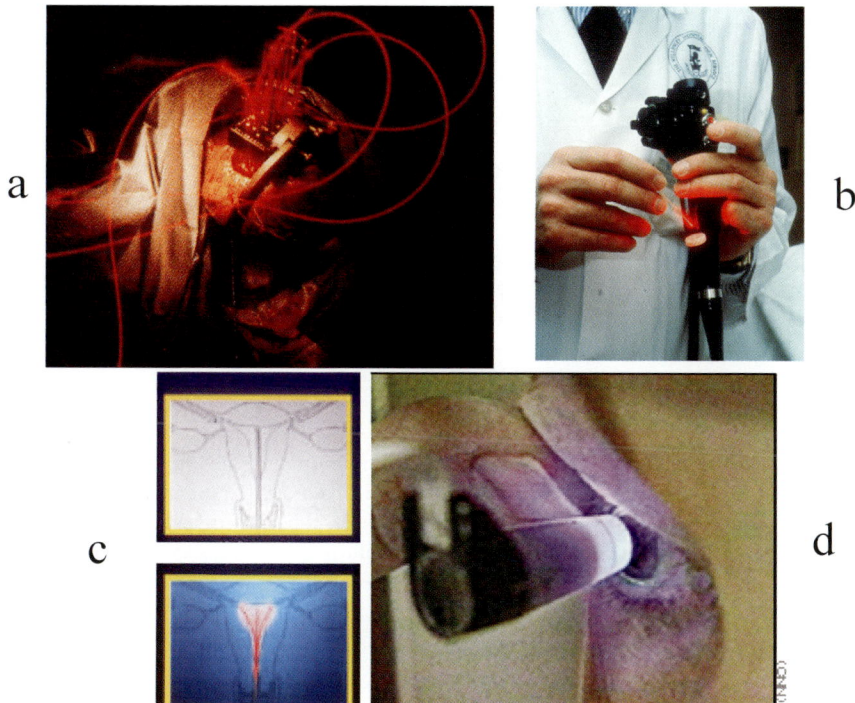

Figure 8. Representative clinical applications using diffusing fiber light delivery. (a) multiple interstitial cylindrical fibers used to treat brain tumors (courtesy Dr T. Origitano, Chicago, USA), (b) cylindrical diffusing fibers being placed into the instrument channel of an endoscope for esophageal PDT (courtesy Dr. N. Marcon, Toronto, Canada), (c) three cylindrical diffusers in an 'umbrella' configuration for endometrial irradiation (courtesy Drs B. Tromberg and Y. Tadir, Irvine, USA), (d) transcorneal irradiation of the retina from a laser diode coupled into a fundus camera (courtesy QLT Inc, Canada).

a simple cut-end fiber and to fill the cavity with a light scattering liquid. Most commonly this is a solution of Intralipid™ (as used for intravenous feeding, diluted to about 1 part per 1000 solids).

For treatments where multiple optical fibers are required simultaneously, with a single laser source, it is necessary to use a beam splitter to divide the light between the fibers. These can involve sets of partially-silvered mirrors to split one beam into two, and these can then be "stacked" to produce four or eight beams. Direct two-way fiber–fiber splitters are also available. A limitation has been that each of the output beams has the same power, or some fixed power ratio. Recently, variable power splitting has become possible, with computer control of the distribution between multiple fibers. These devices are still at the commercial prototype stage. The advantage is that the distribution between the fibers can be adjusted to match the target tissue, and it will be possible in future to modify this dynamically during treatment as changes such as blood flow occur and are monitored (see below).

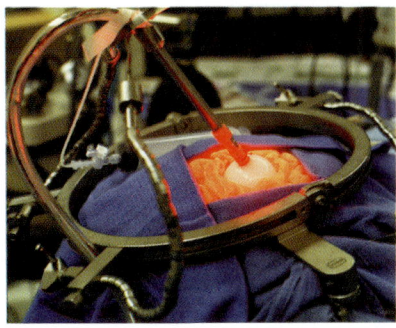
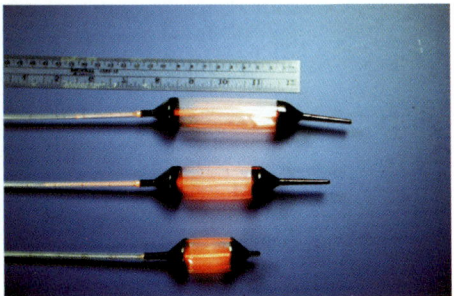

a   b

Figure 9. Balloon applicators for light distribution in hollow cavities (a) brain tumor cavity irradiation post-resection, (b) with centered diffusing fiber for treatment of the esophagus: the balloon is collapsed for passage through the endoscope instrument channel (courtesy Drs M. Panjepour and G. Overholt, Knoxville, USA).

7.3 Measurement of light sources

In order to ensure that the delivery of light to tissue is accurate between treatments, it is necessary to measure one or more of the factors: delivered power and power density, wavelength or wavelength spectrum, and power distribution [3]. We will assume that treatment exposure time can be accurately monitored and recorded, either manually or automatically (if the latter, then any built-in timers should be periodically checked).

Figure 10 shows two of the main ways to check the delivered output power (watts) of a light source: by a flat photodetector, or using an integrating sphere. In the first, a light detector, most commonly a photodiode, is placed in the light field. If the beam is smaller than the photodiode area (e.g. if measuring a laser beam directly or with the tip of a cut-end optical fiber or small fiber bundle close to the detector head), then the power reading represents the total output power. If the beam is larger than the detector area, then the local power density ($W\,cm^{-2}$) can be calculated by dividing the measured power by the detector area. (In some instruments, this reading is given directly as an optional setting.) The total power is then obtained by integrating the power density over the whole beam area. If the power density is not constant everywhere across the light field, then multiple measurements will be needed to "sample" the output distribution. These measurements are typically done in air, prior to and, as a check on the constancy of the source output, immediately following treatment. For interstitial light sources, it is also possible to measure the output (distribution) in water or in a light-scattering medium (e.g. Intralipid™) to simulate the tissue. This would be done, for example, if the output was sensitive to the presence of the tissue.

The second method, which is particularly useful with extended sources such as diffusing optical fibers, is to place the diffuser into a sphere that has a highly reflective inside coating. This randomizes the light by multiple reflections and a known small fraction escapes through an output port to be detected, again usually by

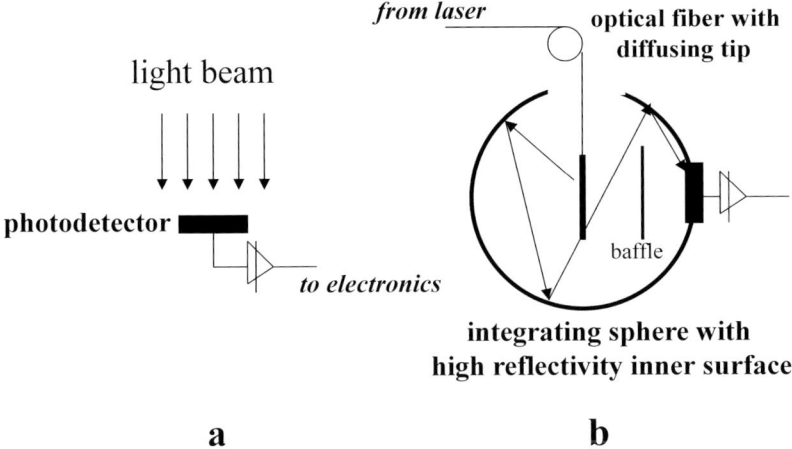

Figure 10. Measurement of delivered output power from light sources, using (a) a calibrated flat photodetector placed in the light beam, (b) a calibrated integrating sphere and photodetector.

a photodiode. Thus, the signal is largely independent of the detailed light output distribution and measures the total power.

In both techniques it is critical that the detector is properly calibrated, so that the power measurement is accurate. This can be achieved by having the instrument traceable to a national standards laboratory (e.g. NIST in the USA). The calibration factor must match the required wavelength and usually the calibration should be checked yearly, as it can drift. The calibration factor (efficiency) for the integrating sphere must also be known at the required wavelength and, unless damaged or modified, should not change over time. Increasingly manufacturers of PDT light sources are providing built-in power meters (including in some cases integrating spheres). However, their accuracy should not be assumed without independently checking against a traceable standard, and they should be periodically re-calibrated.

For wavelength tunable light sources, it will also be necessary to calibrate or check the output wavelength (lasers) or spectrum (lamps, LEDs). Again, a traceable spectrometer should be used and, in cases where the manufacturer has provided a wavelength reading or check, this should be independently confirmed periodically. With LEDs and lamps, it cannot be assumed that the output spectrum will be constant over time, as the device ages. Filters used with lamps may also degrade with time.

At present, there are no established standards for the accuracy of PDT light power and wavelength. However, reasonably achievable values are ± 10% for the power and ± 2 nm for the (peak) wavelength.

7.4 Light distributions

The spatial distribution of light in tissue is critically dependent on the size and shape of the physical tissue volume, on the spatial distribution of the delivered light, and on the tissue optical absorption and scattering properties at the treatment wavelength [4]. The distributions can be accurately predicted if all these factors are known, especially for the cases where the tissue is optically homogeneous, i.e. the absorption and scattering do not vary appreciably within the treatment volume, and the volumes are relatively large compared to the typical light propagation distances (specifically, the effective penetration depth: see below for definition). We will consider these cases first:

7.4.1 Ideal case: large, homogenous tissues

Figure 11(a,b) shows the distribution of fluence (J cm^{-2}) for a broad-beam surface irradiation (where, again, the beam is at least several penetration depths in diameter), for two different cases: high absorption relative to scattering (Figure 11(a)) and vice versa (Figure 11(b)). In the former case, the "depth dose" has a simple exponential form:

$$\phi(z) = \phi_0 \exp(-\mu_a z) \qquad (2)$$

where ϕ is the fluence at depth z below the tissue surface, ϕ_0 is the incident fluence and μ_a is the tissue absorption coefficient (cm^{-1}) at the treatment wavelength. The light fluence is then greatest at the tissue surface, little light is lost by scattering back through the surface, and there is little lateral spreading of the light beyond the incident beam area. This type of situation pertains to, for example, the blue-light treatment of actinic keratosis (Figure 6(b)), since at short visible wavelengths the absorption of tissue is high. Correspondingly, the fluence falls very rapidly with

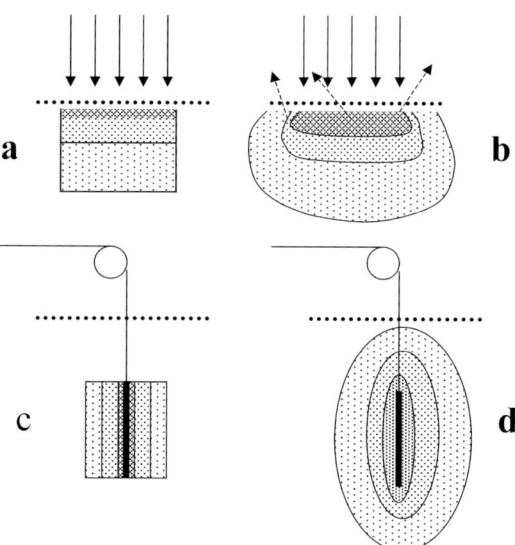

Figure 11. Schematics of light distributions in optically-homogeneous, large-volume tissues (a,b) surface irradiation, (c,d) interstitial irradiation, (a,c) high absorption/scatter ratio, (b,d) high scatter/absorption ratio.

depth: typically at 2 mm below the surface the fluence would be only a few percent of the incident value.

In the more usual case when longer wavelength red or near-infrared light is used, the scattering becomes comparable to or higher than the absorption, and the distribution follows a form as shown in Figure 11(b). There are now several differences, namely a sub-surface peak, a slower depth decrease than for the same absorption value alone, back scattering out of the irradiated surface, and considerable lateral spreading of the beam.

The depth dose is no longer a simple single-exponential form, but rather shows a sub-surface peak and only becomes exponential at some depth beyond this. Mathematically, this can be expressed as:

$$\phi(z) = \phi_0 B(z)\exp(-\mu_{eff}z) \qquad (3)$$

where B is a backscatter factor and μ_{eff} is the so-called effective attenuation coefficient. $B(z)$ depends on the tissue absorption and scattering and on the tissue surface (air or water coupling) in a complex way; suffice to say that it increases with beam size and can be as high as 2–3. The effective penetration depth, $\delta = 1/\mu_{eff}$. This is the *incremental depth* in tissue over which the fluence falls by 63%, i.e. to 37% (1/e) of its previous value (see insert, Figure 2). Over two penetration depths the value falls to approximately 10% (37% of 37%) and over four penetration depths to approximately 1% (10% of 10%). The penetration depth is *not* the depth of PDT treatment, since this depends on the photosensitizer concentration, tissue oxygenation and intrinsic tissue photosensitivity, as discussed below. Typically, for most photosensitizers used clinically to date the effective treatment depth, d_t, is at least 3–5 times δ. The penetration depth is also *not* simply the depth below the surface at which the fluence is 37% of the incident value: this depth is several times greater than δ, because of the backscatter peak.

The penetration depth depends on the tissue absorption and scattering in a well-determined way. To a first approximation:

$$\delta = 1/\mu_{eff} = 1/\sqrt{\{3\mu_a\mu'_s\}} \qquad (4)$$

The factor μ'_s is the transport scattering coefficient of the tissue (which combines the actual scattering probability and the forward nature of light scattering in soft tissues). Some numerical values, presented in Table 1, will clarify this relationship. Note that the attenuation increases, i.e. the penetration depth decreases, equally with increase in absorption or scatter: both are equally important in determining the resulting effective treatment depth.

The loss of backscattered light from the volume can be greater than 50% at long wavelengths for lightly pigmented tissues. This represents a loss of therapeutic light to the target volume. Conversely, in the case of a (partially) closed cavity (bladder, resection cavity), most or all of this light will re-enter the tissue, resulting in a significant elevation in the incident fluence compared to the delivered (primary) value. This increase can be as high as 5–7 times in, say bladder or esophagus [5]. At the same time, the depth distribution gets steeper as the scattering increases, so that the dose is more superficial. A particular issue is that this backscatter factor has been shown to be highly variable from patient to patient, for example in the bladder, and

Table 1. Typical absorption and scattering coefficients of tissues in the red/near-infrared spectral region (600–800 nm) and resulting approximate effective attenuation coefficient and penetration depth. Note that for any tissue type, there is considerable patient-to-patient and, for tumors, lesion-to-lesion variation in these properties

μ_a /cm^{-1}	μ'_s /cm^{-1}	μ_{eff} /cm^{-1}	δ /mm	Typical of tissue with red light treatment
2–7	5–15	5–15	0.5–2	Liver (high pigmentation)
0.2–2	10–30	2.5–10	1–4	Brain white matter (high scattering)
0.2–3	3–8	1.5–5	2–6	Muscle (moderate scatter, low absorption)
–	–	2.5–5	2–4	Non-melanoma solid tumor (moderate scatter and absorption)

this has led to increased normal bladder tissue damage in some cases, where the delivered light dose is not adjusted to take it into account.

The lateral spreading of the light distribution due to scattering increases the effective beam diameter by typically about 1–2 δ, i.e. by a few mm in the case of red activation. This may or may not be important clinically depending on the proximity of adjacent critical normal tissues. A final point is that, if the beam diameter decreases to less than, say, 5δ, the fluence at depth will decrease even for the same incident J cm^{-2}, since there is a reduced volume of tissue for backscatter into the beam.

For interstitial treatments there is an analogous dependence of the spatial distribution of the light on the tissue optical properties. However, geometric spreading of the light causes an additional decrease in the fluence with radial distance, r, from the outer surface of the source. In the case of a point isotropic source (equal in all directions), the fluence distribution is of the form:

$$\phi(z) = E_0 R(r) \frac{\exp(-\mu_{eff} r)}{r} \tag{5}$$

where E_0 is the delivered light energy (Joules) and $R(r)$ depends on the absorption coefficient. Note that the exponent has the same value as in the case of surface irradiation, but there is the extra $1/r$ factor. (For low scattering this factor is more like $1/r^2$, the so-called inverse square law, as seen in photon radiotherapy.) For a cylindrical fiber the dependence is more complex than $1/r$ and depends on the diffuser length, but, qualitatively, the behavior is similar. The shape of the distribution at each end of a diffusing fiber is roughly the same as for a point source, but there is no simple algebraic expression to describe it. These situations are illustrated in Figure 11(c,d).

7.4.2 Complicating factors

In reality, clinical target volumes are neither optically completely homogeneous nor unbounded. Only a few general guidelines can be provided for designing and interpreting clinical studies, and in practice it may be desirable to measure directly the light fluence (rate) at critical points on or in the tissue (see below), rather than relying on the calculated distributions. Since hemoglobin is a significant absorber below 650 nm, major blood vessels in the volume will distort the light distribution, and may cause shadowing behind them. Tissue layers, such as in the skin, will cause local variation in the depth-dose: indeed this is exploited in laser thermal treatment of port wine stain. In the treatment of small tumors, say, by single interstitial fiber insertion, a significant fraction of the light may be lost from the tumor, with a reduced total dose compared to delivering the same light energy to a larger lesion. Computational techniques are available to handle situations such as small tumors or layered tissues once the optical properties are known.

Two further factors may need to be considered that are dynamic, i.e. change during the treatment. The first is that the tissue absorption and, hence, the light distribution may change, primarily due to changes in blood flow caused by the vascular PDT response. This effect is less marked above 650 nm, where the hemoglobin absorption is less, and is more pronounced for vascularly-acting photosensitizers. Typically at 650–700 nm, a $+10\%$ change in blood flow will change μ_a in moderately-pigmented tumors by about $+30\%$, and δ by -30%. Hence, the fluence rate at the base of a tumor, say 10 mm thick, treated from the surface would be decreased by a factor of 2.

With 2nd-generation photosensitizers, the molar extinction coefficients can be very high, typically 10^4–10^5 cm^{-1} M^{-1}. Hence for a tissue concentration of, say, 10 μM the added absorption due to the photosensitizer itself would be ~ 0.01–0.1 cm^{-1}. At longer wavelengths this is significant compared to the intrinsic absorption of the tissue, and so decreases the penetration depth. If the photosensitizer photobleaches (see below), this "self shielding" effect will decrease during the light irradiation.

7.4.3 Treatment planning for PDT

By analogy to radiation therapy, it is possible to envisage creating treatment plans for PDT of individual patients, based on the target geometry and therapeutic intent [6]. The most obvious type of planning is to calculate the distribution of light fluence within the target and adjacent tissues for specific light delivery, and to adjust the delivery to produce the optimal distribution. An example of this is shown in Figure 12. If the photosensitizer concentration in the tissue is known (measured or assumed), then the effective PDT dose can be calculated, and the corresponding boundary of PDT effect (Figure 12(b,c)) determined. Factors such as photobleaching can also be incorporated: these terms will be discussed in more detail below.

The sophistication of such treatment planning can be extended as required by incorporating other factors such as tissue oxygenation and oxygen depletion. This type of computer planning can be used (1) to pre-plan and, thereby, optimize treatments, (2) to study the effects of the various photophysical, photochemical and photobiological factors on the effective treatment region of the tissue (e.g. to determine the sensitivity of the treatment to uncontrolled changes in the treatment parameters) and (3) to analyze, retrospectively, the treatment outcome versus the actual treatment

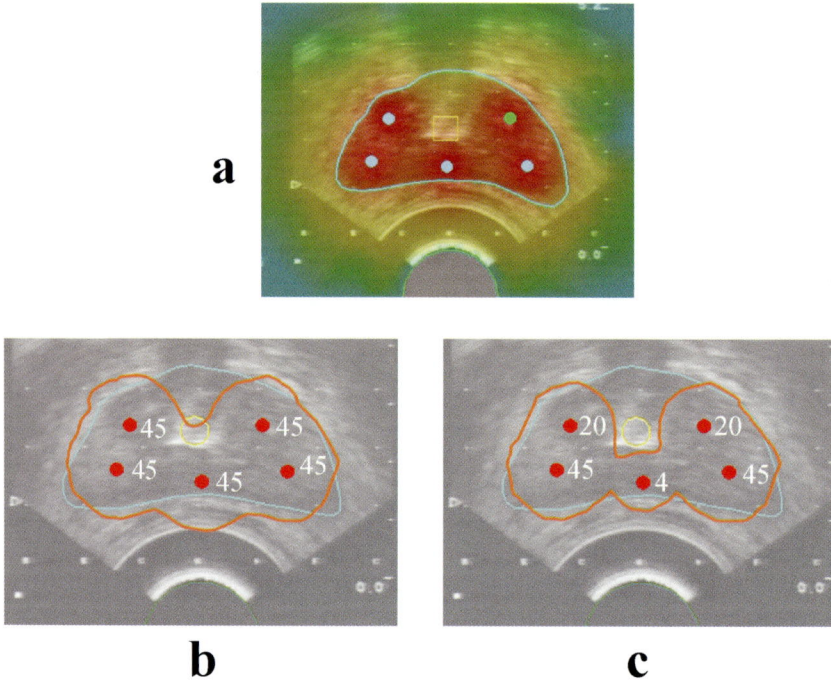

Figure 12. Treatment planning for PDT. This example is for interstitial treatment of prostate cancer, using multiple cylindrical diffusing fiber sources. The treatment volume is defined by trans-rectal ultrasound. (a) light fluence (rate) distribution for five fibers at specific locations and with equal power to each fiber, (b) corresponding threshold-dose boundary, (c) treatment boundary with the light fluences from each fiber adjusted to reduce the dose to the urethra.
(Images courtesy CADMIT Inc, Canada).

delivered, e.g. if the light sources were not placed exactly as planned, or there was a dynamic change in the tissue properties monitored during treatment.

7.4.4 Measurement of light during PDT

A number of techniques and instruments are available to determine the distribution of light on or within the irradiated treatment during PDT. The most widely used are "isotropic" optical fiber probes, illustrated in Figure 13. These are single optical fibers in which the tip has been modified by making it highly light scattering. Hence, photons incident on the tip will be multiply scattered and a fraction will be transmitted along the fiber to a (calibrated) photodetector, as above. The signal is then largely independent of the direction of the incident light (except usually close to the "backwards" direction where the fiber itself shields the light). These probes can be calibrated to read the local fluence rate (W cm^{-2}), and are small enough to insert into tissue, usually through a biopsy needle. Note that the probe response is different in air, in contact with the tissue surface and in the tissue, since the refractive index of

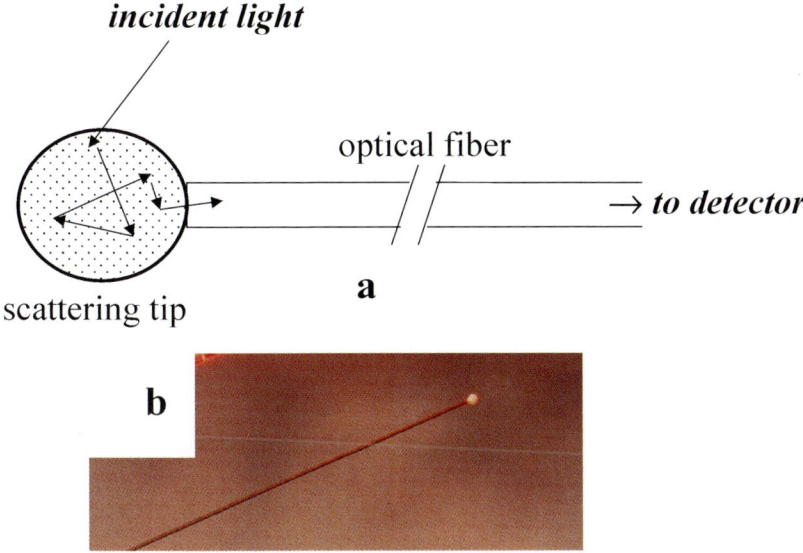

Figure 13. Isotropic light fluence probes. (a) principle of operation, (b) a typical "ball tipped" isotropic probe approximately 1 mm diameter.

tissue alters the fraction of light lost from the scattering tip due to total internal reflection. Hence, it is important to know the conditions under which the probe is calibrated if absolute fluence rate values are required. Procedures for "in-house" calibration of these probes have been published [7].

This type of probe has been used in various ways clinically and preclinically, as shown in Figure 14. For example, several probes may be placed at different points within the target tissue and adjacent normal structures to monitor the local fluence rates. Probes may be placed on the surface of a balloon applicator to monitor the total (primary + backscattered) fluence during endoscopic treatments [8]. Devices incorporating several probes to monitor the uniformity of light delivery and the total fluence in the bladder or bronchus have been reported [5], while probes can also be used to measure the fluence-depth distribution by moving the probe tip along the fiber direction interstitially. This has been extended, mainly in animal studies, to mapping complete light fluence-rate distributions in tissue volumes.

The surface irradiance has been measured clinically in some intraoperative scenarios, for example, during irradiation of the entire peritoneal cavity for ovarian cancer by placing several small photodiodes directly on the tissue surface at representative points. Clearly, such measurements could also be done using fiberoptic probes.

More complex techniques for light measurements in tissues have been demonstrated experimentally or are under development. One example is the use of interstitial probes that have a deliberate directional dependence so that, at a fixed point in tissue, as the probe is rotated circumferentially it samples the local radiance distribution, rather than simply the local fluence rate that is the sum of the radiance over all directions. This gives more information about the light field, especially at

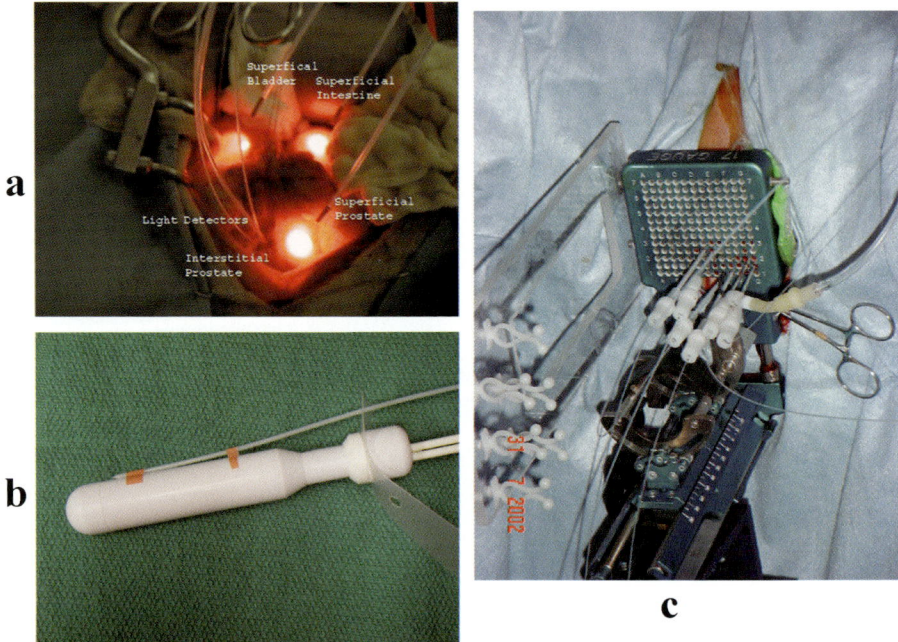

Figure 14. Uses of isotropic optical fiber fluence probes for different applications. (a) interstitial probes placed into dog prostate during an intraoperative experimental procedure (courtesy Drs F. Hetzel and Q. Chen, Denver, USA): source fibers irradiating the prostate surface, bladder and bowel are also seen, (b) a probe on the surface of a transrectal ultrasound applicator to monitor the rectal dose during interstitial PDT of prostate cancer, (c) corresponding interstitial probes placed into the prostate, along with source fibers, through a template (courtesy Dr J. Trachtenberg, Toronto, Canada).

locations near to the light source or to tissue surfaces where the distribution is not completely diffuse [9]. Since it is the fluence (rate) rather than the radiance that determines the drug activation, the purpose is to gain more information on the tissue optical absorption and scattering properties, to allow calculation of the complete fluence distribution. A second example is where multiple interstitial diffusing sources are used but the fibers are set up so as to serve also as light detection probes. By switching between these two functions, more complete information can again be obtained on the light distribution [10].

These detection schemes are necessarily invasive. Non-invasive methods have been developed based on measuring the spatial distribution of light, diffusely reflected from the tissue surface, as illustrated in Figure 15. This can be analyzed to determine μ_a and μ'_s, and thereby to calculate the light distributions: e.g. the fluence-depth/radius curve using Equations (3) and (4) or (5) and (4), or computational models for any given source configuration. Specialized diffuse reflectance probes have been developed, both for accessible body surfaces and for endoscopic application. At present these techniques are not in common use for clinical PDT.

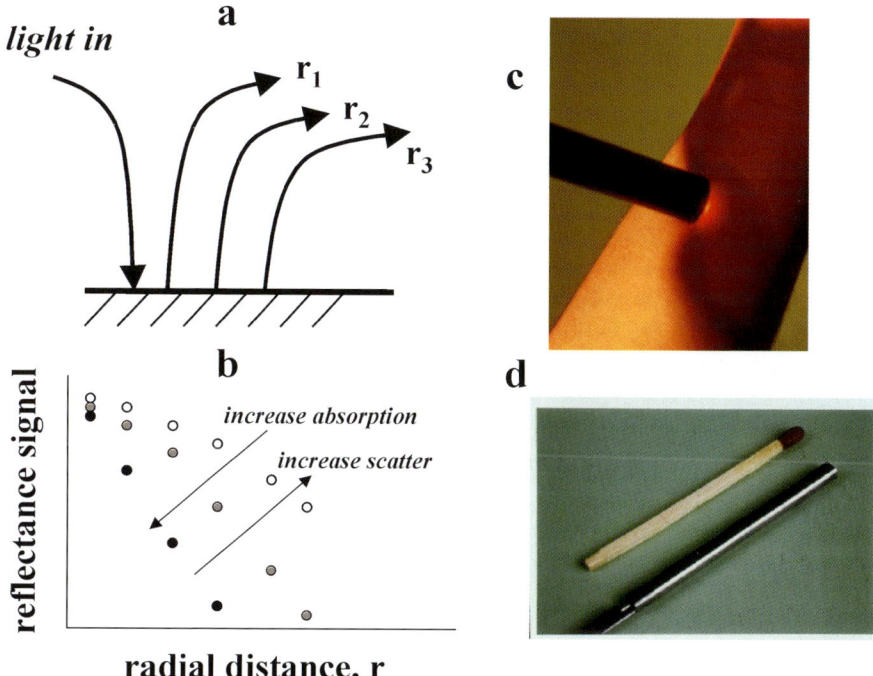

Figure 15. Diffuse reflectance distributions used to measure tissue absorption and scattering properties non-invasively. (a,b) principle of the technique, showing light entering a point on the tissue surface and the measured radial distribution of the diffusely reflected (backscattered) light that depends on the tissue absorption and scattering properties, (c) external surface probe (courtesy Dr M. Patterson, Hamilton, Canada), (d) endoscopic probe (courtesy Dr R. Bays and colleagues, Lausanne, Switzerland): in this case the distribution is measured along the probe from light input at the end, with the probe placed flat on the tissue (e.g. esophagus) surface.

7.5 Photosensitizer measurements and biophysical effects

Several general characteristics of photosensitizers affect their efficacy as PDT agents: photophysical, photochemical and pharmacological. The photophysical/photochemical properties include the absorption (extinction spectrum) in vivo, the quantum efficiency for generating singlet oxygen (or other active photoproducts), the photobleaching rate and the quantum efficiency for fluorescence. The characteristics of particular photosensitizers and the relationship to their molecular structure are discussed in other chapters, as are the tissue uptake and clearance and microlocalization properties. Here, we will focus on the methods, primarily optical, that may be used to measure some of these characteristics in vivo.

7.5.1 Measuring photosensitizer concentration in tissue

Fluorescent drugs can be detected in vivo by monitoring the fluorescence intensity, either on the tissue surface or interstitially. This can be done either with a fiberoptic spectrometer or by an imaging device (fluorescence camera) (examples are shown in Figure 16). These techniques have also been widely investigated for diagnostics, such as early cancer or dysplasia detection [11] and potentially for guiding surgery. For PDT they may be used to monitor photosensitizer pharmacokinetics, to localize the target lesions and to monitor photosensitizer photobleaching.

Problems in fluorescence monitoring arise when quantitative, especially absolute, measurement of the photosensitizer concentration in tissue, C, is required. Consider, for example, Figure 17(a), where a measurement is made at the tissue surface. The incident light used to excite the fluorescence is absorbed and scattered by the tissue. A fraction of the light is absorbed by the photosensitizer and converted into fluorescence; this fraction is proportional to C. The fluorescent light (of longer wavelength) is also absorbed and scattered before a fraction can exit the tissue to be detected. Hence, the detected signal depends on several factors in addition to C. For example, the fluorescence signal will be smaller from a highly scattering or absorbing tissue than for a lightly-pigmented or low scattering tissue, even for the same concentration of photosensitizer.

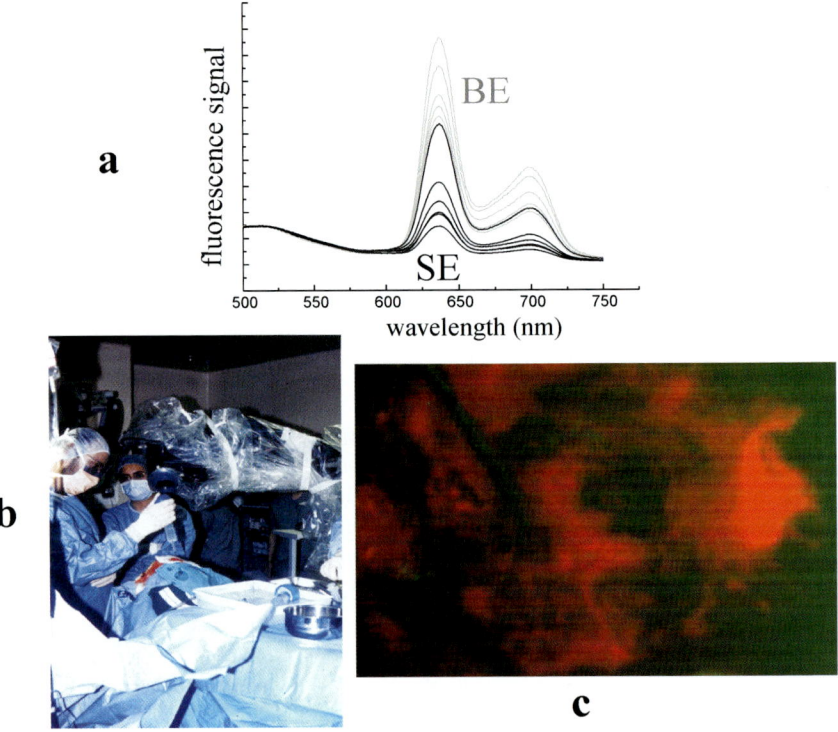

Figure 16. Detection of photosensitizer fluorescence in vivo. (a) fluorescence spectrum of ALA-PpIX fluorescence at different points in the esophagus within Barrett's (BE) and normal squamous epithelium (SE) at 3 h after 10 mg kg^{-1} ALA orally (courtesy Dr N. Marcon, Toronto, Canada), (b) fluorescence camera imaging the photosensitizer distribution in tissue during brain tumor PDT, (c) fluorescence in the resection bed at the end of radical brain tumor resection at 24 h after 2 mg kg^{-1} Photofrin (courtesy Dr P. Muller, Toronto, Canada).

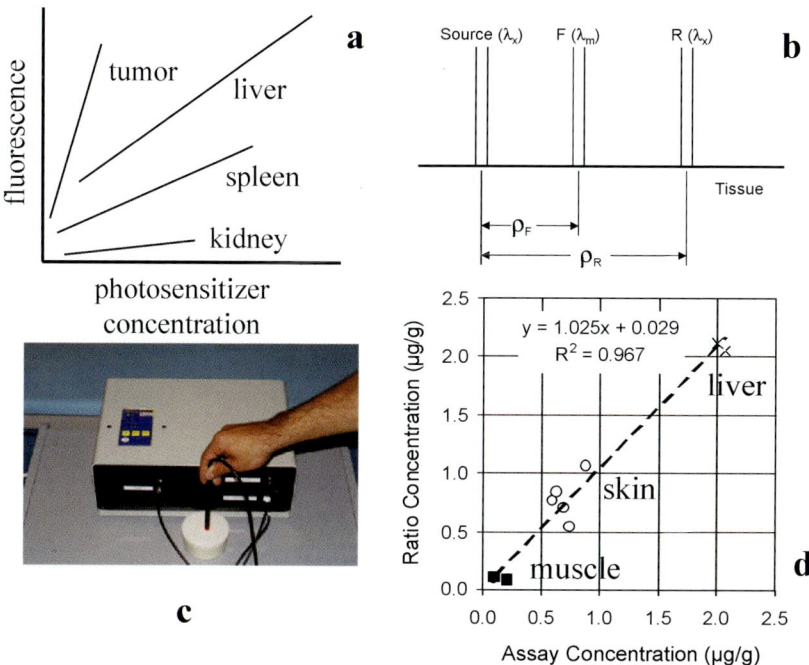

Figure 17. The effects of tissue attenuation on fluorescence measurements. (a) graph showing measured fluorescence versus Photofrin concentration in different tissues (after Panjehpour et al., 1993 [12]), (b–d) correction for attenuation by the ratio of fluorescence to diffuse reflectance at two different radial distances: (b) principle, (c) prototype instrument, (d) photosensitizer concentration measured in vivo versus known concentration by assay on tissues ex vivo for three different tissue types (after Weersink et al., 1997 [13]).

There are several possible solutions to this problem. One option is to measure the tissue optical properties at the excitation and emission wavelengths and correct the signal for the attenuation. A variation on this is to normalize the measured fluorescence signal, F^*, to some other measurement, such as the diffuse reflectance, R, that also depends on the tissue optical properties, so that there is partial cancellation of the attenuation effects in the ratio, F^*/R (this is illustrated in Figure 17(b–d)). A third method is to make the measurement in such a way that it is (relatively) independent of the optical properties. This can be done either by using very small optical fibers that both deliver the excitation light and collect the fluorescence so that the measurement is confined to a very small tissue volume within which the attenuation is minimal, or to separate the excitation and detector fibers at a specific distance where the effects of scatter and absorption roughly cancel each other.

For non-fluorescent photosensitizers, it is possible to measure the concentration in tissue non-invasively using diffuse reflectance spectroscopy. This is based on the fact that, at wavelengths where the drug absorption is high, the diffusely reflected (backscattered) signal from the tissue will be low. Hence, by measuring the reflectance spectrum, and subtracting the tissue background, the absorption

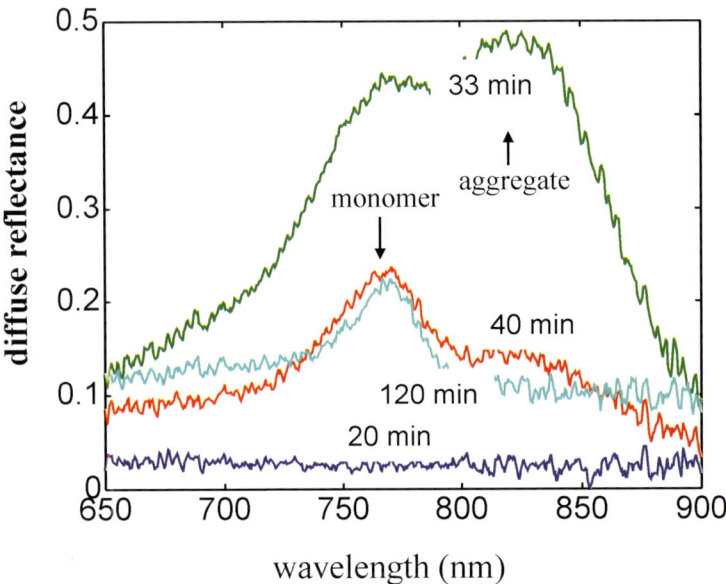

Figure 18. Example of diffuse reflectance spectroscopy in vivo. The reflectance spectrum from normal skin/muscle was measured in an animal model following intravenous administration of a non-fluorescent photosensitizer. The spectra show monomeric and aggregated drug and the changes in these components with time after injection.

coefficient of the photosensitizer, $\mu_{\text{a-ps}}$, ($= A(\lambda)$ in Equation (1)) can be measured, where $C = \mu_{\text{a-ps}}/\varepsilon$. An example is shown in Figure 18. The limitation of this method is that its sensitivity is low compared to fluorescence, so that a relatively high value of $C\varepsilon$ is required to give a good reflectance signal. Nevertheless, it can be quite practical for second-generation photosensitizers with high extinction coefficients.

In measuring photosensitizer fluorescence from the tissue surface, the photosensitizer closest to the surface contributes most to the detected signal. The contribution from greater depths depends on the light attenuation, primarily at the excitation wavelength, since this must be shorter than the emission wavelength and, generally, light penetration increases with wavelength, as previously discussed. If the excitation wavelength is not the same as for the PDT treatment, then the full treatment depth will not be sampled. This is a particular problem with layered tissues, for example tumors with overlying skin, where the drug concentration is different between layers. For interstitial fluorescence measurements of photosensitizer, analogous issues arise, with the effective diameter over which the drug is sampled being dependent on the tissue optical properties.

Many of these problems in making quantitative drug measurements in vivo are avoided by using ex vivo tissue samples where these can be obtained, e.g. by biopsy. Several techniques have been developed, based on either chemical extraction of the photosensitizer or solubilizing the whole tissue [14], followed by fluorescence or absorption spectroscopy or, in some cases, chromatography or atomic absorption spectroscopy. Tissue samples of only a few mg weight can be assayed. An important point is that care must be taken in the exact extraction/assay protocols and in calibrating each technique for specific photosensitizers and specific tissues, in order to obtain accurate and reproducible results. It is also important to rapidly freeze the tissues (e.g. in dry ice or liquid N_2) and to maintain them in the dark, in order to avoid decomposition or photobleaching (see below) of the photosensitizer. Fluorescence microscopy (especially confocal) can also be used to image quantitatively the microdistribution of photosensitizers in tissues (or cells).

7.5.2 Photobleaching

Many PDT agents photobleach, i.e. are destroyed or modified by the light excitation. This may be monitored by measuring the fluorescence as the treatment proceeds. Mathematically, in the simplest case, the fluorescence (or absorption) of the photosensitizer follows a single exponential behavior:

$$C(\phi) = C_0 \exp(-\beta\phi) \tag{6}$$

where C is the concentration (at some point in the tissue) as a function of the light fluence, ϕ (J cm^{-2}), C_0 is the initial concentration, and β is the photobleaching rate (cm^2 J^{-1}). The photobleaching rate depends on the photosensitizer, and may also depend on its tissue localization (binding) and on the tissue oxygenation. Figure 19 shows photosensitizer in tissue pre- and post-treatment. It has been proposed [15] that the photobleaching, as monitored by fluorescence or absorption, may be used to determine the effective PDT dose received, based on the idea that the greater the degree of photobleaching the more singlet oxygen, and hence photodamage, will be produced. Although this may be qualitatively true (e.g. if there is no bleaching of a photobleachable drug at some point in the tissue), then there should be negligible photodynamic effect, or if the photobleaching stops part of the way through irradiation then the remaining light dose should be ineffective. However, in practice there are several factors that may confuse or weaken the relationship between photobleaching and photobiological effect. For example, if the photobleaching is caused by the singlet oxygen, rather than or in addition to direct damage by the initial light absorption, then the simple exponential (first order kinetics) behavior breaks down, since, as the drug is destroyed the rate of singlet oxygen production decreases, slowing the rate of photobleaching. With some photosensitizers, the photobleaching generates photoproducts that themselves may or may not be fluorescent and/or photoactive [16]. Further, the oxygen dependency of the photobleaching may itself depend on how the photosensitizer is bound in the tissue, while there will also be a gradient of photobleaching from the tissue surface (or interstitial fiber source) due to the gradient in fluence rate with depth (radius). Hence, much work remains to be done to determine the utility of photobleaching as a *quantitative* measure of photodynamic dose. It may still be very useful to show regions of the target tissue that have not been adequately irradiated, allowing the opportunity to re-direct light to such areas. Some work has also been done on using the bleaching of the endogenous tissue fluorescence (autofluorescence) for monitoring treatment, based on photodestruction of fluorescent tissue components (e.g. metabolites).

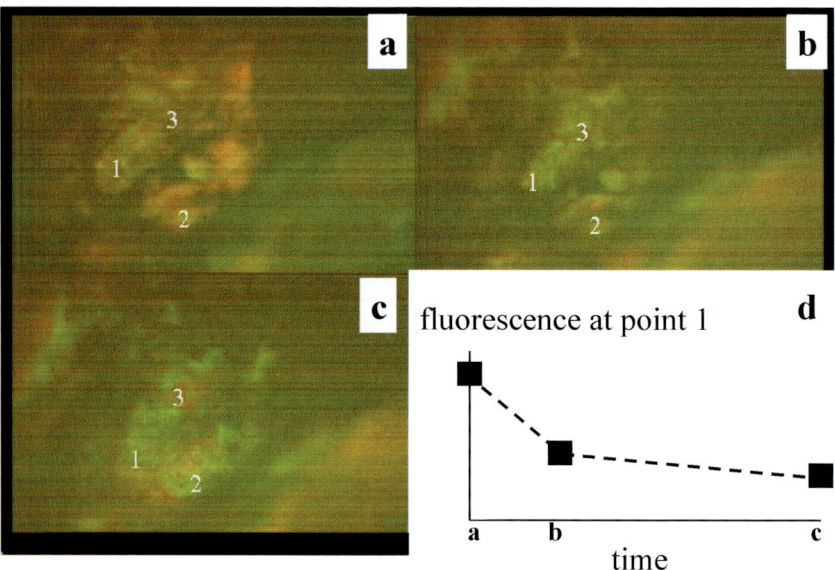

Figure 19. Example of photosensitizer photobleaching, monitored by fluorescence imaging and point spectroscopy in the post-resection bed of a malignant brain tumor during PDT (Photofrin, 2 mg kg^{-1} @ 24 h). (a–c) fluorescence images before, during and after light administration (120 J cm^{-2}), (d) corresponding red fluorescence intensity over a region of tumor at each time point.

7.6 Oxygen monitoring and biophysical effects

Since most clinical photosensitizers are believed to operate through (Type II) singlet oxygen photochemistry, the availability of adequate free oxygen in the tissue is a prerequisite for photobiological effect. The oxygen dependence of PDT has a sharper onset than for ionizing radiation and the onset occurs at lower pO_2.

A number of techniques are available to monitor tissue oxygenation. The most commonly used to date, both in animal studies and in patients, is the polarigraphic micro-electrode, which can be placed interstitially to measure the local pO_2. This is an invasive technique but, if the electrode is moved through the tissue, can provide very valuable information on the distribution of oxygen in the tissue. Alternatively, one or more microelectrodes may be placed in a fixed position to monitor the oxygen levels before, during and after treatment. Non-invasive methods under development include reflectance spectroscopy to monitor the hemoglobin and oxyhemoglobin (an indirect indicator of tissue oxygenation), near-infrared phosphorescence (in which an additional drug or, possibly, the photosensitizer itself is used as an indicator of oxygen level, based on measuring the "quenching" of the phosphorescence emission), and spin-trap electron spin resonance. It is too early to know the usefulness of these methods for clinical PDT, although they have certainly been valuable in pre-clinical research studies.

About a decade ago, it was recognized that, under some circumstances, the oxygen in tissue could become depleted by the photodynamic reaction itself.

Molecular oxygen is used up in the conversion into singlet oxygen so that, if this exceeds the rate of replenishment, the oxygen level will fall, reducing the photo-dynamic reaction rate and, hence, PDT response of the tissue [17]. This has been demonstrated both preclinically and clinically. The photochemical reaction rate at any point in tissue is proportional to the drug concentration and light fluence rate. The main consequence has been the recognition that there is a limit as to how rapidly one can deliver PDT and that a low light fluence rate (mW cm^{-2}) can be more effec-tive than a high fluence rate, for the same total light fluence (J cm^{-2}) delivered.

This photochemical depletion of oxygen in tissue is distinct from changes in pO_2 due to the vascular response of tissue to PDT. Both transient increases or decreases in blood flow in target tissue have been reported during PDT, depending on the photosensitizer, tissue and drug/light doses. Clearly, this has significant biological consequences. From the biophysical perspective, it may increase or decrease the photochemical reaction rate by altering the availability of free oxygen in the tissue. Increased blood flow will also reduce the light penetration in tissue by increasing the absorption by (oxy)hemoglobin, and such effects have been reported. Decreased flow has the opposite effect. These attenuation changes are most pronounced at shorter wavelengths below 650 nm.

The tissue oxygenation during PDT may be increased by hyperoxygenation, i.e. having the patient (or experimental animal) breath carbogen (95% O_2, 5% CO_2) rather than air. This has not been exploited deliberately in clinical studies, but is probably a factor in intraoperative treatments where the patient is under general anesthesia. In animal models hyperoxygenation has been shown to be at least as effective as the more complex technique of hyperbaric oxygen delivery [18].

7.7 Integrated PDT dosimetry

In optimizing PDT treatment for individual patients it is desirable to account, as much as is practical, for variations in light penetration, photosensitizer uptake and tissue oxygenation. These variations may be considerable and, as indicated above, may change dynamically (and unpredictably) during treatment [19]. For example, in esophageal tumors, variations in penetration depth of tissue at 650 nm of a factor of 3 have been reported and in photosensitizer uptake (mTHPC, Scotia, UK) by a factor of >5, while it is well known that the pO_2 of tissues, especially diseased tissues such as tumors, can vary even from point-to-point by more than an order of magnitude. Clearly, such large variations should lead to significant differences in clinical outcome in studies where the delivered doses of drug (mg kg^{-1}) and light (J cm^{-2}) are fixed. On the other hand, implementing meaningful light, drug and oxygen dosimetry for individual patients has not yet become feasible in routine practice in an ergonomic and cost-effective way. This situation should change as the dosimetry technologies become more available and clinically optimized.

Ideally, pre-treatment planning based on predicted true doses would help optimize the treatment delivery, in situ monitoring during treatment would allow compensa-tion for errors in light delivery (e.g. in interstitial fiber position) and compensation for dynamic changes, while recording of the actual doses delivered would allow more meaningful correlation with outcome. This information would then both reduce the size of clinical trials (reduce numbers of patients and dose escalation points) and, ultimately, improve the outcome of individual patients.

Part of the barrier to using these techniques in the clinic is the need to measure multiple factors simultaneously, possibly at multiple points in or on the treated tissue. Hence, there is motivation to develop so-called "implicit" dosimetry techniques [19] that integrate all or some of the factors. One example is the use of photobleaching, discussed above, that effectively uses the destruction of the photo-sensitizer as a surrogate measure of the effective PDT dose. Another is the direct measurement of singlet oxygen generation in the tissue, where detecting the infrared luminescence at 1270 nm has been suggested. Until recently this had not been technically possible in vivo due to the very high reactivity of $^1O_2$ making it very short-lived and hence giving an undetectably low luminescence signal. However, detection of $^1O_2$ emission has now been reported in animal models [20], so that there is hope that this will be feasible clinically in the future.

7.8 Intrinsic photodynamic sensitivity and PDT threshold

The response of tissue to PDT treatment clearly depends on the dose factors of local light fluence (rate), photosensitizer concentration and microlocalization and free oxygen availability. It also depends on the intrinsic photodynamic sensitivity of the tissue. That is, how much of the cytotoxic photoproduct (e.g. $^1O_2$) must be generated locally in the tissue in order to produce a specified biological endpoint, such as apoptosis or necrosis? If this could be known and the various dose factors could be measured during treatment, then the response could be predicted.

The determination of the intrinsic sensitivity is facilitated for those photosensi-tizers and tissues for which there is a clear boundary in the tissue (at some depth for surface irradiation or at some radius from the fiber source for interstitial treatment) between the undamaged and PDT-responding regions. There are numerous, examples where such sudden transitions are observed in the case of tissue necrosis, as illustrated in Figure 20. A sharp boundary implies that there is a minimum local concentration of $^1O_2$ required to produce the biological effect. There is a correspon-ding "threshold" for the number of light photons absorbed by the photosensitizer per unit volume of tissue, and a number of preclincal studies have determined these T-values for different tissues and photosensitizers [21]. Generally, these are in the order of 10^{17}–10^{19} photons per cm^3. If there is unlimited molecular oxygen available and the photosensitizer has a high quantum yield for $^1O_2$, then this would be approx-imately the concentration of singlet oxygen required for necrosis. Limited studies to date suggest a lower threshold for apoptosis, although the generality of this is not yet known.

In terms of PDT dosimetry, an important conclusion arises from this threshold concept, independent of the particular T value. In the simplest case of surface irradi-ation of a uniform tissue volume, the depth of necrosis, d_n, can be written as

$$d_n = \delta \ln(\varepsilon C\phi/T_n) \qquad (7)$$

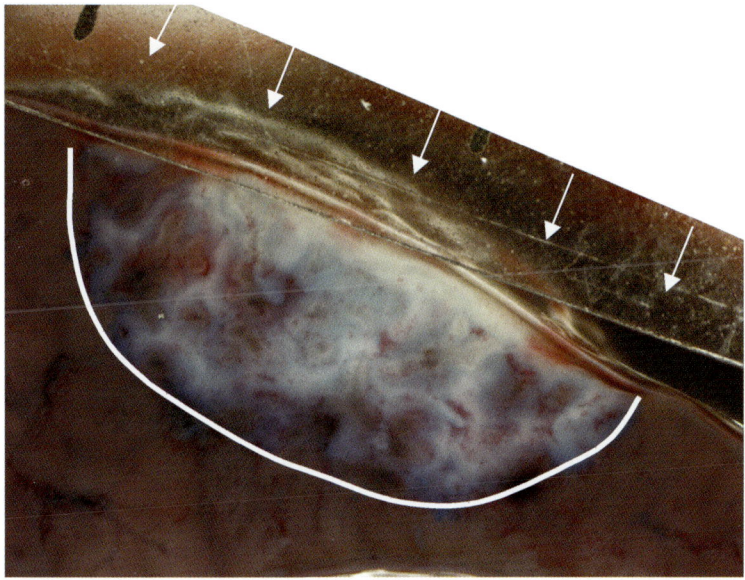

Figure 20. Example of the sharp 'PDT threshold' boundary in tissue (in this case, normal liver treated with a second-generation photosensitizer). The surface light irradiation is indicated, and the contour follows the necrosis boundary, defined by Evans Blue dye exclusion.

The important point is that d_n is proportional to the optical penetration depth, δ, but depends only logarithmically on the photosensitizer concentration, light fluence and threshold value. Hence, doubling any one of the latter factors does not double the effective treatment depth: a 7-fold increase is required. This is a limitation, in that it makes it more difficult to extend the treatment depth, but conversely it means that these factors do not have to be known as accurately as they would if d_n depended linearly on them. (Note that Equation (7) must be modified to include effects like photobleaching and oxygen depletion, but the essential conclusions remain valid.)

7.9 Treatment response monitoring

An alternative or adjunct to in situ PDT dosimetry is to monitor the PDT-induced changes to the target (and adjacent) tissues. Several methods have been investigated to date, although there are few extensive systematic studies for any of these, and there is a particular paucity of clinical reports. The methods include different forms of radiological imaging: computed tomography, magnetic resonance imaging (and spectroscopy), ultrasound, and radionuclide imaging (both conventional gamma-ray imaging and positron emission tomography). CT, MRI and ultrasound imaging have demonstrated imaging of the zone of PDT-induced damage following treatment. In particular, it has been possible to see the sharply-delineated boundary of damage in some tissues, corresponding to the threshold zone. Differentiation of necrotic from edematous changes post treatment remains a problem in some tissues, depending on how soon after treatment the imaging is performed. Gamma-ray scintigraphy and positron emission tomography have shown local functional/metabolic tissue responses.

In treatment of tumors in hollow organs (GI and urogenital tracts), endoscopic imaging during and post treatment may be useful for monitoring immediate and short-term tissue responses. For superficial lesions, particularly skin, blood flow changes may be monitored by ultrasound or laser Doppler imaging, and have been shown to correlate with outcome. In PDT of macular degeneration, fluorescein angiography is used before treatment to identify the retinal target area and post-PDT to monitor the completeness of neovascular shut-down.

There are several methods still under development that have potential for response monitoring. High-frequency ultrasound imaging has been shown to be capable of imaging specifically apoptotic cell death in vivo with high spatial resolution, but limited depth of imaging. Optical coherence tomography is the optical analogue of ultrasound imaging, and studies are in progress to determine whether it is also sensitive to necrotic and/or apoptotic responses. Electrical impedance spectroscopy, a point measurement technique in which the (complex) electrical properties of tissue are measured as a function of the frequency (kHz–MHz range) of a small applied electrical voltage, has recently been shown to be sensitive to PDT-induced changes in tissue and may differentiate necrosis and apoptosis. Finally, new imaging methods are appearing that may provide specific molecular information on the tissue response, such as the use of fluorescently-tagged reporter molecules or bioluminescence imaging using luciferase gene transfection. These will be invaluable for pre-clinical research, even if they do not eventually reach the clinic.

References

1. L.H. Murrer, J.P. Marijnissen, W.M. Star (1997). Improvements in the design of linear diffusers for photodynamic therapy. *Physics in Medicine and Biology*, **42**,1461–1464.
2. L.H. Murrer, J.P. Marijnissen, W.M. Star (1998). Monte Carlo simulations for endobronchial photodynamic therapy: the influence of variations in optical and geometrical properties and of realistic and eccentric light sources. *Lasers in Surgery and Medicine,* **22**,193–206.
3. B.C. Wilson, D.R. Wyman (2000). Dosimetry and thermal monitoring. In: R.W. Wynant (Ed.), *Lasers in Medicine* (pp. 47–84). CRC Press, Boca Raton.
4. W.M. Star (1997). Light dosimetry in vivo. *Physics in Medicine and Biology*, **42**, 763–787.
5. J.P. Marijnissen, W.M. Star, H.J. Zandt, M.A. D'Hallewin, L. Baert (1993). In situ light dosimetry during whole bladder wall photodynamic therapy: clinical results and experimental verification. *Physics in Medicine and Biology*, **38**, 567–582.
6. B.C. Wilson, W. Whelan, S.R.H. Davidson, R. Weersink, M.D. Sherar (2002). Treatment planning platform for photodynamic therapy: architecture, function and validation. *Proceedings of the Society of Photo-Optical Instrumentation Engineers*, **4612**, 85–92.
7. J.P. Marijnissen, W.M. Star (1996). Calibration of isotropic light dosimetry probes based on scattering bulbs in clear media. *Physics in Medicine and Biology*, **41**, 1191–1208.

8. M. Panjehpour, B.F. Overholt, J.M. Haydek (2000). Light sources and delivery devices for photodynamic therapy in the gastrointestinal tract. *Gastrointestinal Endoscopy Clinics of North America*, **10**, 513–532.

9. O. Barajas, A.M. Ballangrud, G.G. Miller, R.B. Moore, J. Tulip (1997). Monte Carlo modelling of angular radiance in tissue phantoms and human prostate: PDT light dosimetry. *Physics in Medicine and Biology*, **42**, 1675–1687.

10. T. Johansson, M.S. Thompson, M. Stenberg, C. af Klinteberg, S. Andersson-Engels, S. Svanberg, K. Svanberg (2002). Feasibility study of a system for combined light dosimetry and interstitial photodynamic treatment of massive tumors. *Applied Optics*, **41**, 1462–1468.

11. G.A. Wagnieres, W.M. Star, B.C. Wilson (1998). In vivo fluorescence spectroscopy and imaging for oncological applications. *Photochem. and Photobiol.*, **68**, 603–632.

12. M. Panjehpour, R.E. Sneed, D.L. Frazier, M.A. Barnhill, S.F. O'Brien, W. Harb, B.F. Overholt (1993). Quantification of phthalocyanine concentration in rat tissue using laser-induced fluorescence spectroscopy. *Lasers in Surgery and Medicine*, **13**, 23–30.

13. R.A. Weersink, J.E. Hayward, K.R. Diamond, M.S. Patterson (1997). Accuracy of non-invasive in vivo measurements of photosensitizer uptake based on a diffusion model of reflectance spectroscopy. *Photochem. and Photobiol.*, **66**, 326–335.

14. L. Lilge, C. O'Carroll, B.C. Wilson (1997). A solubilization technique for photosensitizer quantification in ex vivo tissue samples. *Journal of Photochem. and Photobiol.*, **B39**, 229–235.

15. W.R. Potter, T.S. Mang, T.J. Dougherty (1987). The theory of photodynamic therapy dosimetry: consequences of photo-destruction of sensitizer. *Photochem. and Photobiol.*, **46**, 97–101.

16. E.F. Gudgin-Dickson, R.H. Pottier (1995). On the role of protoporphyrin IX photoproducts in photodynamic therapy. *Journal of Photochem. and Photobiol.*, **B29**, 91–93.

17. J.C. Finlay, D.L. Conover, E.L. Hull, T.H. Foster (2001). Porphyrin bleaching and PDT-induced spectral changes are irradiance dependent in ALA-sensitized normal rat skin in vivo. *Photochem. and Photobiol.*, **73**, 54–63.

18. Q. Chen, Z. Huang, H. Chen, H. Shapiro, J. Beckers, F.W. Hetzel (2002). Improvement of tumor response by manipulation of tumor oxygenation during photodynamic therapy. *Photochem. and Photobiol.*, **76**, 197–203.

19. B.C. Wilson, M.S. Patterson, L. Lilge (1997). Implicit and explicit dosimetry in photodynamic therapy: A new paradigm. *Lasers in Medical Science*, **12**, 182–199.

20. M. Niedre, M.S. Patterson, B.C. Wilson (2002). Direct near-infrared luminescence detection of singlet oxygen generated by photodynamic therapy in cells in vitro and tissues in vivo. *Photochem. and Photobiol.*, **75**, 382–391.

21. T.J. Farrell, B.C. Wilson, M.S. Patterson, M.C. Olivo (1998). Comparison of the in vivo photodynamic threshold dose for Photofrin, mono- and tetra-sulfonated aluminum phthalocyanine using a rat liver model. *Photochem. and Photobiol.*, **68**, 394–399.

Chapter 8

Fluorescence bronchoscopy for early detection of lung cancer

Dr. G. Sutedja

Table of Contents

Summary

Due to the often advanced stage of lung cancer at the time of detection, the 12% cure rate of patients has remained dismally low for the last few decades. The stage shift hypothesis, i.e., detecting and treating cancer at an early stage, has propelled the interest to screen the population at risk, as early lung cancer is known to be only several millimeters in size. Together with the feasibility of minimal invasive treatment techniques, the applications of less morbid diagnostic and treatment approaches may benefit many individuals at risk.

Autofluorescence bronchoscopy (AFB) is a diagnostic imaging method based on the difference of fluorescence characteristics between malignant tissue and normal bronchial mucosa. AFB can, therefore, be used to assist clinicians to localize previously invisible pre-cancerous lesions in the bronchial mucosa. AFB has been shown to be more sensitive than the conventional fiberoptic bronchoscopy in the detection of pre-cancerous and early cancer lesions. The number of suspicious lesions observed by AFB was found to be predictive of the chances of developing squamous cell cancer. AFB allows accurate delineation of tumor border for performing bronchoscopic treatment such as photodynamic therapy. The higher image resolution of AFB enables accurate sampling of specimens to study the molecular genetic changes and the natural history of pre-cancerous lesions. AFB, in combination with minimal invasive bronchoscopic treatment, seems cost-effective for the early management of cancerous lesions in the central airways.

8.1 Introduction

It is widely acknowledged that lung cancer is an important global epidemic leading to severe morbidity and high mortality for the individuals at risk [1]. Ten percent of heavy smokers can be expected to suffer from lung cancer. Despite many new treatment approaches applied in the past, the cure rate remains low at 12%. This is due to the fact that the majority of lung cancer patients are diagnosed at an advanced stage, precluding any curative treatment attempt. Even for those who underwent radical lung resection (< 25% of the total), tumor recurrence (30%) and the high chance of developing subsequent cancers (4% cumulative new incidence per year) remain a serious threat [2–5].

The resurgence of interest for lung cancer screening is supported by the appreciation that early stage lung cancer, i.e., when cancerous tissue is only several millimeters in size [6–9], can now be better managed with less invasive therapeutic modalities [10]. The applications of less invasive new techniques for early management of lung cancer have raised controversial issues because screening by conventional chest X-rays and sputum cytology in previous randomized clinical trials failed to reduce lung cancer mortality [11–14].

The use of molecular genetic markers [7], autofluorescence bronchoscopy (AFB) [15], etc. in combination with minimally invasive therapy such as intraluminal bronchoscopic treatment (IBT) [10] may provide clinicians with new modalities for early intervention. Smoking related morbidities such as emphysema and poor cardiovascular condition have raised important limitations to appropriately treating lung cancer patients. The aim of AFB is to allow clinicians to detect early and accurately localize pre-cancerous lesions. Cost-effectiveness of a lung cancer screening, however, depends on the proper selection of the population and the optimal application of these new techniques. The clinical perspective on the use of AFB for the early management of lung cancer in the target population will be discussed.

8.2 The target population

Many Japanese centers have been conducting sputum cytology in (ex-) heavy smokers (combined with chest X-rays and spiral, high resolution computer tomography for the detection of early adenocarcinoma in the lung parenchyma) for decades [16,17]. Data have been consistent and in accordance with previous randomized trials in showing that early detection of stage I–II cancers, mainly early stage squamous cell cancer (ESSCC), has led to a higher resection rate and improved survival [18,19]. This seems in contrast to subjects with sputum positive adenocarcinoma, who were shown to have a poor survival because >82% of them were already in stage IIIA–IV [20]. Data from randomized clinical trials performed by the three institutions — Mayo Clinic, Johns Hopkins, Memorial Sloan Kettering — showed that especially in the group of subjects with ESSCC [14], more early cancers were detected and survival rate after resection was significantly better compared to the group of patients who refused surgery, 63–76% versus <19%, respectively [19]. Despite many controversies, detection of ESSCC appeared to be of some benefit when combined with surgical resection.

Extended follow-up data, however, reconfirmed earlier findings that no true mortality benefit exists [14]. This is because lung cancer was not the ultimate cause of death in the majority of the individuals in the screening arm. Smokers' mortality from all causes of death is about three-fold compared to non-smokers. Mortality due to lung cancer was ultimately only 7% in both arms of the Mayo study, and ESSCC formed only a relatively minor part. Many individuals died due to other causes, including ischemic heart disease and COPD. The absence of mortality reduction has raised controversies about the potential value of screening. Overdiagnosis is considered an important problem due to the detection of relatively indolent cancers which would never have become clinically relevant [21]. Some individuals are destined — despite early detection and treatment — not to die from lung cancer *per se*. In the control arm of a screening study, the presence of lung cancer may remain undetected [22]. Two percent of lung cancers were missed, out of the 153 primaries found in a postmortem study [23]. Death in the screened arm from other causes may have been falsely attributed to lung cancer. Traditionally, health care workers and the individuals at risk strongly perceive that screening can save lives, because lung cancer is regarded as a fatal disease [6]. The inherent risks for death due to COPD and cardiovascular diseases withhold for lung cancer deaths as well [9,14]. In retrospect, early lung cancer treatment may prove to be treating a pseudo disease. Screening may also lead to additional morbidity and mortality (for example finding an aortic aneurysm), not always expressed correctly in the number of disease specific mortalities [21,22]. All these factors have led to different interpretations and controversies regarding the potential benefit of a lung cancer screening.

As it is currently impossible to accurately predict the cause of death for each individual, it is ethically difficult for a clinician to refuse treatment in case an early lung cancer is diagnosed. The lack of mortality benefit in previous randomized clinical trials, the issues of 'overdiagnosis' and treatment of 'pseudo-disease' may all be epidemiologically interesting, but offer no solution for clinical practice. In previous decades, the advanced disease at presentation has been blamed as the major reason for the dismal cure rate, so a strategy of no treatment in early detected lung cancers to investigate the issue of overdiagnosis will be difficult to defend. The next best approach is to compare treatment with a group of patients who refuse treatment, however, such a strategy is obviously not free from biases [18,19].

About 90% of (ex-)smokers will not die of lung cancer, so, using a highly specific diagnostic test may allow us to focus our efforts on the ~10% group who possess the greatest risk. Molecular epidemiology for studying cancer susceptibility, e.g., CYP1A1, CYP2D6 and GSTM polymorphisms looking for individuals' capacity to detoxify carcinogens have initially shown that the presence of null GSTM1 and mutant CYP1A1 genotypes are associated with a 41 times higher risk of developing lung cancer [24]. However, recent meta-analysis by Kiyohara et al. showed that only myeloperoxidase and microsomal epoxide hydrolase exon 4 polymorphisms are found to be associated with a higher lung cancer risk [25].

It is important to realize that FEV1 as % predicted ventilatory impairment is a prognostic factor for death from ischemic heart disease [26]. COPD with an FEV1 <70% predicted showed a 10-year cumulative percentage for lung cancer death of 8.8% versus 2% for those with an FEV1 > 85% predicted [27]. The difference in survival at 10 years is significant, 74% versus 91.1%, respectively. The higher relative risk for lung cancer in COPD smokers was significantly and proportionally related to the degree of airway obstruction. The presence of ventilatory impairment was associated with a 6.44 times greater risk for lung cancer [28]. Hence the approach of selecting primarily COPD individuals in a lung cancer screening program [9,29]. Paradoxically, screening of this population may fuel more controversies regarding the treatment of pseudo-diseases, because of the relatively higher risk for COPD death. It remains to be seen whether genetic susceptibility studies and focusing on COPD patients may help us to achieve a more economical screening program, instead of repeat testing and enlisting all (ex-) heavy smokers with or without pulmonary function impairment for a screening program. The population of (ex)-smokers who have a 10% chance of developing lung cancer is indeed very large. For peripheral early stage adenocarcinoma, a lobectomy and mediastinal lymph nodes dissection is the current standard of treatment [30,31]. In contrast, intraluminal bronchoscopic treatment (IBT) provides an alternative to surgery for patients with central type ESSCC [10,32,33]. IBT can be conducted in a matter of minutes in an outpatient setting [10]. Especially in this group of individuals, long term local control without surgical resection may suffice to preserve quality of life. Therefore, the application of less invasive and morbid treatment methods has to be put into proper perspective. The efficacy of IBT allows us to treat non-operable candidates, who previously were not enlisted in a screening program [14,34]. Long term survival after IBT and the fact that recurrences can be retreated successfully, indicates the slow progression rate of early cancer from millimeters in size towards advanced local disease [35]. This is evidence of the potential benefit of early detection in combination with IBT regarding central type ESSCC.

The combination of accurate diagnostic tests and minimally invasive treatment are crucial elements for cost-effective screening [36,37]. Screening related anxiety, morbidity and mortality may increase the burden and costs. The efficacy of minimal invasive treatment, i.e., IBT as an alternative to surgery warrants the inclusion of the population with severe COPD and poor general health, despite their inherent risk of dying from non-lung cancer related causes [35]. Paradoxically, high local control for ESSCC patients using a less invasive treatment method exposes them longer to the risk of death from non-lung cancer related causes [9,14,35]. This will lead to an epidemiological quagmire, if one insists that lung cancer screening has to result in the reduction of the overall mortality figure [38]. Currently, the large number of COPD and (ex-) smokers form a major hurdle to the realization of a cost-effective screening program.

8.3 Early stage cancer in the central airways

Sputum cytology mainly detects ESSCC [5,9,14,16]. Treatments offered have been both surgical and bronchoscopic (IBT, e.g., photodynamic therapy) [18,19,32,33]. Many phase II studies have been dealing with photodynamic therapy as the most popular treatment method. Screening with spiral CT has shown significant stage shift for the detection of stage I/II adenocarcinoma in >80% of the cases [8,17,39]. However, the use of spiral CT screening is controversial for the same reasons mentioned above [11,14,21,22]. Also, >50% of the individuals were found to have one or more non-calcified nodule(s) at baseline examination. This leads to additional diagnostic steps with increased morbidity and costs [8,39,40]. Compared to low dose spiral CT, the 'false positive' rate of early detection programs using sputum cytology is favorably low [16]. Recent data from the SPORE trial in COPD (ex-) smokers have shown that 103 (4.2%) individuals developed cancer among the 2441 individuals screened, which is a seven-fold increase over the incidence of the normal USA population (Hirsch et al. ASCO abstract no. 1201, 2002).

Molecular abnormalities in the bronchial mucosa may be present prior to the emergence of overt cancer [41,42]. However, senescence and inherent dynamism due to continuous exposure of bronchial mucosa to irritants may lead to changes that are not necessarily related to carcinogenesis *per se*. Not a single set of biomarkers has yet emerged as an accurate predictor for the development of lung cancer in any particular individual. Early molecular changes have also been studied — in terms of field carcinogenesis [43] — by looking at 'normal' specimens collected from areas remote to where the primary tumor was located [44]. Changes were detected in about 50–60% of the individuals, underscoring our limited knowledge about genetic changes in the bronchial mucosa related to carcinogenesis.

Studies on archived sputum have shown sensitivities from malignant associated changes (MAC seen using DNA texture analysis) and immunostaining with hnRNP A2/B1 of 78 and 91%, respectively [45,46]. Both methods seemed to predict early carcinogenesis, years in advance of early lung cancer emerging. Semi-automated MAC has the advantage of being reproducible and consistent, in contrast to the subjective classification of pre-cancerous lesions by the pathologists due to intra- and inter-individual variability [47].

Step-wise morphological changes are believed to occur from normal through the various grades of pre-cancerous lesions: squamous metaplasia, mild-, moderate-, severe dysplasia and CIS, before the lesion becomes micro-invasive squamous cell cancer [48–51]. The carcinogenetic pathways of adenocarcinoma and neuro-endocrine tumors have not been fully elucidated and are beyond the scope of this chapter, as these tumors are mainly located in the lung parenchyma [48]. Auerbach in 1961 [49,50], investigated the relation of smoking to changes in the bronchial epithelium. Post-mortem step sections of the entire tracheobronchial mucosa of 402 male patients was carried out, up to 208 sections per individual. The true presence especially of CIS was shown to correlate with smoking history and the presence of cancer. CIS lesions were found in 23 (2.4%) of the 956 sections, showing atypical cells in the entire specimen, with an average depth of five cells only (range 4–38 cells). In ESSCC, 61% of the lesions had abnormalities to a depth of ≥ 6 cells. So, early ESSCC is indeed < 1 mm thick. Non-smokers did not seem to harbor any CIS. CIS was present in 75% (27/36) of the subjects who smoked ≥ 2 packs/day and in 83% (52/63) of lung cancer subjects. The presence of multiple CIS strongly supported field carcinogenesis [43], which was shown in 11.4% of ≥ 2 packs/day smokers and in 15% of lung cancer patients.

However, the exact location of CIS using conventional fiberoptic bronchoscopy (FB) could only be determined in 29% of the cases [5]. Two-thirds of these 'radiographically occult' lung cancers were shown to be only a few millimeters thick. Sato et al. reported that in 180 patients with 200 occult cancers, 527 repeat FB sessions were required to ultimately locate CIS and early stage cancer lesions [52]. An average of three lengthy FB sessions of up to 45 minutes were performed with separate brushings of all bronchial segments. In retrospect, 175 lesions were found proximal to the sub-segmental bronchi within the visible range of FB. Tedious and repeat FB examinations led to an average delay of 29.2 months before the exact location within the tracheobronchial tree was determined. This indicates the failure of current diagnostic procedure for finding the exact location of early cancerous lesions. This will be counterproductive for the stage shift hypothesis if ever any advanced sputum cytology screening method becomes available.

8.4 Autofluorescence bronchoscopy for localizing pre-cancerous lesions

Based on the relatively selective uptake of photosensitizers in cancerous tissue similar to photodynamic therapy [32,33], photodynamic diagnosis has been initially developed to increase sensitivity in finding early cancerous lesions [53]. However, the lack of drug selectivity, complex pharmacokinetics and skin photosensitivity raise issues as to whether photodynamic diagnosis will be cost-effective and practical to screen large number of individuals [54,55].

Laser-induced fluorescence endoscopy (AFB-LIFE, Xillix®, Richmond, BC Canada) uses a Helium–Cadmium laser of λ 442 nm for tissue excitation [56,57]. The emission spectrum is captured by two sensitive CCD cameras and is processed through a fluorescence collection sensor and an optical multi-channel analyzer. Real time digitized images are obtained using the ratios of red to green (λ 630–520 nm) fluorescence emissions for correcting distance and breathing movements during in vivo bronchoscopic examinations. Special imaging algorithms have been generated, based on the data from in vivo–in vitro analysis of pre-cancerous tissues and Monte Carlo simulation modeling [58]. Digitized images reflect the in vivo fluorescence signals collected, exploiting the differences in autofluorescence spectra between suspicious (high grade dysplasia and CIS) lesions and normal bronchial mucosa. The suspicious lesions appear dark red-brownish due to the lower green reflectance while normal mucosa appears green. Other systems use the fluorescence–reflectance system in their imaging algorithms, such as the D-light system (Storz®, Tüttlingen, Germany) [53]. Non-coherent ultraviolet to blue 300-W Xenon filtered lamp (λ 380–460 nm) is used to excite a broad emission spectra of the different chromophores (Flavins, NADH) in cancerous tissues [59]. In combination with some reflectance of blue light, the D-light produces bluish images for normal areas and darker images for high grade dysplasia and CIS. The SAFE® 1000 autofluorescence system (Pentax®, Asahi Optical Tokyo, Japan) also uses a Xenon-lamp (λ 420–480 nm) and the camera contains a fluorescence filter as well as an image intensifier [60]. Other new fluorescence–reflectance imaging systems (DAFE®, Wolf, Knittlingen, Germany and ONCO-LIFE® Xillix, Richmond, BC Canada) will soon enter clinical trial, with the aim of reducing the complexity of equipment and cost. Exploiting both quantitative and qualitative imaging algorithms may improve sensitivity. The use of digitized fluorescence ratio, i.e. quantitative imaging, projected on the monitor screen in addition to the subjective classification of the images has improved the CIS detection rate to 100% [61]. It is now obvious as to why the achievements of autofluorescence detection systems have pushed further interest in photodynamic diagnosis into the background.

AFB-LIFE programs have so far included the largest number of individuals and thousands of lesions have been investigated in several large multicenter trials [15,62,63]. A panel of pathologists reviewed all biopsy specimens for proper WHO/IASLC classification, as intra- and inter-individual variability of histology classification has been shown to be quite significant [64]. The commonly accepted WHO/IASLC classification is clearly not a gold standard. The reproducibility of classifying pre-cancerous lesions showed an intra-observer agreement of 0.71 and an inter-observer agreement of 0.55 [47]. In classifying the CIS materials in the initial study period, <50% was reconfirmed by the panel to be true CIS [64,65]. Significant intra- and inter-observer variability is an important hurdle to consistent analysis of data from carcinogenesis studies. Especially when looking for genetic markers in pre-cancerous lesions, the use of internet communication for a more accurate morphological classification is a key issue [66]. A κ value of 0.9 has been obtained.

It is also important to realize that sensitivity, specificity and accuracy remain a relative issue because of the unknown total number of pre-cancerous lesions present at that particular moment in the entire bronchial mucosa of a certain individual [15,49,50]. Smoking history [49–51], gender distribution with relative paucity of pre-cancerous lesions in women [67], the expertise of the bronchoscopist [62,63] the variability of morphological classification [47,66], are all factors to be taken into account.

Nevertheless, AFB has been shown to be significantly more sensitive than conventional FB alone for the detection of pre-cancerous lesions [65,68], especially as the sequence of examination, either using AFB first or FB first, did not influence the results because of the image memory bias of the bronchoscopists.

Unfortunately, biopsy sampling using AFB-LIFE may remove entire patches of clonal cells [69]. Many lesions were <1.5 mm in diameter and about 50% of these lesions were smaller than the biopsy forceps. Twenty-seven of the 69 paired biopsies obtained at 6-month intervals showed one or more molecular changes in the initial biopsy specimens, 86% had no abnormality after re-biopsy and 24% lacked the initial changes found after repeat biopsy. So, the natural history of minute lesions cannot be studied because of complete mechanical removal during baseline sampling.

Spectroscopic images capture the entire spectral images of the target tissue within a wavelength of interest [70,71]. Optical coherence tomography (OCT) can record two-dimensional cross-sectional images of cell layers [72]. Confocal micro-endoscopy provides sub-cellular images used in the study of the process of carcinogenesis [73]. The use of OCT and confocal techniques may enable us to study the natural history of pre-cancerous lesions in vivo without the necessity of taking biopsy.

In conclusion, the difference between autofluorescence spectrum of suspicious (moderate dysplasia, severe dysplasia and CIS) and normal bronchial mucosa seems to be caused by:

(1) Relative thickening of dysplastic mucosa and CIS, causing reduction of green light reflectance leading to a red-brownish appearance of the lesion [56,58].
(2) Increased microvascularity or 'angiogenic squamous dysplasia' [68,74]. The presence of capillary sized blood vessels juxtaposed to and projecting into dysplastic epithelium has been shown in dysplastic lesions compatible with the pattern of tortuous capillaries shown with high magnification broncho-video-scope indicating increased angiogenesis [75].
(3) Dysplastic lesions are known to contain higher concentrations of chromophores, such as Flavins and NADH, contributing to the differences in autofluorescence signals between normal and potentially malignant cells [59].

The size of current bronchoscopes, however, does not allow inspection beyond the 5th–6th order of the bronchial segmentation. So, AFB can only be applied to the inspection of the central airways' mucosa, focusing on high grade dysplasia, CIS and microinvasive squamous cell cancer.

So far, AFB-LIFE® programs have analyzed the largest number of individuals and thousands of lesions have been studied [15,62–64], and specimens have been reviewed by a panel of pathologists for proper WHO/IASLC classification. The results of several years' clinical experiences in many tertiary institutions have shown that:

(1) AFB-LIFE examination is well tolerated by the patient, extending the duration of FB session by <15 min. AFB can be performed under local anesthesia in an outpatient setting [57,65].

(2) AFB-LIFE is 2–4 times more sensitive than conventional FB for the detection of previously invisible pre-cancerous lesions with 80–100% detection rate [64,68]. Increased sensitivity is irrespective of the sequence of examinations (no memory bias: AFB or conventional FB first) or the expertise of the broncho-scopists. Nevertheless, sensitivity, specificity and accuracy remain relative issues as has been described before, as the absolute number of pre-cancerous lesions in the entire tracheobronchial tree of an individual at a certain moment cannot be determined [49,50].

(3) Proper classification of pre-cancerous lesions according to WHO/IASLC criteria by a panel of pathologists is essential [66]. The intra- and inter-individual variability in classifying pre-cancerous lesions by pathologists can be quite significant [47,64].

(4) AFB-LIFE enables the study of the 'natural history' of pre-cancerous lesions [76–78]. A progression rate toward malignancy of 6–17% has been reported so far. Many pre-cancerous lesions seem to regress 'spontaneously'.

(5) All CIS ultimately progress to micro-invasive ESSCC [79]. Unforeseen occult synchronous lesions in the central airways have been found in ~10% of individuals [80].

(6) AFB-LIFE assists accurate delineation of the tumor border [37]. Only 25% of these lesions have been found to be occult, thereby justifying intraluminal bronchoscopic treatment (IBT) with curative intent [81].

(7) AFB-LIFE enables longitudinal sampling of pre-cancerous lesions. The presence or persistence of genetic alterations (e.g., p53+, FHIT) in AFB suspi-cious lesions is predictive for ESSCC invasion [82].

(8) AFB-LIFE biopsies show minute pre-cancerous lesions <1.5 mm in size, 50% of which are smaller than the biopsy forceps [69]. This may explain the 'sponta-neous regression' rate, especially regarding low-grade dysplastic lesions [76–78].

(9) The number of AFB-LIFE suspicious lesions at baseline are found to be predic-tive of the development of ESSCC in the central airways [83].

8.5 How to treat early stage lesions

The most important determinant for outcome is accurate staging [10,36,37,81]. ESSCC should be assessed with regard to tumor margins and depth of tumor invasion in the bronchial wall. It is important to recognize the correlation between tumor size and nodal disease metastasis [83–86]. Involvement of lymph nodes precludes any curative treatment attempt with IBT alone [10,81,87]. Endobronchial ultrasonography (EBUS) has been used to exactly determine the bronchial wall invasion of early cancer lesions, but its definite role needs to be determined [88,89]. Also, the role of PET-scan for accurate staging to exclude nodal disease needs further study [90].

Many phase II studies have dealt with the efficacy of IBT [10,32,33,54]. Ideally, phase III studies comparing surgery versus various IBT modalities versus a no treatment arm are required. However, the relative paucity of early stage SCC — as many are detected by chance — and the ethical dilemma described previously make such a study protocol unworkable. The choice of IBT modality is of less importance provided nodal metastasis has been ruled out [81,83,84,90], so IBT is an alternative to surgery and is a parenchyma conserving technique. All IBT modalities, including photodynamic therapy, Nd–YAG laser and electrocautery seem to be equally effective, and electrocautery is the most cost-effective [10,35,81,91]. Any IBT modality prior to surgery may also enable a less extensive surgical resection [92–94]. IBT as the initial treatment strategy for intraluminal ESSCC is a prudent approach to preserve healthy lung tissue and quality of life, and is found to be very cost-effective compared to surgical resection [95].

8.6 Cost-effectiveness

We have calculated the cost of AFB-LIFE bronchoscopy for early detection in the target population consisting of individuals with previously resected stage I–II lung and head and neck cancers, and those with positive sputum cytology. Twenty-one lesions have progressed to cancer so far (171 dysplastic lesions and 429 bronchoscopies). The tariff of one bronchoscopy examination in the Netherlands is € 81. The endoscopic cost for early detection per cancer lesion was € 1653. Treatment with electrocautery was € 380 [10]. The costs for early bronchoscopic intervention per ESSCC was € 2033.

We recently analyzed the cost-effectiveness of IBT for inoperable individuals with ESSCC in comparison to matched controls of surgically resected stage IA, T1N0 cancer patients after a follow up period of 2–10 years. Patients were matched with regard to various co-morbidities, tumor type and age to individuals who underwent surgery for a T1N0 cancer. So far, the average cost per individual for early management of IBT, including long term follow up has been € 6547 versus € 22 173 for each surgical patient. Despite the worse initial health status of patients treated with IBT, actual survival rate and expenses for early intervention underscores the superior cost-effectiveness of a minimally invasive diagnostic and therapeutic approach for early intervention in properly selected individuals. Both autofluorescence bronchoscopy and intraluminal bronchoscopic treatment, i.e., electrocautery are clearly less invasive, less morbid and can be applied effectively with acceptable costs.

8.7 Future perspective

New, less invasive diagnostic and therapeutic techniques enable us to detect and follow lesions that were previously invisible and study the molecular genetics of lung cancer. If a significant squamous cell cancer type exists in the target population, more benefit can be drawn from early management such as the combined use of autofluorescence bronchoscopy and intraluminal bronchoscopic treatment. In the Netherlands, for example, squamous cell type comprises ~ 35% of all histology types. A cure rate of ~ 70% due to stage shift will improve the total cure rate to ~ 24%. However, the compliance rate of a screening trial has been shown to be only ~ 75% [14]. An effective early intervention program for only the squamous cell type group may, therefore, improve the absolute cure rate to ~ 18% (from the current cure rate of 12% only). One should remember that the so-called improved outcome of any interventional strategy in the past decades was mainly determined by better selection and staging, the so-called Will Rogers phenomenon. It did not cure more people because in daily practice, 75–80% of new lung cancer patients are still at the non-curable stage. Improved survival rate was fallacious, because advanced stage cases have been shifted out from the many studies of population cohorts. Therefore, the total cure rate for lung cancer patients has remained relatively stable at 12%. So the use of new and minimally invasive techniques is a promising prospect, as improved survival will be based on stage shift, by increasing substantially the percentage of early stage cancers detected combined with cost-effective therapeutic techniques that preserve quality of life.

References

1. R.T. Greenlee, M.B. Hill-Horman, T. Murray, et al. (2001). Cancer statistics. *CA Cancer J. Clin.*, **51**, 15–36.
2. R. Feld, L.V. Rubinstein, T.H. Weisenberger and the Lung Cancer Study Group (1984). Sites of recurrence in resected stage I Non-Small-Cell Lung Cancer. A guide for future studies. *J. Clin. Oncol.*, **2**, 1352–1358.
3. P.C. Pairolero, D.E. William, E.J. Bergstrahl, et al. (1984). Postsurgical stage I bronchogenic carcinoma: Morbid implications of recurrent disease. *Ann. Thor. Surg.*, **38**, 331–338.
4. N. Martini, M.S. Bains, M.E. Burt (1995). Incidence of local recurrence and second primary tumors in resected stage I lung cancer. *J. Thorac. Cardiovasc. Surg.*, **109**, 120–129.
5. L.B. Woolner, R.S. Fontana, D.A. Cortese (1984). Roentgenographically occult lung cancer: Pathologic findings and frequency of multicentricity during a 10-year period. *Mayo Clin. Proc.*, **59**, 453–466.
6. G. M. Strauss, R.E. Gleason, D.J. Sugarbaker (1997). Screening for lung cancer: Another look, a different view. *Chest*, **111**, 754–768.
7. F.R. Hirsh, W.A. Franklin, A.F. Gazdar, P.A. Bunn (2001). Early detection of lung cancer: clinical perspectives of recent advances in biology and radiology. *Clin. Cancer Res.*, **7**, 5–22.
8. J.R. Jett (2001). Pro/Con editorials spiral computed tomography screening for lung cancer is ready for prime time. *Am. J. Respir Crit. Care Med.*, **163**, 812–815.
9. T.L. Petty (2000). Screening strategies for early detection of lung cancer. The time is now. JAMA, **284**, 1977–1980.
10. T.J. van Boxem, B.J. Venmans, G. Sutedja (1999). Curative endobronchial therapy in early-stage non-small cell lung cancer. Review, *J. of Bronchol.*, **6**, 198–206.

11. E.F. Patz, P.C. Goodman, G. Bepler (2000). Screening for lung cancer. Review article. *NEJM*, **343**, 1627–1633.
12. M.R. Melamed, B.J. Flehinger, M.B. Zaman (1987). Impact of early detection on the clinical course of lung cancer. *Surg. Clin. North Am.*, **67**, 909–924.
13. A. Kubik, D.M. Parkin, M. Khlat, et al. (1990). Lack of benefit from semi-annual screening for cancer of the lung: Follow-up report of a randomized controlled trial on population of high-risk males in Czechoslovakia. *Int. J. Cancer*, **45**, 26–33.
14. P.M. Marcus, E.J. Bergstrahl, R.M. Fagerstrom, et al. (2000). Lung cancer mortality in the Mayo lung project: impact of extended follow-up. *J. Natl. Cancer Inst.*, **92**, 1308–1316.
15. G. Sutedja, B.J. Venmans, E.F. Smit, et al. (2001). Fluorescence bronchoscopy for early detection of lung cancer. A clinical perspective. *Lung Cancer*, **34**, 157–168.
16. Y. Hayata, H. Funatsu, H. Kato (1982). Results of lung cancer screening programs in Japan: Early detection and localization of lung tumors in high-risk groups. In: P.R. Band (Ed.), *Recent Results of Cancer Research* (Vol. 82, pp. 179–186). Heidelberg, Springer Berlin.
17. M. Kaneko, K. Eguchi, H. Ohmatsu, et al. (1996). Peripheral lung cancer: Screening and detection with low-dose spiral CT versus radiography. *Radiology*, **201**, 798–802.
18. T. Koike, M. Terashima, T. Takizawa, et al. (1999). The influence of lung cancer mass screening on surgical results. *Lung Cancer*, **24**, 75–80.
19. B.J. Flehinger, M. Kimmel, M.R. Melamed (1992). The effect of surgical treatment on survival from early lung cancer: The impact of screening. *Chest*, **101**, 1013–1018.
20. H. Miura, C. Konaka, N. Kawate, et al. (1992). Sputum cytology-positive, bronchoscopically negative adenocarcinoma of the lung. *Chest*, **192**, 1328–1332.
21. W.C. Black (2000). Overdiagnosis: An underrecognized cause of confusion and harm in cancer screening. *J. Natl. Cancer Inst.*, **92**, 1280–1282.
22. H.G. Juffs, I.F. Tannock (2002). Screening trials are even more difficult than we thought they were. Editorial. *J. Natl. Cancer Inst.*, **94**, 156–157.
23. M.J. McFarlane, A.R. Feinstein, C.K. Wells (1986). Clinical features of lung cancer discovered as a postmortem "surprise". *Chest*, **90**, 520–523.
24. K. Nakaichi, K. Imai, S. Hayashi (1993). Polymorphisms of the CYP1A1 and Glutathione S-transferase genes associated with susceptibility to lung cancer in relation to cigarette dose in a Japanese population. *Cancer Res.*, **53**, 2994–2999.
25. C. Kiyohara, A. Otsu, T. Shirakawa, et al. (2002). Genetic polymorphisms and lung cancer susceptibility: A review. *Lung Cancer*, **37**, 241–256.
26. H.J. Schuneman, J. Dorn, B.J. Grant, et al. (2000). Pulmonary function is a long-term predictor of mortality in the general population. 29-year follow-up of the Buffalo Health Study. *Chest*, **118**, 656–664.
27. D.M. Skillrud, K.P. Offord, R.D. Miller (1986). Higher risk of lung cancer in chronic obstructive pulmonary disease. *Ann. Int. Med.*, **105**, 503–507.
28. M.S. Tockman, N.R. Anthonissen, E.C. Wright, et al. (1987). Airways obstruction and the risk for lung cancer. *Ann. Int. Med.*, **106**, 512–518.
29. T.C. Kennedy, S.P. Proudfoot, W.A. Franklin, et al. (1996). Cytopathological analysis of sputum in patients with airflow obstruction and significant smoking histories. *Cancer Res.*, **56**, 4673–4678.
30. R.J. Ginsberg, L.V. Rubinstein (1995). Randomized trial of lobectomy versus limited resection for T1N0 non-small cell lung cancer. Lung Cancer Study Group. *Am. Thorac. Surg.*, **60**, 6615–6623.
31. Y. Ichinose, T. Yano, H. Yokoyama (1984). The correlation between tumor size and lymphatic vessel invasion in resected peripheral stage I non-small cell lung cancer. *J. Thorac. Cadiovasc. Surg.*, **108**, 684–686.

32. Y. Hayata, H. Kato, K. Furuse, et al. (1996). Photodynamic therapy of 169 early stage cancers of the lung and oesophagus: a Japanese multi-centre study. *Laser Med. Sci.*, **11**, 255–259.
33. D.A. Cortese, E.S. Edell, J.H. Kinsey (1997). Photodynamic therapy for early stage squamous cell carcinoma of the lung. *Mayo Clin. Proc.*, **72**, 595–602.
34. R.J. Van Klaveren, H.J. de Koning, J. Mulshine, et al. (2002). Lung cancer screening by spiral CT. What is the optimal target population for screening trials. *Lung Cancer*, **38**, 243–252.
35. A. Vonk Noordegraaf, P.E. Postmus, G. Sutedja (2003). Bronchoscopic treatment of patients with intraluminal microinvasive radiographically occult lung cancer not eligible for surgical resection: a follow-up study. *Lung Cancer*, **39**, 49–53.
36. G. Sutedja, R.P. Golding, P.E. Postmus (1996). HRCT in patients referred for intra-luminal bronchoscopic therapy with curative intent. *Eur. Resp. J.*, **9**, 1020–1023.
37. G. Sutedja, H. Codrington, E.K. Risse, et al. (2001). Autofluorescence bronchoscopy improves staging of radiographically occult lung cancer and has an impact on therapeutic strategy. *Chest*, **120**, 1327–1332.
38. W.C. Black, D.A. Haggstrom, H.G. Welch (2002). All-cause mortality in randomized trials of cancer screening. *J. Natl. Cancer Inst.*, **94**, 167–173.
39. C.L. Henschke, D.I. McCauley, D.F. Yankelevitz, et al. (1999). Early lung cancer action project: Overall design and findings from baseline screening. *Lancet*, **354**, 99–105.
40. S.J. Swenson, J.T. Jett, J.A. Sloan, et al. (2002). Screening for lung cancer with low-dose spiral computed tomography. *Am. J. Resp. Crit. Care Med.*, **165**, 508–513.
41. L. Thiberville, P. Payne, J. Vielkinds, et al. (1995). Evidence of cumulative gene losses with progression of premalignant epithelial lesions to carcinoma of the bronchus. *Cancer Res.*, **55**, 5133–5139.
42. I. Wistuba, C. Behrens, S. Milchgrub, et al. (1999). Sequential molecular abnormalities are involved in the multistage development of squamous cell lung carcinoma. *Oncogene*, **18**, 643–650.
43. D. Slaughter, H.W. Southwick, W. Smeijkal (1953). "Field cancerization" in oral strati-fied squamous epithelium. *Cancer*, **6**, 963–968.
44. S.A. Ahrendt, J.T. Chow, L.H. Xu, et al. (1999). Molecular detection of tumor cells in bronchoalveolar lavage fluid from patients with early stage lung cancer. *J. Natl. Cancer Inst.*, **91**, 332–339.
45. P.W. Payne, T.J. Sebo, A. Doudkine, et al. (1997). Sputum screening by quantitative microscopy: A re-examination of a portion of the National Cancer Institute Cooperative Early Lung Cancer Study. *Mayo Clin. Proc.*, **72**, 697–704.
46. M.S. Tockman, J.L. Mulshine, S. Piantadosi, et al. (1997). Prospective detection of preclinical lung cancer: results from two studies of heterogeneous nuclear ribonucleo-protein A2/B1 overexpression. *Clin. Cancer Res.*, **3**, 2237–2246.
47. A.G. Nicholson, L.J. Perry, P.M. Cury, et al. (2001). Reproducibility of the WHO/IASLC grading system for pre-invasive squamous lesions of the bronchus: A study of inter-observer and intra-observer variation. *Histopathology*, **38**, 202–208.
48. K.M. Kerr (2001). Pulmonary preinvasive neoplasia. *J. Clin. Pathol.*, **54**, 257–271.
49. O. Auerbach, A.P. Stout, C. Hammond, L. Garfinkel. Changes in bronchial epithelium in relation to cigarette smoking and in relation to lung cancer. *N. Engl. J. Med.*, **265**, 253–268.
50. O. Auerbach, C. Hammond, L. Garfinkel (1979). Changes in bronchial epithelium in relation to cigarette smoking 1955–1960 VS. 1970–1977. *N. Engl. J. Med.*, **300**, 381–386.
51. G. Saccomano, V.E. Archer, O. Auerbach, R.P. Saunderss (1974). Development of carci-noma of the lung as reflected in exfoliated cells. *Cancer*, **33**, 256–270.

52. M. Sato, Y. Saito, K. Usuda, et al. (1998). Occult lung cancer beyond bronchoscopic visibility in sputum cytology positive patients. *Lung Cancer*, **20**, 17–24.
53. K. Haussinger, J. Pichler, F. Stanzel, et al. (2000). Autofluorescence bronchoscopy: the D-light system. In: C.T. Bolliger, P.N. Mathur (Eds). *Interventional bronchoscopy. Prog. Respir. Res.* (Vol. 30, pp. 243–252). Basel, Karger.
54. G. Sutedja, S. Lam, J.C. LeRiche, et al. (1994). Response and pattern of failure after photodynamic therapy for intraluminal stage I lung cancer. *J. Bronchology*, **1**, 295–298.
55. T. Kawaguchi, K. Furuse, M. Kawahara, et al. (1998). Histological examination of bronchial mucosa after photodynamic therapy showing no selectivity of effect between tumor and normal mucosa. *Lasers Med. Sci.*, **13**, 265–270.
56. J. Hung, S. Lam, J.C. LeRiche, et al. (1991). Autofluorescence of normal and malignant bronchial tissue. *Lasers Surg. Med.*, **11**, 99–105.
57. S. Lam, C. MacAulay, J. Hung, et al. (1993). Detection of dysplasia and CIS with a lung imaging fluorescence endoscopy device. *J. Thorac. Cardiovasc. Surg.*, **105**, 1035–1040.
58. J. Qu, C. MacAulay, S. Lam, et al. (1995). Laser-induced fluorescence spectroscopy at endoscopy: tissue optics, Monte Carlo modelling and in vivo measurements. *Optical Engineering*, **34**, 3334–3343.
59. J.D. Pitts, R.D. Sloboda, K.H. Dragnev, et al. (2001). Autofluorescence characteristics of immortalized and carcinogen-transformed human bronchial epithelial cells. *J. Biomed. Optics*, **6**, 31–40.
60. M. Kakihana, K.I. Kim, T. Okunaka, et al. (1999). Early detection of bronchial lesions using system of autofluorescence endoscopy (SAFE) 1000. *Diagnostic and Therapeutic Endoscopy*, **5**, 99–104.
61. A. McWilliams, C. MacAulay, A.F. Gazdar, et al. (2002). Innovative molecular and imaging approaches for the detection of lung cancer and its precursor lesions. *Oncogene*, **21**, 6949–6959.
62. S. Lam, T. Kennedy, M. Unger, et al. (1998). Localization of bronchial intraepithelial neoplastic lesions by fluorescence bronchoscopy. *Chest*, **113**, 696–702.
63. L. Thiberville, G. Sutedja, P. Vermijlen (1999). A multicenter European study using the light induced fluorescence endoscopy system to detect pre-cancerous lesions in high-risk individuals. *Eur. Resp. J.*, **14**, 2475.
64. B.J. Venmans, J.C. van der Linden, G. Sutedja, et al. (2000). Observer variability in histopathological reporting of bronchial biopsy specimens: influence on the results of AFB in detection of bronchial neoplasia. *J. of Bronchol.*, **7**, 210–214.
65. B.J. Venmans, T.J. van Boxem, G. Sutedja, et al. (1999). Results of two years experience with fluorescence bronchoscopy in detection of preinvasive bronchial neoplasia. *Diagnostic and Therapeutic Endoscopy*, **5**, 77–84.
66. F.R. Hirsch, A.F. Gazdar, E. Gabrielson, et al. (2000). Histopathologic evaluation of premalignant and early malignant bronchial lesions: an interactive program based on internet digital images to improve WHO criteria for early diagnosis of lung cancer and for monitoring chemoprevention studies – A SPORE collaborative project. *Lung Cancer*, **29**, suppl 2, 209.
67. S. Lam, J.C. LeRiche, Y. Zheng, et al. (1999). Sex-related differences in bronchial epithelial changes associated with tobacco smoking. *J. Natl. Cancer Inst.*, **91**, 691–696.
68. F.R. Hirsch, S.A. Prindiville, Y.E. Miller, et al. (2001). Fluorescence versus white-light bronchoscopy for detection of preneoplastic lesions: a randomized study. *J. Natl. Cancer Inst.*, **93**, 1385–1391.
69. A. Gazdar, I. Park, S. Sood, et al. (2000). Clonal patches of molecular changes in smoking damaged respiratory epithelium. *Lung Cancer*, **29**, S7.

70. T. Vo-Dinh, P.N. Mathur (2000). Optical diagnostic and therapeutic technologies in pulmonary medicine. In: C.T. Bolliger, P.N. Mathur (Eds), *Interventional bronchoscopy. Prog. Respir. Res.* (Vol. 30, pp. 267–279). Basel, Karger.
71. G.A. Wagnieres, W.M. Star, B.C. Wilson (1998). Invited review. In vivo fluorescence spectroscopy and imaging for oncological applications. *Photochem. Photobiol.*, **68**, 603–632.
72. J.G. Fujimoto (2000). Optical coherence tomography for the diagnosis of neoplasia. *Lung Cancer*, **29**, S2, 89.
73. C.E. MacAulay, M. Guillaud, J. LeRiche, et al. (2000). 2D and 3D quantitative microscopy for preinvasive lung cancer. *Lung Cancer*, **29**, S1, 252.
74. R.L. Keith, Y.E. Miller, R.M. Genmill, et al. (2000). Angiogenic squamous dysplasia in bronchi of individuals at high risk for lung cancer. *Clin. Cancer Res.*, **6**, 1616–1625.
75. K. Shibuya, H. Hoshino, M. Chiyo, et al. (2002). Subepithelial vascular patterns in bronchial dysplasias using a high magnification bronchovideoscope. *Thorax*, **57**, 902–907.
76. S. Bota, J.-B. Auliac, C. Paris, et al. (2001). Follow-up of bronchial pre-cancerous lesions and carcinoma in situ using fluorescence endoscopy. *Am. J. Respir. Crit. Care Med.*, **164**, 1688–1693.
77. R.H. Breuer, E.H. Elzinga, G. Sutedja, et al. (2002). The natural course of preneoplastic lesions in bronchial epithelium – A longitudinal study. *Am. J. Resp. Crit. Care Med.*, **165**, 8, Abstract 403.
78. S. Lam, C. MacAulay, J. LeRiche, et al. (2002). A randomized phase Ib trial of anethole dithiolethione in smokers with bronchial dysplasia. *J. Natl. Cancer Inst.*, **94**, 1001–1009.
79. B.J. Venmans, T.J. van Boxem, G. Sutedja, et al. (2002). Outcome of bronchial CIS. *Chest*, **117**, 1472–1576.
80. P. Pierard, P. Vermijlen, T. Bosschaerts, et al. (2000). Synchronous roentgenographically occult lung carcinoma in patients with resectable primary lung cancer. *Chest*, **117**, 779–785.
81. G. Sutedja, A.J. van Boxem, P.E. Postmus (2001). Comprehensive review. The curative potential of intraluminal bronchoscopic treatment for early-stage non-small-cell lung cancer. *Clinical Lung Cancer*, **2**, 264–270.
82. G. Sozzi, M. Oggionno, L. Alasio, et al. (2002). Molecular changes track recurrence and progression of bronchial pre-cancerous lesions. *Lung Cancer*, **37**, 267–270.
83. A. Vonk Noordergraaf, A. Pasic, R. Brever, et al. (2003). Multiple suspicious lesions detected by autofluorescence bronchoscopy predict malignant development in the bronchial mucosa in high risk patients. *Lung Cancer*, in press.
84. N. Nagamoto, Y. Saito, S. Ohta, et al. (1989). Relationship of lymph node metastasis to primary tumor size and microscopic appearance of roentgenographically occult lung cancer. *Am. J. Surg. Pathol.*, **13**, 1009–1013.
85. K. Usuda, Y. Saito, N. Nagamoto, et al. (1993). Relation between bronchoscopic findings and tumor size of roentgenographically occult bronchogenic squamous cell carcinoma. *J. Thorac. Cardiovasc. Surg.*, **106**, 1098–1103.
86. E. Akaogi, I. Ogawa, K. Mitsui, et al. (1994). Endoscopic criteria of early squamous cell carcinoma of the bronchus. *Cancer*, **74**, 3113–3117.
87. H. Nakamura, N. Kawasaki, Hagiwara et al. (2001). Endoscopic evaluation of centrally located early squamous cell carcinoma of the lung. *Cancer*, **91**, 1412–1147.
88. Y. Miyazu, T. Miyazawa, N. Kurimoto, et al. (2002). Endobronchial ultrasonography in the assessment of centrally located early-stage lung cancer before photodynamic therapy. *Am. J. Respir. Crit. Care Med.*, **165**, 832–837.

89. M. Baba, Y. Sekine, M. Suzuki, et al. (2002). Correlation between endobronchial ultra-sonography (EBUS) images and histologic findings in normal and tumor-invaded bronchial wall. *Lung Cancer*, **35**, 65–71.

90. G.J. Herder, R.H. Breuer, E.F. Comans, et al. (2001). Positron emission tomography scans can detect radiographically occult lung cancer in the central airways. *J. Clin. Oncol.*, **19**, 4271–4272.

91. T.J. van Boxem, J. Westerga, B.J. Venmans, et al. (2001). Photodynamic therapy, Nd-YAG laser and electrocautery for treating early-stage intraluminal cancer. Which to choose? *Lung Cancer*, **31**, 31–36.

92. H. Kato, C. Konaka, J. Ono, et al. (1985). Preoperative laser photodynamic therapy in combination with operation in lung cancer. *J. Thorac. Cardiovasc. Surg.*, **90**, 420–429.

93. S. Shankar, P.J. George, M.R. Hetzel (1990). Elective resection of tumors of the trachea and main carina after endoscopic laser therapy. *Thorax*, **45**, 493–495.

94. G. Sutedja, G. Baris, F. Zoetmulder, et al. (1992). High dose rate brachytherapy improves resectability in squamous cell lung cancer. *Chest*, **102**, 308–309.

95. H. Kato, T. Okunaka, T. Tsuchida, et al. (1999). Analysis of the cost-effectiveness of photodynamic therapy in early stage lung cancer. *Diagnostic and Therapeutic Endoscopy*, **6**, 9–16.

Chapter 9

Fluorescence diagnosis and photodynamic therapy in dermatology: an overview

C. Fritsch, K. Lang, K.W. Schulte, W. Neuse, T. Ruzicka and P. Lehmann

Table of contents

Summary

The topical cutaneous application of δ-aminolevulinic acid (ALA) induces porphyrin formation in skin, preferentially in tumor tissues. Irradiation of the porphyrin-enriched tumor tissue with Wood's light leads to emission of a typical brick-red fluorescence. This principle may be used as a diagnostic procedure which is termed fluorescence diagnosis with ALA-induced porphyrins (FDAP) or photo-dynamic diagnosis (PDD). In FDAP, tumors and precancerous lesions of the skin reveal a homogenous intensive fluorescence. Psoriatic lesions also show a strong but non-homogenous porphyrin fluorescence. FDAP in preoperative planning is a valuable method to determine the peripheral borders of a given tumor. The histopathological extensions of the tumors correlate well with the borders detected by the specific fluorescence. The main indications of FDAP are the delineation of clinically ill-defined skin tumors and the control of the efficacy of other tumor therapies.

Photodynamic therapy (PDT) with various photosensitizers has been proven to be highly efficient in the treatment of skin tumors. However, most valid data are available for ALA only, which has been shown to effectively sensitize solar keratoses and superficial basal cell carcinomas with porphyrins. Initial squamous cell carcinomas also show good response to ALA-PDT. Treatment of psoriasis by PDT is a subject under investigation. During the last decade numerous studies on PDT for dermatological diseases have been published, the more important ones are reviewed here.

List of abbreviations

ALA	δ-Aminolevulinic acid
BCC	Basal cell carcinoma
BD	Bowen's disease
BpD	Benzoporphyrin derivative
CR	Complete response
DHE	Dihematoporphyrin ether/ester
EPP	Erythropoietic protoporphyria
FDAP	Fluorescence diagnosis with ALA-induced porphyrins
Hp	Hematoporphyrin
HpD	Hematoporphyrin derivative
Met	Metastasis
MM	Malignant melanoma
Npe6	Monoaspartyl chlorine e 6
PD	Paget's disease
PDD	Photodynamic diagnosis
PDL	Pumped dye laser
PDT	Photodynamic therapy
PR	Partial response
SCC	Squamous cell carcinoma
SK	Solar keratoses
TPPS	Tetrasodium-meso-tetraphenylporphyrinsulfonate

9.1 Introduction

In this chapter we will discuss the significance of a novel pre-surgical method, the fluorescence diagnosis with ALA-induced porphyrins (FDAP) of tumors and pre-cancerous lesions and the present state of photodynamic therapy (PDT) techniques in dermatology.

Policard used the characteristic brick-red fluorescence of porphyrins, e.g. hemato-porphyrin, for tumor detection for the first time in 1924 [1]. The predominant porphyrin fluorescence in tumor tissue was confirmed by several investigators in humans and animals [2,3]. Hematoporphyrin derivative (HpD) consisting of porphyrin derivatives was shown to be even more preferentially stored in carcinomas [3]. However, the systemic administration of photosensitizers induces generalized phototoxicity. Since the introduction of ALA-PDT in 1990, interest in PDT has also risen among dermatologists [4–10]. In particular, solar keratoses have been shown to be highly sensitive to topical ALA-PDT [8,11]. The PDT, and in particular the fluorescence diagnosis, initiate interesting perspectives, which have been discussed increasingly at photobiological conferences and in dermatological journals.

Apart from the therapeutic application of porphyrins, tumor tissues can be made visible on the basis of porphyrins' typical brick-red fluorescence under Wood's light. This principle has been termed fluorescence diagnosis with ALA-induced porphyrins (FDAP) or photodynamic diagnosis (PDD).

At present, contrary to other dermatological diagnostic techniques, such as the photopatch test or photoprovocation testing, there are neither standardized test protocols nor established sources of light or test substances available for FDAP, so that there is still a high demand for more research.

We examined ALA-induced porphyrin fluorescence in various dermatological disorders, particularly skin tumors. In the case of epithelial tumors, correlation was made between the clinically detectable fluorescence extension and tumor margins as examined histopathologically [12].

9.2 Technique of fluorescence diagnosis with ALA-induced porphyrins (FDAP)

9.2.1 Pretreatment and application of ALA

It is recommended to remove any scabs (e.g. with 3–5% acidum salicylicum in vaselinum album and a sharp spoon) and to clean fatty surfaces (e.g. with an alcoholic solution). In the case of ALA treatment in facial or intertriginous areas a pretreatment with an antibiotic-containing cream (e.g. erythromycine 1%) for several days may be useful to improve the result of FDAP. Exemplarily, the following useful pretreatments are recommended:

(A) Solar keratoses on the capillitium with thick hyperkeratoses or
(B) Basal cell carcinoma on the trunk with (hemorrhagic) scabs:
    1. Abrade the skin with an alcoholic (e.g. Dibromol®)-solution

2. Remove visible and gropable scabs with e.g. a sharp spoon
3. If there is bleeding, stop it by means of gauze-compression
(C) Basal cell carcinoma, solid type, e.g. in the face:
    1. Abrade the skin with an alcoholic (e.g. Dibromol®)-solution
    2. Roughen or clear away the exophytic part of the tumor (debulking)
    3. If there is bleeding, stop it by means of gauze-compression

In general, ALA-ointments are used [8], although cream mixtures are also effective. In tumor-bearing mice, ALA lotion applied on the skin overlying the tumor induced higher accumulation of tumoral porphyrins than cream [3]. We use ALA mixed in an ointment vehicle (e.g. Neribas®, Schering; 10–20%; 20–40 mg ALA cm^{-2}). The mixture is applied to the cutaneous lesions under occlusive foil (e.g. Tegaderm®, 3M) and tape to enhance tissue penetration and to avoid photobleaching. In the case of a wide application, occlusive handling does not necessarily have to be performed. For light protection, waiting in a darkened room is sufficient or, alternatively, aluminium foil and gauze may be applied. After an application period of 3–4 h tape and ointment are removed. In a completely darkened room illumination of the tissue with Wood's light is performed (Figures 1–5) [7,12,14].

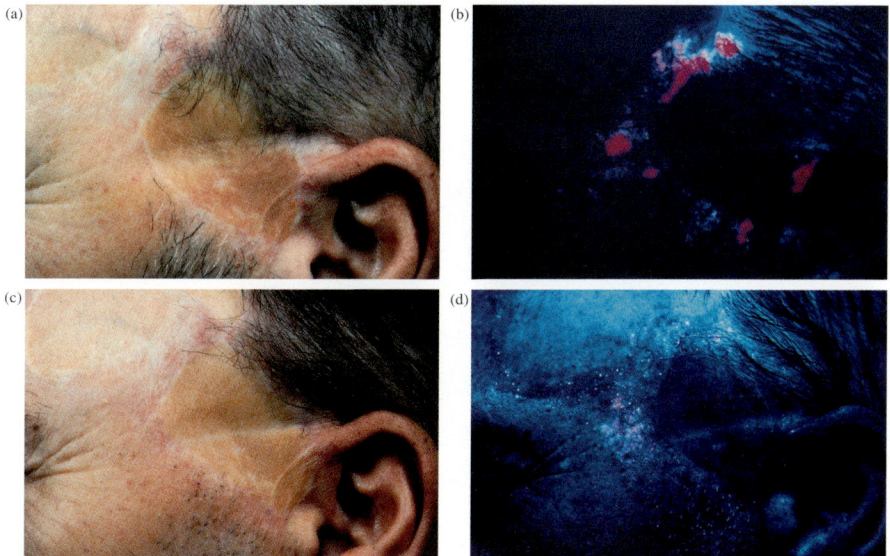

Figure 1. (a) Patient with history of more than 200 BCCs which were surgically treated over a period of 10 years. Tissue defects were partly covered by full thickness skin grafts. The picture shows a preauricularly localized skin graft. The patient was presented by the department of plastic surgery to detect any novel or regrowing tumor tissue. Clinically, there was no hint of any tumor tissue, except for eczematous areas on the upper part of the skin graft and on the front of the upper part of helix. (b) FDAP (ALA 20%, 6 h, Wood's light, 370–405 nm), however, showed deep red porphyrin fluorescence in 6 skin areas. Fluorescence allowed a sharp demarcarion of the tumor tissues. Lesions were marked according to the fluorescence and excised. All lesions were histopathologically proven to be BCCs. (c) Nine months after FDAP-guided surgery. Neither clinically nor (d) by FDAP is there any tumor tissue.

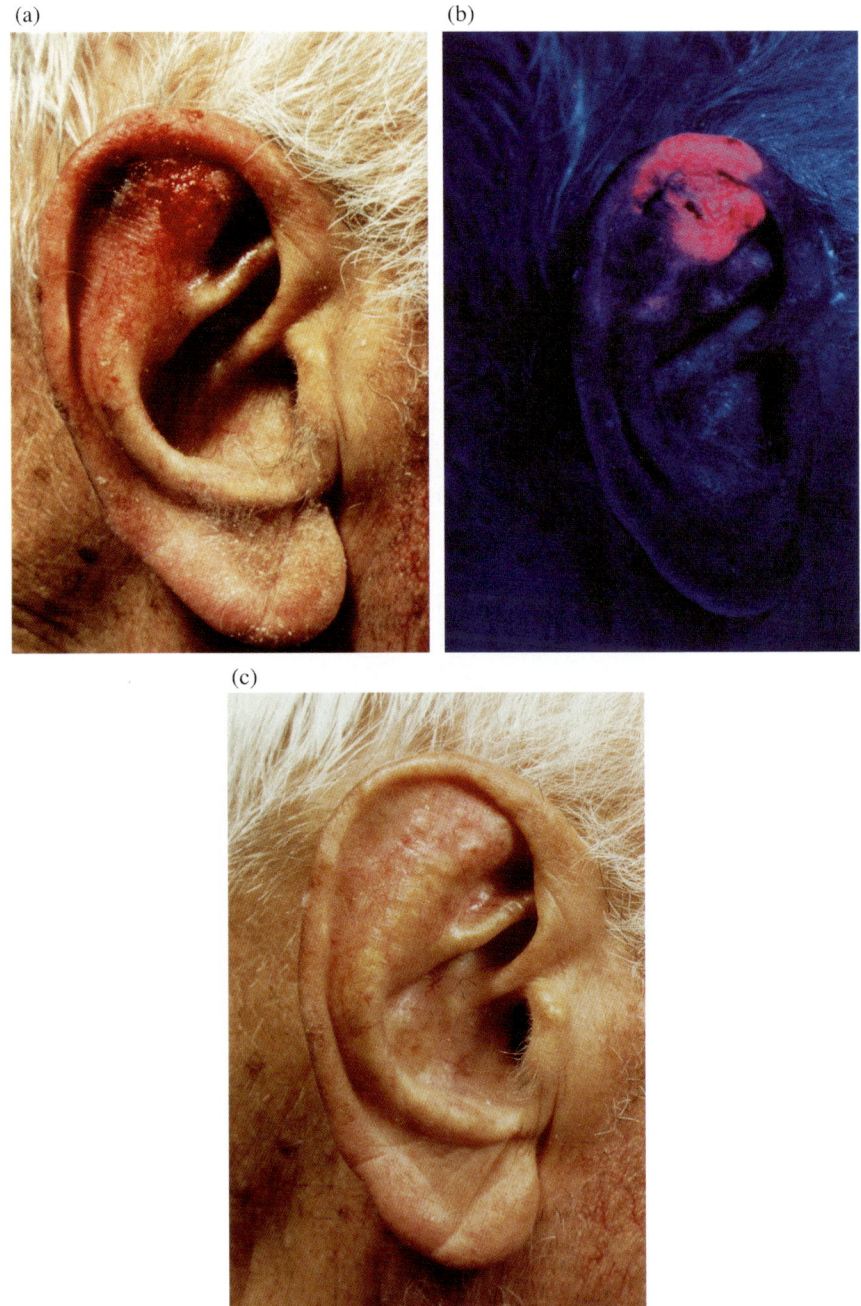

Figure 2. (a) Squamous cell carcinoma on the ear helix. The lesion is very ill-defined. (b) FDAP clearly demarcates the extent of the lesion. (c) The ear after 3 (monthly repeated) PDT sessions (20% ALA, 180 J cm^{-2} red light; PDT 700, Waldmann, Germany) with an excellent cosmetic result. Complete response still maintained after 4 years of follow-up.

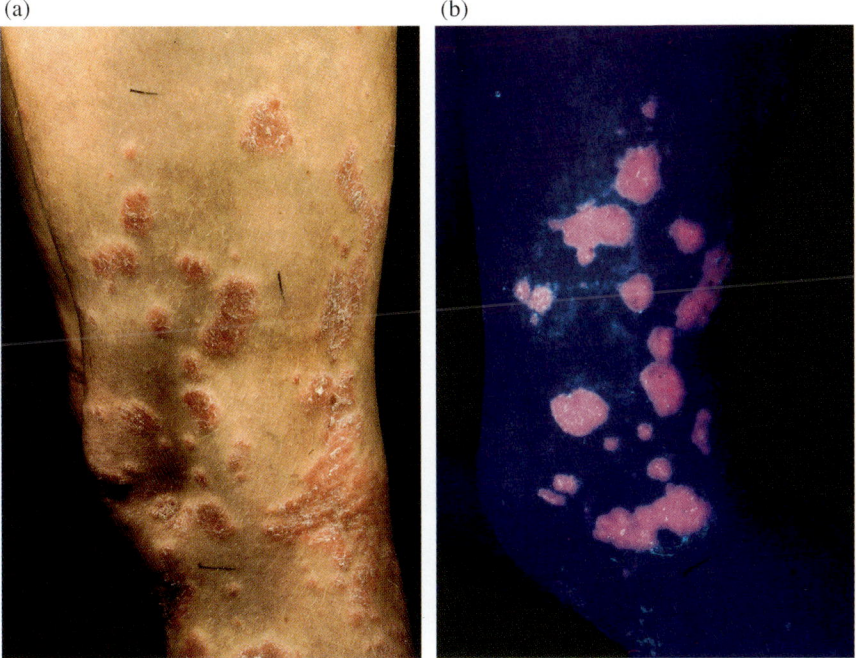

Figure 3. (a) Psoriatic lesions on the right leg. (b) All treated lesions reveal bright red porphyrin fluorescence upon topical ALA treatment. The fluorescence is sharply limited to the lesional skin.

For FDAP, suitable light sources must emit an intensity that includes 405 nm (Figure 1) – the absorption maximum of the porphyrin molecules (so-called Soret band). In general, coherent (lasers) and incoherent light sources have been tested for tumor detection. In dermatology, above all, incoherent lamp systems, so-called Wood's light systems (long-wave UVA light) in the form of handsets are used (named after the American physicist Wood 1868–1955). In the past only one radiation source, which was principally developed for textile testing, met the necessary criteria in order to make the formed porphyrins effectively visible (Fluotest® forte; Atlas Material Testing Technology; Gelnhausen, Germany; 180 W; maximum with 366 nm). Recently, the German light systems company Saalmann (Herford) has developed a more powerful and very selective irradiation source especially for the use in FDAP [Fluolight®, 10–20 mW cm^{-2}, 370–400 nm, cutaneous and ocular threshold time (ACGIH): 3 h (10 cm distance)].

9.3 Results of FDAP in different tissues

FDAP was performed in precancerous areas, neoplasms, and inflammatory diseases of the skin. The most important data are summarized in the Tables 1 and 2. These results show that malignant and inflammatory skin diseases can show a greater

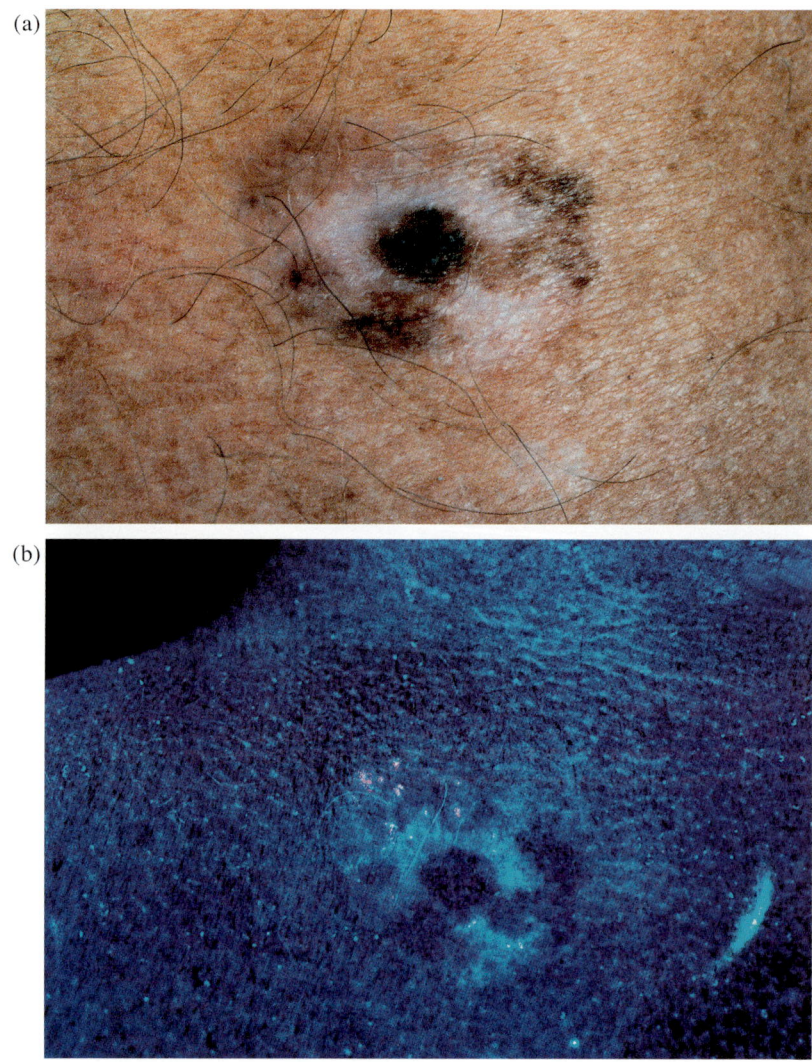

Figure 4. (a) Malignant melanoma (Level III, 1.5 mm) on the back. (b) FDAP: Except for some follicular-bound fluorescences (normal) there is no tumor-specific fluorescence.

porphyrin formation after exogenous ALA application as compared to the (lesion-adjacent) "normal" skin.

Epidermal neoplasms such as basal cell carcinoma (Figures 1, 5, 7), squamous cell carcinoma (Figure 2), Bowen's disease, solar keratoses (Figure 6) and extra-mammary Paget's disease show an intensive uniform red fluorescence. All other melanocytic or amelanotic, benign or malignant tumors, such as malignant melanoma (Figure 4), lentigo senilis, verruca seborrhoica, nevus cell nevus, demon-strate no or only minimal fluorescence. In all verrucae vulgares no fluorescence can be detected. Psoriatic lesions also reveal bright fluorescence (Figure 3) which is,

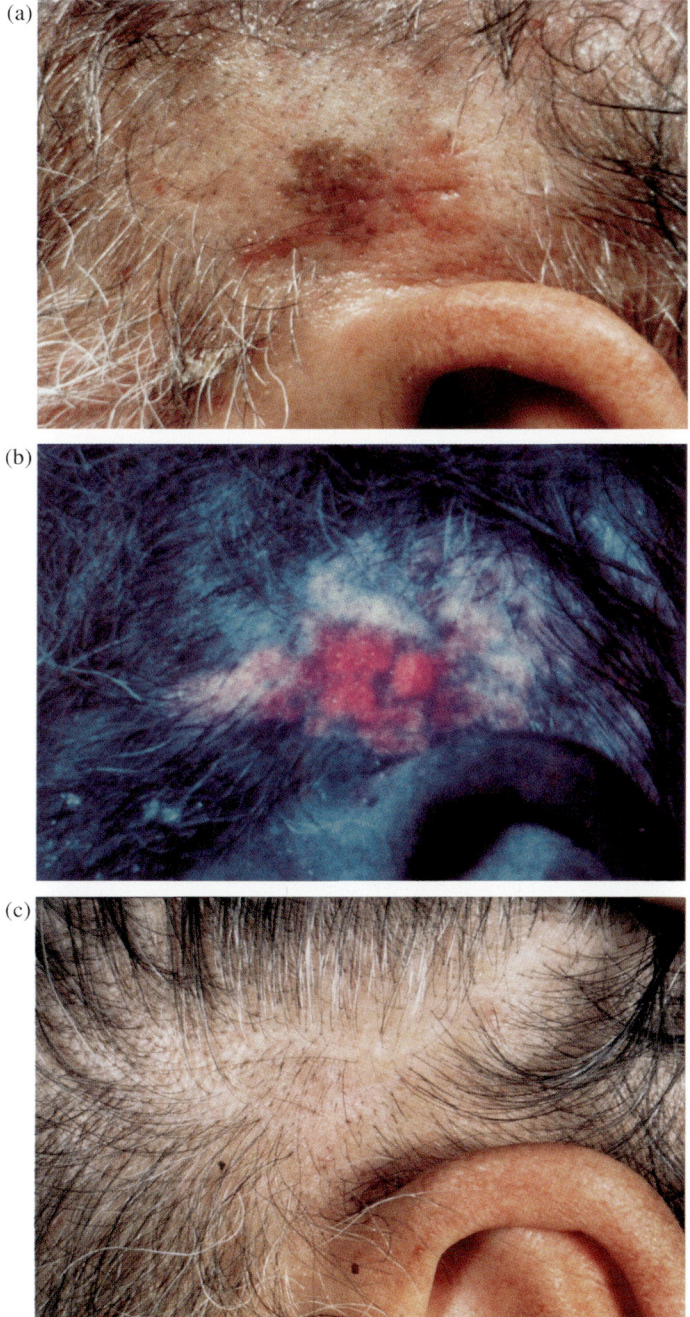

Figure 5. (a) BCC, in partum nodular, in the left supraauricular area. (b) FDAP (20% ALA, 6 h): The lesion shows a relative sharply bordered fluorescence pattern. (c) Result after 3 monthly PDT sessions (180 J cm^{-2} red light, 570–750 nm) six months after the last PDT session.

Table 1. Fluorescence pattern in skin diseases and normal skin

Type of Tissue	Intensity	Fluorescence Homogeneity		Demarcation	
		Yes	No	Sharp	ϕ Sharp
BCC: superficial	1 2 3 4	x		x	
BCC: sclerodermiformic	1 2 3 4		x		x
BCC: nodular	1 2 3 4	x		x	
SCC: superficial	1 2 3 4	x		x	
SCC: nodular	1 2 3 4	x		x	
SCC: ulcerated	1 2 3 4		x		x
Keratoacanthoma	1 2 3 4	x		x	
Paget's disease	1 2 3 4	x		x	
Bowen's disease	1 2 3 4	x		x	
Solar keratosis A	1 2 3 4	x		x	
Solar keratosis B	**1** 2 3 4		x	x	
Pseudocarcinomatous Epithelhyperplasia	1 2 3 4		x	x	
Mycosis fungoides	1 **2 3** 4	x		x	
Kaposi's sarcoma	**1 2 3** 4	x			x
MM – metastasis	**1 2 3** 4				x
Lupus erythematosus	1 **2 3** 4		x	x	
Psoriasis: guttata	1 2 3 **4**	x		x	
Psoriasis: plaques A	1 2 3 **4**	x		x	
Psoriasis: plaques B	1 **2 3** 4		x	x	
Atopic eczema	1 **2 3** 4		x		x
Acne vulgaris	1 2 3 4		x		x
Dyskeratosis follicularis	1 2 3 4		x	x	
Normal skin: face	1 2 3 4		x	none	
Normal skin: trunk	**1** 2 3 4		x	none	
Normal skin: intertrigin	**1** 2 3 4		x	none	

Fluorescene intensity is given semiquantitatively according to a fluorescence standard. A = without hyperkeratoses; B = with intensive hyperkeratotic crusting.

however, often inhomogeneous and sometimes absent. In plaques of mycosis fungoides an intermediate fluorescence intensity is demonstrable [12].

9.4 Kinetics of fluorescences in epithelial tumors and psoriatic plaques

The best fluorescence contrast between tumor and normal tissue is obtained if the tumor accumulates porphyrins as much as possible and the normal tissue accumulates very few porphyrins. Fluorescence studies show that superficial lesions of BCC exhibit a maximum porphyrin fluorescence 6 h post ALA application, whereas in nodular lesions the observed intensity is highest after 2–4 h [15]. Detailed biochemical analyses prove that the respective fluorescence intensity correlates well with the extracted quantity of porphyrins [16–19]. In epithelial skin tumors, the porphyrin maxima are achieved 1–4 h after ALA application. In normal skin, porphyrin

Table 2. Skin diseases without any fluorescence in FDAP

Lentigo maligna
Nevus
Verruca seborrhoica
Verruca vulgaris
Lentigo maligna melanoma
Malignant melanoma

---

enrichment is slower and weaker. On the other hand, slightly increased porphyrin values are measured in normal skin up to 24 h after ALA application, indicating photosensitivity for this period of time. The best fluorescence yield in the tumor is thus expected after 2–4 h. Murine skin revealed maximal porphyrin accumulation in the tumor 3 h after ALA application in both cream and lotion preparations [3]. PDT efficacy in a subcutaneous mammary carcinoma was reported to be optimal if irradiation was performed between 2 and 3 h after topical ALA application [20].

On the face, on the capillitium, as well as inguinal, the fluorescence demarcation of tumors against the healthy skin may be more difficult since a medium strength fluorescence can be present in the healthy skin also (Figures 1, 5, 7). Nevertheless, in this case neoplastic tissue can be distinguished from healthy skin, since fluorescence is substantially more intense in the tumor.

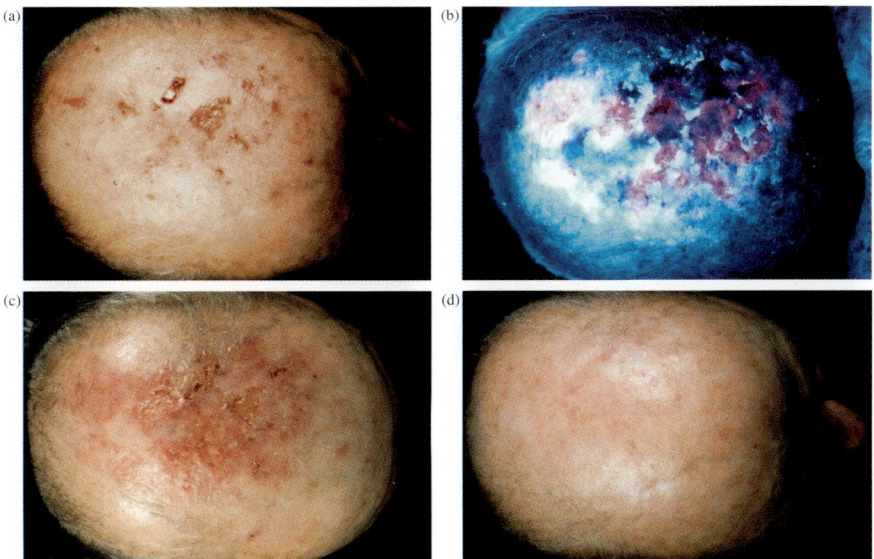

Figure 6. (a) Multiple solar keratoses on the capillitium. Some lesions are already transforming into SCC. (b) FDAP (10% ALA) clearly demarcates all neoplastic tissue. (c) Capillitium 2 days after PDT (green light 30 J cm^{-2} Saalmann, Germany). Treated area is presenting signs of inflammation. (d) 10 days after PDT. The skin is healed with an excellent cosmetic result and no symptoms of remaining neoplastic tissue.

9.5 Value of autofluorescence and effect of bacteria in FDAP

Uninvolved skin shows different fluorescence intensities depending on the anatomical area and ALA application time. The higher fluorescence intensity in the face, axilla or groin as compared to the trunk or extremities is probably due to the presence of an increased bacterial flora (propionibacteria) also producing relatively high porphyrin levels [21]. In these body areas, fluorescence in normal skin may interfere with that in neoplastic skin, but differentiation is still possible due to the higher fluorescence intensity in tumor tissue. The fluorescence intensity in normal skin is increased by a longer ALA exposure time [22], and preliminary data show that shorter application times of ALA, e.g. 3 h, or pretreatment with an erythromycin-containing cream facilitate and improve the differentiation between tumor and normal tissue fluorescence in bacterially-enriched skin areas.

Without ALA application, mucosal tissues such as the mouth [23,24], gastrointestinal tract [25] or bronchial system [26] may show high "falsely positive" fluorescences,

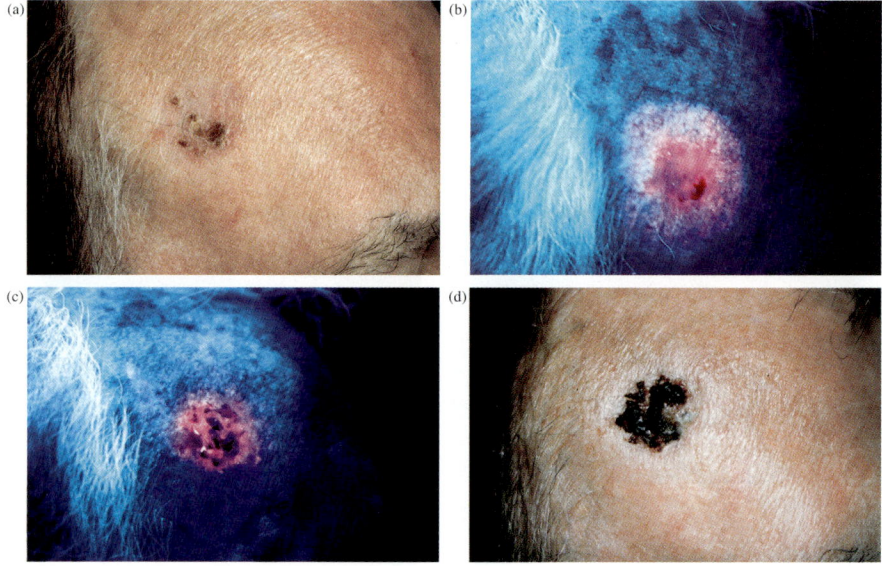

Figure 7. (a) Patient with a multicentric BCC on the front right. The clinical picture does not allow a detailed delineation of the lesion. (b) FDAP (ALA 20%, 6 h). The lesion shows a bright red porphyrin fluorescence. The tumor can be much better demarcated than by the clinical picture. However, the surrounding normal tissue also shows quite intensive porphyrin fluoresence. This phenomenon is very typical for the facial and intertriginous skin and might be partly explained by the presence of follicle-bound bacteria which also produce relative high amounts of porphyrins upon ALA-treatment. (c) One day later, FDAP was repeated using an ALA-methylester (Metvix®, Photocure, Oslo, Norway) for 3 h. FDAP shows a fluorescence pattern limited to the tumor tissue, indicating that the ALA-ester induces a much more selective intralesional porphyrin accumulation as compared to the free ALA. (d) Three days after PDT (Metvix® 20%; 150 J cm⁻², red light; Photocure, Oslo, Norway). The treated lesion is covered by hemorrhagic crusting.

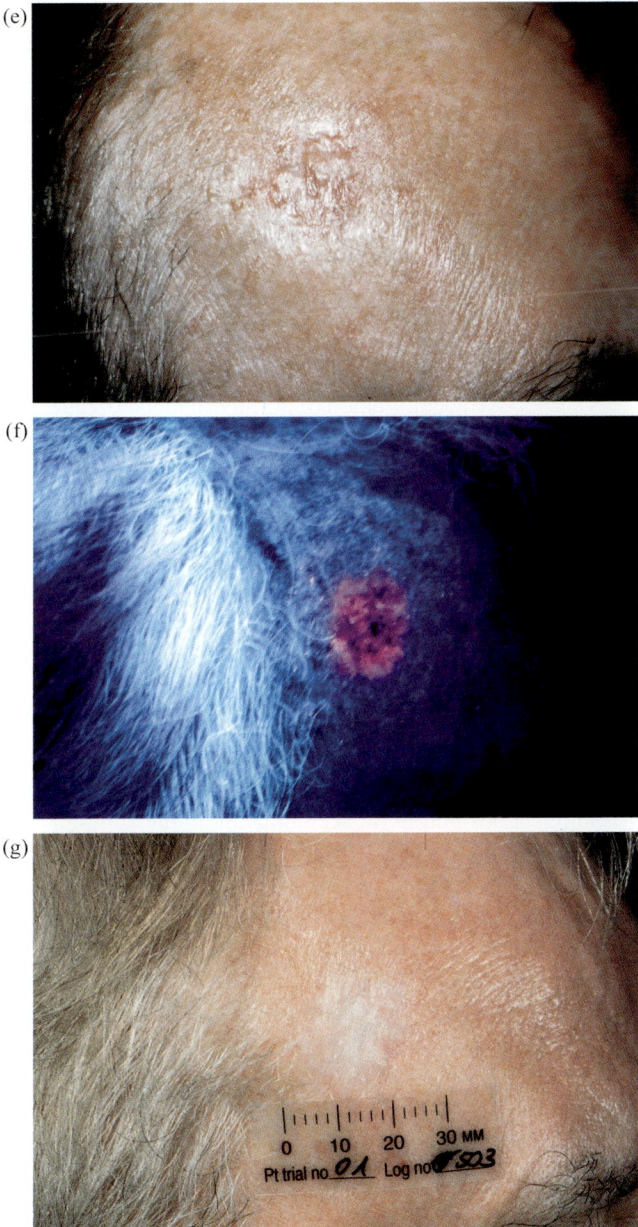

Figure 7. (*cont.*) (e) Three weeks after PDT. There is already a dramatic improvement of the BCC. However, multiple skin colored papules indicates remaining tumor tissue. (f) FDAP (Metvix® 20%) at this time point still shows an intensive sharply bordered fluorescence proving the presence of tumor tissue. PDT was repeated as described above. (g) Complete response three months after the third PDT session, showing an excellent cosmetic result with discrete hypopigmentation (cicatrix).

caused by the presence of bacteria in the mucous membranes [27]. In mucosal areas, tumors partly show high ground levels of porphyrins, a phenomenon that workers tried to use for tumor detection. In dermatology, autofluorescence proved to be of little help since skin tumors do not show tissue autofluorescence. Only skin diseases that are accompanied by an intensified bacterial settlement (e.g. erythrasma, acne) can specifically fluoresce [21]. In, e.g., erythrasma, *Corynebacterium minutissimum* forms high quantities of porphyrins.

9.6 Intralesional porphyrin enrichment

Our experiences with FDAP underline an increased ALA-induced porphyrin biosynthesis in neoplastic, hyperplastic and also inflamed tissues. It is postulated that ALA treatment induces a selective accumulation of porphyrin metabolites in tumors [19,28]. The mechanism of preferential intratumoral uptake of photosensitizers (or their precursors) is still not fully understood. In the case of ALA, active transport is the most likely explanation, but passive diffusion may be operative as well. Enzymatic differences between normal and neoplastic tissue such as a lower activity of ferrochelatase, which in erythropoietic protoporphyria (EPP) leads to an accumulation of protoporphyrin [29], seem to be less effective in ALA-induced porphyrin sensitization. The level of synthesized porphyrins depends mainly on the amount of ALA that penetrates through the skin into neoplastic cells. The efficiency of FDAP or PDT depends on an homogeneous distribution of ALA(-induced porphyrins) into the skin. In the case of BCC, ALA-induced porphyrin accumulation is limited up to 0.75–0.81 mm [19,30] and can be intensified up to 1.25 mm by addition of dimethyl sulfoxide (DMSO) or ethylendiaminetetraacetic acid (EDTA). Orally administered ALA has been postulated to induce a more homogeneous porphyrin accumulation in the depth of BCC [19]. In 13 patients with BCC, who ingested ALA (10, 20 or 40 mg kg^{-1} single doses), protoporphyrin fluorescence peaked in tumors before normal adjacent skin after from 1 to 3 h. Gross fluorescence revealed a greater protoporphyrin fluorescence in tumor than in normal skin, and fluorescence microscopy showed distinct, full-thickness fluorescence only after ingestion of 40 mg kg^{-1} ALA [31].

Thus, we assume a reduced ALA penetration or uptake in lesions such as verruca vulgaris, thick solar keratosis, or psoriatic lesions due to the hyperkeratotic areas which may limit penetration of ALA and, in addition, of the exciting light. The weak fluorescence in Kaposi's sarcoma or the lack of fluorescence in melanoma or naevi is probably mainly due to the high pigmentation of these tissues and the consequent lower absorption as well as emission of light.

9.7 FDAP – state of the art

9.7.1 FDAP in the detection of various tumors

FDAP is already of major use for the detection of cervix neoplasms, bladder tumors [32], malignant gliomas [33] and precanceroses as well as carcinomas of the tongue, oral mucosa, larynx, esophagus [34] and colon.

9.7.2 FDAP in the detection of cutaneous tumors

In the dermatological department of the Heinrich-Heine-University, Duesseldorf FDAP has been routinely used, at the time of writing, for the detection and demarcation of epithelial tumors for six years. In light of this experience the preoperative use of FDAP is particularly recommended, in order to simplify the surgical procedure. The histopathological processing proved that tumors excised on the basis of the fluorescence expansion could be generally excised *in toto* [12,14]. Investigations of other groups point to the same conclusion [35,36]. In current histopathological studies we are comparing the clinical picture as well as the fluorescence pattern in FDAP with the exact extent of epithelial tumors.

9.7.3 FDAP for preoperative planning and control of tumor therapy

In general, all deeply fluorescing areas probably represent neoplastic tissues, as proven by histopathology (Figures 1, 7). The clinical fluorescence corresponds well with the histological borders of the tumor [7,14]. FDAP is very sensitive, so that the smallest neoplastic areas can be visualized, areas which would not have been detected by the clinical aspect alone. In anatomically difficult sites such as the nose or the ear, and especially in pretreated skin, the detection of (regrowing) tumor areas can be facilitated by using FDAP (Figures 1, 2, 5, 7). FDAP has shown itself a helpful diagnostic tool in the control of any tumor therapy [7,8].

On the one hand, FDAP serves to verify that porphyrins are enriched plentifully in tissue prior to a planned PDT [37]. On the other hand, surgical excision [7,14], cryosurgical handling [38] or therapy with the CO_2-laser [39] can be easily controlled by FDAP and performed in a sufficiently effective way. ALA-FDAP allows delineation of clinically ill-defined tumors and the detection of tumor relapses or new tumors which are clinically not detectable [14].

9.8 Technique of photodynamic therapy (PDT)

To perform photodynamic therapy (PDT) in skin tumors, the most often used substance is ALA. The porphyrin precursor is topically applied under occlusive foil as described above. Irradiation should be performed when the optimal ratio of photosensitizer levels between tumor and normal tissue is reached (in the case of ALA 2–6 h after application; Figures 2, 5, 7) [16]. The type of light source (laser or incoherent light) and the required fluence depend on the photosensitizer used as well as on the type and localization of the lesion.

9.9 Photosensitizers and light sources used in PDT

The main classes of photosensitizers used in dermatological PDT are porphyrin derivatives, chlorins, phthalocyanines and porphycenes [8]. The data on PDT efficacy of the latter are still experimental in nature.

Topical application of ALA esters induced higher fluorescence intensities in mouse skin [40] and higher intralesional selectivity of porphyrins in solar keratoses than using ALA (Figure 7) [41]. In ALA-PDT, protoporphyrin is the predominant porphyrin metabolite [42]. Tetrasodium-meso-tetraphenylporphyrinsulfonate (TPPS) is a lipophilic compound which has proved to be highly effective in topical PDT of super-ficial BCCs, but reveals prolonged local photosensitivity. Benzoporphyrin derivative mono-acid ring A (BpD), a reduced porphyrin, was effective in treating skin tumors [43] and psoriatic lesions [44] and has further been tested in murine lupus erythe-matosus [45]. Porphycenes, synthesized porphyrins, and phthalocyanines induced regression of experimental tumors better than did porfimer sodium [46]. After as early as 6 h incubation, the fluorescence induced by topically applied porphycene (ATMPn) was up to the deep dermis in BCC. In lesion-surrounding tissue fluorescence, penetra-tion, however, was restricted to the upper parts of the epidermis [47]. PDT with mono aspartyl chlorin e6 (Npe6) led to a complete response rate (CR) of approximately 50% in skin and oropharynx cancers [48]. In 11 lesions, including recurrent adenocarci-noma of the breast, BCC, and SCC, iv injection of Npe6 and 664 nm laser light caused immediate tissue bleaching followed by a marked necrosis of the tumor mass. However, the CR was not that encouraging and a temporary generalized skin photo-sensitivity was noted [49]. Meso tetra(hydoxyphenyl)chlorin (mTHPC) was primarily tested for diffuse interstitial tumors and has been reported to be very effective in the treatment of BCC [Richter; meeting of the international photodynamic association, Nantes 7–9 July 1998]. The cutaneous phototoxic reaction induced by intravenous injection of mTHPC ($0.1–0.3$ mg kg^{-1}) has been clinically evaluated in 23 patients undergoing photodynamic therapy. These tests have shown that the duration of the skin photosensitization induced by a typical therapeutic dose of mTHPC (0.15 mg kg^{-1}) is less than with Photofrin (2 mg kg^{-1}) [50].

In dermatological applications, the most widely used light systems are the argon ion pumped dye laser (argon-PDL) (630, 635 nm), the gold vapor laser (628 nm) and incoherent light sources emitting close to the sensitizer's absorption peaks (e.g. for porphyrins: 505, 540, 580, and 635 nm) [8,11,51]. In the treatment of large skin lesions incoherent light devices are superior to laser systems due to larger irradiation fields. In early clinical studies incandescent lamps or slide projectors were used. The wavelength 635 nm was revealed as the most effective in ALA-PDT [52]. Reduction of fluence rate or fractionated irradiation may increase PDT efficacy due to increased singlet oxygen levels in regions of spare capillary density and re-accumulation of porphyrins in pretreated and sensitizer-bleached tissue.

9.10 Side effects in FDAP and ALA-PDT

Topical application of ALA in amounts from 0.05 to 0.2 g cm^{-2} (total: 0.05 to 7.0 g ALA) does not lead to measurable systemic porphyrin levels in humans [53]. Therefore no systemic side effects are observed in topical ALA application techniques. However, in ALA-treated mice, liver, kidney, spleen and blood porphyrins also increased from basal levels, showing that ALA and/or ALA-induced porphyrins reach all tissues in this species even after topical application [20].

In PDT with systemically administered ALA, transient rises in serum aspartate aminotransferase, nausea, vomiting, headache, circulatory failure, and prolonged photosensitivity have been reported [8,19]. The increase of hepatic transaminases might be partly due to the prolonged increase in hepatic protoporphyrin levels which was shown to be 50-fold in hamsters treated with 500 mg kg^{-1} ALA iv although ALA and porphyrins were cleared rapidly from the blood and the skin [28]. On the other hand, it has been postulated that ALA, up to 60 mg kg^{-1}, is associated with minimal side effects and no patient developed cutaneous phototoxicity or abnormal neurologic function [54]. Using mTHPC, a patient with an esophageal cancer suffered a second-degree burn on the index finger with subsequent loss of the nail [55].

Since 1994, we have performed topical FDAP in approximately 3000 cases and topical PDT in approximately 1500 cases. We could not find serious unwanted effects, except local dysaesthesia and hypersensitivity. During the application of ALA, a little pruritus can be present. During Wood's light irradiation or red light treatment a slight tingling (FDAP) up to a severe burning sensation (PDT) may occur. Irradiation of porphyrin-sensitized skin areas cause erythema that persists for up to some hours after FDAP and up to 25 days after PDT. After performance of FDAP, the treated skin must be covered with a light-protecting tape until the evening (Figure 7), since otherwise (in particular in the summer) phototoxic reactions can develop. Severe burning and pain, particularly on the face, may occur during PDT; they can last for several hours after irradiation, with decreasing severity [8]. Crusting is common in treated areas and disappears after a few days. Healing occurs within 10 to 14 days with cosmetic results equal or superior to other treatment modalities such as cryosurgery or surgery. Hyperpigmentation is common but subsides within several months in all patients [4,7,8].

A case report on the development of a malignant melanoma at the ALA-PDT exposed site on the scalp seems to be coincidental but has to be taken seriously [56].

9.11 Literature data concerning PDT in dermatology

In general, surgical excision is the most effective and preferential treatment epithelial skin tumors. However, alternative modalities are necessary for extensive or multiple disseminated lesions such as superficial BCC and solar keratoses to improve functional and cosmetic results. In general, the outcomes of clinical studies on PDT treatment of skin tumors are difficult to compare since different specifications of PDT were used. The following discussion will review mainly data on topical ALA-PDT. The trend of the effectiveness of PDT will be given for different dermatological disorders. Detailed information is summarized in Tables 3–8 below.

9.11.1 Basal cell carcinoma (BCC) (Figures 1, 5, 7)

For systemically administered porphyrins, CR from 31–100% were reported [57–63] (Table 3). Topical ALA-PDT resulted in CR between 10 and 100% [5–10,52,64–72] depending on the size and the thickness of the BCCs and was proportional to the

Table 3. Clinical studies on PDT efficacy in basal cell carcinomas

Treated lesion	n	Photosensitizer Type (Intravenous)	Dose/ mg kg⁻¹	Time/h	Light source Type	λ /nm	Intensity/ mW cm⁻²	Fluence/ J cm⁻²	CR[a] (%)	Follow up (mo)	Ref.
BCC	3	Photofrin	5	96	Xenon arc	600–700	100	120	100	7	[64]
BCC	6	Photofrin	1–2	24/72	Argon-PDL	630	29–90	40–60	100	3	[63]
BCC	15	DHE	2	72	Cop. Vap-PDL	628	n.d.	50	93	6	[62]
BCC	21	HpD	5	72	n.d.	630	n.d.	30	0	6	[61]
BCC	36	Photofrin	2	48–144	Argon-PDL	630	11–110	40–144[b]	31	6	[60]
BCC	7	DHE	1.5–2	72	Argon-PDL	630	n.d.	50–100	50	6	[58]
BCC	3	NPe6	0.5	4–8	Argon-PDL	664	50–400	25–200	33	n.d.	[57]
BCC	67	Photosan 3	2	48	Argon-PDL	630	100	100	97	54	[64]
BCC	149	Photofrin	1	48–72	Argon-PDL	630	150	72–288	88[c,d]	20–43	[72]
BCC[e]	13	SnET2	1.2	24	n.d.	n.d.	150	200	100	6	[74]
BCC	94	mTHPC	0.1, 0.15	96	Argon-PDL	652	100	5–20	93	3–24	[73]

[a] Complete response rate is generally based on the results after one treatment.
[b] Treatment was given either by surface irradiation or interstitially.
[c] CR of lesions was dependent on the body site.
[d] Nodular, morphea-like and superficial BCC were examined; however, CR was not given for the different histologic types.
[e] 26-year-old man with basal cell nevus syndrome.

Table 4. Clinical studies on PDT efficacy in basal cell carcinomas

Treated lesion	n	Photosensitizer Type Topical	Time/h	Dose (%)	Light source Type	λ/nm	Intensity/ mW cm^{-2}	Fluence/ J cm^{-2}	CR[a] (%)	Follow up (mo)	Ref.
BCC	292	TPPS	3+6+24	2	Argon-PDL	645	20–200	120/150	94	n.d.	[67]
BCC	98	ALA	>3	20	Cop. Vap-PDL	630	100–150	50–100	96	3	[70]
BCC	27	ALA	4	20	Xenon arc	615–645	20–86	150	59	1	[66]
BCC	19	ALA	4–6	20	Nd:YAG-PDL	635	<110	60	42[b]	0.5–36	[75]
BCC	55	ALA	6	10	Halogen-lamp	400–800	200	240	85	12	[77]
BCC nodular	10	ALA	4–8	20	Slide projector	>570	50–100	90	10	3–12	[10]
BCC nodular	24	ALA	4–6	20	Nd:YAG-PDL	630	100	150	64	6–14	[69]
BCC nodular	22	ALA+Desferal	20	20	PTL-Penta	570–680	150–250	<300	32	20	[6]
BCC nodular	30	ALA	6–8	20	Argon-PDL	630	100	60–80	80	29	[5]
BCC nodular	10	ALA	3	20	Xenon arc	620–670	125–166	75–100	20	6	[71]
BCC pigmented	4	ALA	6–8	20	Argon-PDL	630	100	60–80	0	29	[5]
BCC superficial	80	ALA	3–6	20	Slide projector	>600	150–300	54–540	90	2–3	[9]
BCC superficial	30	ALA	4–6	10–40	n.d.	630	n.d.	75–200	100	3	[68]
BCC superficial	37	ALA	4–8	20	Slide projector	>570	50–100	90	97	3–12	[10]
BCC superficial	8	ALA	3	20	Slide projector	red light	19–44	100	50	1–3	[65]
BCC superficial	55	ALA	4–6	20	Nd:YAG-PDL	630	100	150	100	6–14	[69]
BCC superficial	34	ALA+Desferal	20	20	PTL-Penta	570–680	150–250	<300	88	20	[6]
BCC superficial	23	ALA	6–8	20	Argon-PDL	630	100	60–80	100	29	[5]
BCC superficial[c]	1	ALA	6	20 mg cm^{-2}	PDT 700	570–750	150	180	"95"	9	[7]
BCC superficial	40	ALA	6	20	PDT 700	570–750	150	180	0–83[d]	12–24	[8]
BCC superficial	157	ALA	3	20	Xenon arc	620–670	125–166	75–100	92	6	[71]
BCC superficial	95	ALA	4	20	Slide projector	>515, UVA	50–100	18–131	86	3–60	[78]

[a] Complete response rate is generally based on the results after one treatment.
[b] CR after 3 PDR sessions.
[c] Large BCC of 60 cm^2 on the breast. Performance of 3 PDT sessions achieved a tumor remission of 95%.
[d] CR was dependent on the size of lesion and was improved by repetition of PDT (maximum 3 sessions): superficial BCC 33–100%; BD 75%; nodular SCC 50%; superficial SCC 100%.

Table 5. Clinical studies on PDT efficacy in squamous cell carcinomas, solar keratoses and Bowen's disease

Treated lesion	n	Photosensitizer Type Intravenous	Time/h	Dose/ mg kg^{-1}	Light source Type	λ/nm	Intensity/ mW cm^{-2}	Fluence/ J cm^{-2}	CR[a] (%)	Follow up (mo)	Ref.
BD	3	Photofrin	24	2	Argon-PDL	630	80–194	40–60	100	3	[63]
BD	87	DHE	72	2	Cop. Vap-PDL	628	n.d.	25	85	4–6	[62]
BD	3	Photofrin	48–144	2	Argon-PDL	630	11–110	40–144[b]	66	6	[60]
BD	50	DHE	72	1.5–2	Argon-PDL	630	n.d.	25	50–100	n.d.	[58]
BD	8	Photofrin	48	1	Argon-PDL	630	150	185–250	100	12–24	[80]
SCC	32	HpD	72	5	n.d.	630	n.d.	30	<50	6	[61]
SCC	5	Photofrin	48–144	2	Argon-PDL	630	11–110	40–144[b]	40	6	[60]
SCC	2	NPe6	4–8	1	Argon-PDL	664	50–400	50–100	100	n.d.	[57]
SCC	7	Photosan 3	48	2	Argon-PDL	630	100	100	86	54	[64]
SCC	9	mTHPC	96	0.1, 0.15	Argon-PDL	652	100	5–20	100	3–24	[73]

[a] Complete response rate is generally based on the results after one treatment.
[b] Treatment was given either by surface irradiation or interstitially.

Table 6. Clinical studies on PDT efficacy in squamous cell carcinomas, solar keratoses and Bowen's disease

Treated lesion	n	Photosensitizer Type Topical	Time/h	Dose (%)	Light source Type	λ/nm	Intensity/ mW cm⁻²	Fluence/ J cm⁻²	CRa (%)	Follow up (mo)	Ref.
BD	36	ALA	3–5	0.05 g cm⁻²	Cop. Vap-PDL	630	<150	125–250	89	17	[4]
BD	10	ALA	4–6	20	Nd:YAG-PDL	630	100	150	90	6–14	[69]
BD	10	ALA+desferal	20	20	PTL-Penta	570–680	150–250	<300	30	20	[6]
BD	6	ALA	6–8	20	Argon-PDL	630	100	60–80	100	29	[5]
BD	8	ALA	6	20	PDT 700	570–750	150	180	50b	12–24	[8]
BD	18	ALA	3	20	Xenon arc	620–670	125–166	75–100	61	6	[71]
BD	3	ALA	3	20	Slide projector	400–700	150	125	>90	3	[83]
BD	20	ALA	4	20	Xenon arc	600–660	70	125	75	2	[82]
SCC	8	ALA	3–6	20	Slide projector	>600	150–300	54–540	80	2–3	[9]
SCC nodular	2	ALA	3	20	Slide projector	Red light	19–44	100	0	17	[65]
SCC nodular	6	ALA	6–8	20	Argon-PDL	630	100	60–80	67	29	[5]
SCC nodular	4	ALA	6	20	PDT 700	570–750	150	180	0b	12–24	[8]
SCC superficial	6	ALA	4–8	20	Slide projector	>570	50–100	90	83	3–12	[10]
SCC superficial	3	ALA	3	20	Slide projector	Red light	19–44	100	67	17	[65]
SCC superficial	12	ALA	6–8	20	Argon-PDL	630	100	60–80	92	29	[5]
SCC superficial	10	ALA	6	20	PDT 700	570–750	150	180	60b	12–24	[8]
SCC superficial	35	ALA	4	20	Slide projector	>515	50–100	5–180	54	3–47	[78]
SK	10	ALA	3–6	20	Slide projector	>600	150–300	54–540	90	2–3	[9]
SK	9	ALA	4–8	20	Slide projector	>570	50–100	100	100	3–12	[10]
SK	43	ALA+desferal	20	20	PTL-Penta	570–680	150–250	<300	81	20	[6]
SK	50	ALA	6–8	20	Argon-PDL	630	100	60–80	100	29	[5]
SK	52	ALA	6	10	PDT 700	570–750	120	144	86–95b	12–24	[8]
SK	36	ALA	6	10	PDT 1200	580–740	160	150	0–71c	1	[85]
SK	218	ALA	3	10–30	Argon PDL	630	150	10–150	0–91c	2	[84]

a Complete response rate is generally based on the results after one treatment. b CR was dependent on the size of lesion and was improved by repetition of PDT (maximum 3 sessions): superficial BCC 33-100%; BD 75%; nodular SCC 50%; superficial SCC 100%. c CR was dependent on the body site location and the size of lesion.

Table 7. Clinical studies on PDT efficacy in metastatic and other skin tumors

Treated lesion	n	Photosensitizer Type Intravenous	Time/h	Dose/ mg kg⁻¹	Light source Type	λ/nm	Intensity/ mW cm⁻²	Fluence/ J cm⁻²	CRᵃ (%)	Follow up (mo)	Ref.
Kaposi's sarcoma	1	Photofrin	72	2.5	Xenon	600–700	100	120	100	24	[59]
Kaposi's sarcomaᵇ	5	Photofrin	48–72	2	Argon-PDL	630	50–130	50–200ᶜ	60	3	[95]
Kaposi's sarcomaᵈ	289	Photofrin	48	1	Argon-PDL	630	150	100–400	33	<5dᵉ	[96]
MM	1	Photofrin	96	3.8	Argon-PDL	635	n.d.	60–140	50–80	5	[59]
MM	27	Photofrin	48–144	2	Argon-PDL	630	11–110	40–144ᶜ	7	6	[60]
Recurrent Tu (Met.)	34	Photofrin	48–72	1–2	Argon-PDL	630	40–172	25–100	47	3–5	[97]
Breast Ca	174	Photofrin	48–144	2	Argon-PDL	630	11–110	40–144ᶜ	4	6	[60]
Breast Ca (Met.)	37	Photofrin	48–72	1–2.5	Argon-PDL	630	120–200	104–244	14	1	[98]
Recurrent Tu.	86	TEEPurlytin	24	1.2	Diode-L	664	150	200	92	6	[65b]
Papillary Ca (Met.)	8	NPe6	4–8	0.5–1	Argon-PDL	664	50–400	50–200	75	n.d.	[57]
Papillary Ca	6	NPe6	4–8	1	Argon-PDL	664	50–400	50	0	n.d.	[57]
Paget's disease	1	Photofrin	24–48	1	Argon-PDL	630	150	200–250	100	6–12	[117]
Verrucous Ca	1	NPe6	4–8	0.5	Argon-PDL	664	50–400	100	100	n.d.	[57]
Adeno-Ca (Met.)	50	Photofrin	72	1.5–2	Cop. Vap-PDL	630	5–1500	52–81ᶠ	52	n.d.	[99]
Adeno-Ca (Met.)	13	SnET2	24	1.2	Laser	664	150	200	100	6	[101]
SCC (Met.)	20	Photofrin	72	1.5–2	Cop. Vap-PDL	630	5–1500	150	25–74	n.d.	[99]

ᵃ Complete response rate is generally based on the results after one treatment.
ᵇ Oral lesions of AIDS-related Kaposi's sarcoma.
ᶜ Treatment was given either by surface irradiation or interstitially.
ᵈ AIDS-related Kaposi's sarcoma.
ᵉ Response to PDT was controlled by histopathology and immunohistochemistry. n.d. = no data available.
ᶠ Light irradiation was performed interstitially.

Table 8. Clinical studies on PDT efficacy in metastatic and other skin tumors

Treated lesion	n	Photosensitizer Type Topical	Time/h	Dose (%)	Light source Type	λ/nm	Intensity/ mW cm^{-2}	Fluence/ J cm^{-2}	CR[a] (%)	Follow up (mo)	Ref.
Adeno-Ca (Met.)	6	ALA	3–5	50 mg cm^{-2}	Cop. Vap-PDL	630	<150	125–250	83	17	[4]
Breast Ca (Met.)	4	ALA	3–6	20	Slide projector	>600	150–300	54–540	0	2–3	[9]
Breast Ca (Met.)	9	TPPS4	n.d.	0.15–0.3 mg cm^{-2}	Argon PDL	630	312–680	150	30	n.d.	[100]
MM (Met.)	8	ALA	4–8	20	Slide projector	>570	50–100	90	0	3–12	[10]
SCC (Met.)	6	ALA	3–5	50 mg cm^{-2}	Cop. Vap-PDL	630	<150	125–250	0	17	[4]
T-Cell-lymphoma	n.d.	ALA	4–6	2–10	n.d.	630	n.d.	75–200	100	n.d.	[68]
T-Cell-lymphoma	4	ALA	4–6	20	Nd:YAG-PDL	630	100	150	50	6–14	[69]
Mycosis fungoides	2	ALA	4–6	20	Slide projector	>570	44	40	100	3–6	[94]
Keratoacanthoma	4	ALA	6–8	20	Argon-PDL	630	100	60–80	100	29	[5]

[a] Complete response rate is generally based on the results after one treatment.

light intensity [8]. Fifteen patients with a total of 93 BCCs in the head and neck area and a mean follow up of 15 months (ranging 3–24 months) were treated by iv mTHPC and incoherent light of 652 nm (520 J cm^{-2}). Within several days tumor necrosis appeared followed by wound healing within 4–8 weeks, leaving only minor scars behind. Eighty-nine (93%) showed a CR with an excellent cosmetic outcome and only seven tumors responded partially due to low light dosage [73]. Thirteen lesions in a 26-year-old man with basal cell nevus syndrome treated by SnET2 and red light showed no evidence of recurrence in the 6-month follow-up period [74]. BCCs localized in periocular skin only showed partially good responses after three treatment sessions with ALA and 635 laser light: CR: 42%, PR: 42 % [75]. Thus, PDT can eradicate near-eye non-melanoma skin malignancies without compromising the function of the eyelids. However, further development of the method is needed to reach the same cure rates as those of the conventional treatment modalities (Table 4). ALA 20% + EDTA 2% + DMSO 2% in water in an oil cream base was applied to 31 BCCs. CRs were achieved after one-to-three ALA-PDT treatments in 84% of cases [76]. Some 55 BCCs treated by a novel nanocolloid ALA-lotion and irradiation with visible light from an unfiltered halogen lamp resulted in 85% CR [77].

In contrast to early enthusiastic clinical reports [63], histological and long-term follow-up studies showed a less favorable outcome, especially in large tumors – probably due to insufficient and inhomogeneous ALA penetration and inhomogeneous light irradiation [5,30,61]. In a retrospective study of 95 superficial BCCs and 35 superficial SCCs treated by topical ALA-PDT, primary tumor responses of 86% for superficial BCC and 54% for superficial SCC were revealed. However, 36 months after therapy, the projected disease-free rate was limited to 50% for BCC vs. 8% for SCC. Histopathological studies revealed a significant increase of fibrosis in the dermis after ALA-PDT and the appearance of a sharp border between fibrotic and nonfibrotic tissues [78]. Although primary surgery remains the treatment of choice in large tumors, pretreatment with PDT may improve the cosmetic and functional outcome in difficult locations [7]. Special pretreatment including an initial debulking procedure and topical application of DMSO in order to enhance penetration of ALA (20% in cream) was shown to dramatically increase the CR [79]. Fifty-eight patients with 119 nodular BCCs were successfully treated (95% CR) with ALA-PDT (follow up: 1 year).

9.11.2 Bowen's disease (BD)

In several patients successful treatment of BD with porfimer sodium and an argon-PDL has been described [58,80,81] (Table 5). BD also showed a good initial response to PDT with topical ALA, but long-term results varied considerably (CR: 30–100%) (Table 6) [4,6,69,80]. Even repeated PDT treatments yielded a CR of only 50–75% [6,8]. The incomplete response of BD may be due to the thickened epithelial layer with reduced ALA penetration [4,5,8,58,60,63,69,71,80–83]. Complete clearance was achieved in one patient with Bowen's disease of the penis using ALA 20% + EDTA 2% + DMSO 2% and VersaLight® [76].

9.11.3 Solar keratoses (SK) (Figure 6)

SK represent one of the best indicators for (ALA-) PDT in dermatology at present (Tables 5, 6). In most clinical studies, a CR of 80 to 100% was achieved using 10 to 20% ALA [5,6,8–10,84]. The "burning" pain experienced by most patients during irradiation of multiple extended lesions on the scalp can be significantly reduced by using 10% ALA only [8] and irradiating with green instead of red light [11]. SK on the arms and hands respond to a lesser extent to ALA-PDT than lesions located on the scalp [85].

9.11.4 Squamous cell carcinoma (SCC) (Figure 2)

With systemic photosensitizers 40–100% remission of SCCs could be obtained [57,60,61,64] (Table 5). In topical ALA-PDT the CR was 60 to 92% for superficial SCCs [5,8,10,65] and 0–67% for nodular SCCs [5,8,65]. ALA 20% + EDTA 2% + DMSO 2% in a water in oil cream base applied to five superficial SCC led to CRs after one-to-three ALA-PDT treatments of 80% [76]. The initial stages of SCCs can be effectively treated by topical ALA-PDT, and even in nodular SCCs promising remission rates can be obtained after repeated PDT [5,8]. All four SCC treated by iv mTHPC and incoherent light of 652 nm (5–20 J cm^{-2}) showed a CR with an excellent cosmetic outcome [73] (Table 6).

9.11.5 Other neoplastic skin disorders

PDT also seems to be a promising modality to treat premalignant epithelial lesions and SCCs of the oral mucosa [86]. Twelve patients, who had been suffering from leukoplakia of the oral mucosa for several years, were treated with 20% ALA cream for two hours and light activated at 630 nm (100 mW cm^{-2}, 100 J cm^{-2}). Five patients showed CR to the treatment and four patients showed a PR [87]. There were also initial enthusiastic reports on the efficacy of PDT in genital precancerous stages such as erythroplasia of Queyrat [88], an intraepithelial carcinoma in situ affecting the mucosal surfaces of the penis, with a significant risk of invasion and metastasis. One of two patients with limited disease achieved a long-term CR (36 months) and the other developed a recurrence at 18 months after initial CR. Two further patients with more extensive disease showed a significant improvement, allowing easier treatment by laser vaporization. Although topical ALA-PDT offers the advantages of tumor specificity, preservation of function and a good cosmetic result, more extensive erythroplasia of Queyrat appears less responsive to this new therapeutic modality using current treatment parameters [88]. Actinic cheilitis [89], tumors in xeroderma pigmentosum [90] and nevus sebaceous [91] are other tumors that show response to PDT.

9.11.6 Malignant melanoma (MM) (Figure 4)

So far, there is little information on the efficacy of ALA-PDT in the treatment of primary and metastatic MM, and the results are contradictory (Table 7) [10,59,60].

The strong pigmentation of melanoma tissues may be the limiting factor by inhibiting light penetration. Combined cobalt-60 gamma radiation and argon laser irradiation using injected merocyanine (MC540) as a photosensitizer exerts a significant therapeutic effect on rapidly growing pigmented and non-pigmented Bomirski hamster melanoma growing in the eye [92]. In mice bearing a subcutaneously transplanted B1 melanoma, PDT efficacy was enhanced by pretreatment with 1064 nm light, which caused a selective breakdown of melanosomes [93].

9.11.7 Mycosis fungoides

Topical ALA application and subsequent exposure to polychromatic or laser light was used effectively to treat plaque-stage cutaneous T-cell lymphoma [68,69,94a]. However, the apparent clinical cure was not confirmed histologically [94b] (Table 8).

9.11.8 Kaposi's sarcoma

Classical and AIDS-related oral Kaposi's sarcoma lesions have been successfully treated showing early and late CR [59,95] (Table 7). Twenty-five patients with AIDS-related Kaposi's sarcoma received 1 mg kg^{-1} Photofrin 48 h before exposure to 100–400 J cm^{-2} of 630 nm light. Of the 289 treated lesions 33% had CR, 63% had PR and 4% were treatment failures. There was a strong correlation between response and light dose [96].

9.11.9 Cutaneous and subcutaneous metastases

Some clinical PDT studies focused on the treatment of breast cancer and other metastases (systemic PDT: 4–75% CR [57,97–99] topical PDT with ALA or TPPS 0–83% CR [4,5,9,100] (Tables 7, 8). PDT was shown to offer an excellent local control rate of chest wall recurrence of breast cancer (after mastectomy, radiation therapy, and chemotherapy), in general, raising a therapeutic dilemma. A total of 86 lesions were treated in eight patients who had biopsy-proven chest wall recurrence despite surgery, chemotherapy, and radiation therapy. Each patient underwent a single PDT with 1.2 mg kg^{-1} of the drug tin ethyl etiopurpurin (Purlytin) and laser light treatment at 660 nm. CR was 92% and PR 8%. Three patients with 13 biopsy-proven metastatic adenocarcinoma of the skin were effectively treated (CR: 100%) with SnET2 and laser light at 664 nm [101] (Tables 7, 8).

9.11.10 Lichen sclerosus et atrophicus

Twelve women with vulvar lichen sclerosus et atrophicus, a pseudosclerodermatic disease, were treated by 635 nm light (argon ion-pumped dye laser, 40–70 mW cm^{-2}, 80 J cm^{-2}) 45 h after topical application of 10 ml ALA (20% solution). There was need of intravenous opioids during the procedure in 3 patients, due to burning sensations. Minimal local toxicity included vulvar erythema but no necrosis, sloughing, or scarring. Six to eight weeks after photodynamic therapy, pruritus significantly improved in 83% of cases [102].

9.11.11 Lichen planus

Treatment of lichen planus of the penis by 20% ALA (for 4 h) and incoherent light (~630 nm) induced curing of the lesions [103].

9.11.12 Capillary vascular malformation

Eleven facial lesions of port wine stain (PWS) disappeared completely without any scarring after red light PDT [104]. PDT has strong skin-penetrating ability and irradiation just after iv injection of photosensitizing drugs is effective to embolize the malformative vasculature in pink lesions, especially in dark purple lesions and purple lesions with proliferation [104]. A review of 118 patients with PWS reveals that 98% responded to PDT with varying degrees of success after one single treatment [105]. In addition, hypertrophic scars, permanent hyperpigmentation, and hypopigmentation were not seen based on proper parameters. To obtain an objective assessment of the curative effectiveness of PDT for PWS, the relationship between the microvascular perfusion changes of PWS and the bleaching of the lesions before and after PDT was investigated in 24 patients (28 lesions; face and neck) [106]. After iv injection of HpD, the copper vapor laser was adopted as light source and the lesions of PWS were irradiated. All the lesions showed a remarkable decrease of tissue perfusion after PDT. The colors of lesions were correlated with a decrease of microcirculatory perfusion, and became lightened close to normal skin color without causing any scarring. PDT seems to be an effective modality for PWS. The microcirculation perfusion can reflect the degrees of PWS objectively. The curative effectiveness of PDT for PWS is due to tissue microcirculation response.

9.11.13 Psoriasis (Figure 3)

Systemic and topical sensitization with Hp followed by visible light irradiation resulted in clinical improvement of psoriatic plaques and palmopustular psoriasis [107,108]. Topical ALA-PDT was speculated to be comparable with dithranol in psoriasis therapy. Compared to PUVA the risk of malignancy is lower for ALA-PDT [109]. In 8 out of 10 patients with plaque psoriasis topical ALA-PDT using a broadband visible radiation (3 times per week, with a maximum of 12 treatments, light dose 8 J cm^{-2}, 15 mW cm^{-2}) achieved clinical response. Out of 19 treated sites, 4 cleared, 10 responded but did not clear and 5 showed no improvement [81]. In another study, ALA-PDT induced a CR of approximately 18% only [110] probably due to the non-homogeneous distribution of ALA-induced porphyrins in psoriatic plaques [12,111]. A case report proved a CR of chronic psoriatic plaques refractory to conventional therapy; however, this happened in combination with topical corticosteroids and calcipotriol [112].

Systemic administration of ALA may result in a more homogeneous and selective accumulation of porphyrins in psoriatic lesions as already shown in BCCs [19] but pharmacokinetic and toxicity issues are not yet settled. Weekly PDT sessions using intravenously administered BpD (8 mg m^{-2}) and incoherent red light in the treatment of plaque-stage psoriasis led to a good response of most lesions after only 2–5 weeks [113]. Bathing in a PS-containing solution is another interesting procedure. Red light emitting cabins or high dose UVA1 may prove superior to presently available lamps for whole body PDT.

9.11.14 Wound healing

The effect of PDT on wound healing was tested in full-thickness incisional wounds in 24 hairless Sprague Dawly rats [114]. Cytokines, specifically TGF-beta, are believed to be instrumental in sustaining the fibrotic process, which leads to scarring. Rats were injected with 0.25 or 0.5 mg kg^{-1} BPD-MA or 5 or 10 mg kg^{-1} CASP, 3 and 24 h prior to irradiation with light (1–20 J cm^{-2}), respectively. However, there was no apparent influence of PDT on either the rate or final appearance of wound healing.

9.11.15 Viral-induced diseases

In condylomata acuminata, gross and microscopic fluorescence spectra suggested selectivity (68%) of PpIX formation after topical ALA application [115]. The greatest lesional to normal skin fluorescence ratios occurred after 2 h. A patient with epidermodysplasia verruciformis (types of human papilloma virus were proven in lesional skin) presenting wart-like lesions on the hands, lower arms and forehead was treated using 20% ALA for 6 h and incoherent light 580–740 nm, 160 mW cm^{-2}, 160 J cm^{-2}, with an excellent cosmetic result [116]. In situ hybridization was positive for HPV type 8 in skin which was clinically and histologically normal. Twelve months after PDT a few lesions had recurred on the hands.

9.12 Perspectives of FDAP and PDT in dermatology

The clinical detection of the borders of BCCs and SCCs, particularly in anatomically difficult sites such as the face, is a frequent problem. Multiple surgical procedures can become necessary for complete tumor removal. FDAP was demonstrated to be very effective in detecting and also demarcating clinically ill-defined tumor tissue (Figure 5) [111]. Thus, we recommend FDAP as a useful easy technique to visualize and detect the extent of the tumor preoperatively.

PDT with ALA is very effective in the treatment of SK, small superficial BCCs, and superficial SCCs. Results of PDT in epithelial skin tumors should, however, be viewed critically due to the methodological shortcomings of many studies. Histological examination demonstrated tumor tissue in a large proportion of tumors despite clinical regression after a single PDT cycle, and tumor recurrence was common after long-term follow-up [5,8]. In conclusion, follow-up periods in most published studies were too short. The lack of topical PDT efficacy in more deeply localized and remaining tumor parts may be due to the limited photophysical properties of PDT, the limited permeability for ALA due to overlying normal skin, and the presence of encapsulated tumor cell islands resistant to ALA permeability [8,19]. Thus, retreatment is useful before normal skin covers possible tumor remnants in deeper tissue layers.

Future progress will be achieved with the development of more effective light sources, fractionated light irradiation, the use of new promising compounds such as esterified ALA derivatives or second generation photosensitizers; systemic administration of ALA may enhance FDAP and PDT efficacy. In addition, debulking of the superficial tumor parts optimizes FDAP and PDT efficacy especially in nodular tumors.

Comparative controlled trials and long-term follow-up studies must be performed to establish PDT in dermatology.

Acknowledgement

We thank Mrs K. Kleinert for her help in preparing the manuscript.

References

1. A. Policard (1924). Etude sur les aspects offerts par des tumeurs expérimentales examinées à la lumière de Wood. *CR Soc. Biol.*, **91**, 1423–1424.
2. F.H.J. Figge, G.S. Weiland, L.O.J. Manganiello (1948). Cancer detection and therapy: affinity of neoplastic, embryonic, and traumatized tissues for porphyrins and metalloporphyrins. *Proc. Soc. Exp. Biol. Med.*, **68**, 640–641.
3. H.G. Gregorie Jr, E.O. Horger, J.L. Ward (1968). Hematoporphyrin-derivate fluorescence in malignant neoplasms. *Ann. Surg.*, **167**, 820–828.
4. F. Cairnduff, M.R. Stringer, E.J. Hudson, D.V. Ash, S.B. Brown (1994). Superficial photodynamic therapy with topical 5-aminolevulinic acid for superficial primary and secondary skin cancer. *Br. J. Cancer*, **69**, 605–608.
5. P.G. Calzavara-Pinton (1995). Repetitive photodynamic therapy with topical δ-aminolaevulinic acid as an appropriate approach to the routine treatment of superficial nonmelanoma skin tumors. *J. Photochem. Photobiol. B Biol.*, **29**, 53–57.
6. S. Fijan, H. Hönigsmann, R. Ortel (1995). Photodynamic therapy of epithelial skin tumors using delta-aminolevulinic acid and desferrioxamine. *Br. J. Dermatol.*, **133**, 282–288.
7. C. Fritsch, P.M. Becker-Wegerich, K.W. Schulte, W. Neuse, P. Lehmann, T. Ruzicka, G. Goerz (1996). Photodynamische Therapie und Mamillenplastik eines großflächigen Rumpfhautbasalioms der Mamma. Effektive Kombinationstherapie unter photodynamischer Diagnostik. *Hautarzt*, **47**, 438–442.

8. C. Fritsch, G. Goerz, T. Ruzicka (1998). Photodynamic therapy in dermatology. A review. *Arch. Dermatol.*, **134**, 207–214.

9. J.C. Kennedy, R.H. Pottier, D.C. Pross (1990). Photodynamic therapy with endogenous protoporphyrin IX: basic principles and present clinical experience. *J. Photochem. Photobiol.*, **6**, 143–148.

10. P. Wolf, E. Rieger, H. Kerl (1993). Topical photodynamic therapy with endogenous porphyrins after application of 5-aminolevulinic acid: an alternative treatment modality for solar keratoses, superficial squamous cell carcinomas, and basal cell carcinomas? *J. Am. Acad. Dermatol.*, **28**, 17–21.

11. C. Fritsch, S. Stege, G. Saalmann, G. Goerz, T. Ruzicka, J. Krutmann (1997). Green light is effective and less painful than red light in photodynamic therapy of facial solar keratoses. *Photodermatol. Photoimmunol. Photomed.*, **13**, 181–185.

12. C. Fritsch, K. Lang, W. Neuse, T. Ruzicka, P. Lehmann (1998). Photodynamic diagnosis and therapy in dermatology. *Skin Pharmacol. Appl. Skin Physiol.*, **11**, 358–373.

13. A. Casas, H. Fukuda, A.M. Batlle (1999). Tissue distribution and kinetics of endogenous porphyrins synthesized after topical application of ALA in different vehicles. *Br. J. Cancer*, **81**(1), 13–28.

14. C. Fritsch, P.M. Becker-Wegerich, M. Menke, T. Ruzicka, G. Goerz, R.R. Olbrisch (1997). Successful surgery of multiple recurrent basal cell carcinomas guided by photo-dynamic diagnosis. *Aesthetic Plast Surg.*, **21**, 437–439.

15. C. Klinteberg, A.M. Enejder, I. Wang, S. Andersson-Engels, S. Svanberg, K. Svanberg (1999). Kinetic fluorescence studies of 5-aminolaevulinic acid-induced protoporphyrin IX accumulation in basal cell carcinomas. *J. Photochem. Photobiol. B*, **49**, 120–128.

16. C. Fritsch, P. Lehmann, W. Stahl, K.W. Schulte, E. Blohm, K. Lang, H. Sies, T. Ruzicka (1999). Optimum porphyrin accumulation in epithelial skin tumors and psoriatic lesions after topical application of δ-aminolaevulinic acid. *Br. J. Cancer*, **79**, 1603–1608.

17. E.W. Grant, C. Hopper, A.J. MacRobert, P.M. Speight, S.G. Bown (1993). Photodynamic therapy of oral cancer: photosensitisation with systemic aminolevulinic acid. *Lancet*, **342**, 147–148.

18. Z. Hua, S.L. Gibson, T.H. Foster, R. Hilf (1995). Effectiveness of δ-aminolevulinic acid-induced portoporphyrin as a photosensitizer for photodynamic therapy in vivo. *Cancer Res.*, **55**, 1723–1731.

19. Q. Peng, T. Warloe, J. Moan, H. Heyerdahl, H.B. Stehen, J.M. Nesland, K.E. Giercksky (1995). Distribution of 5-aminolevulinic acid-induced porphyrins in noduloulcerative basal cell carcinoma. *Photochem. Photobiol.*, **62**, 906–913.

20. A. Casas, H. Fukuda, R. Meiss, A.M. Batlle (1999). Topical and intratumoral photo-dynamic therapy with 5-aminolevulinic acid in a subcutaneous murine mammary adeno-carcinoma. *Cancer Lett.*, **141**, 29–38.

21. L.C. Lucchina, N. Kollias, R. Gillies, S.B. Phillips, J.A. Muccini, M.J. Stiller, R.J. Tranick, L.A. Drake (1996). Fluorescence photography in the evaluation of acne. *J. Am. Acad. Dermatol.*, **35**, 58–63.

22. C. Fritsch, W. Stahl, K.W. Schulte, H. Sies, G. Goerz, P. Lehmann, T. Ruzicka (1998). Optimum porphyrin formation in skin tumors and psoriatic lesions after topical applica-tion of δ-aminolevulinic acid. *J. Invest. Dermatol.*, **110**, 673(A).

23. J.M. Nauta, O.C. Speelman, H.L. van Leengoed, P.G. Nikkels, J.L. Roodenburg, W.M. Star, M.J. Witjes, A. Vermey (1997). In vivo photo-detection of chemically induced premalignant lesions and squamous cell carcinoma of the rat palatal mucosa. *J. Photochem. Photobiol. B*, **39**, 156–166.

24. A. Leunig, C.S. Betz, M. Mehlmann, H. Stepp, S. Arbogast, G. Grevers, R. Baumgartner (2000). Detection of squamous cell carcinoma of the oral cavity by imaging

5-aminolevulinic acid-induced protoporphyrin IX fluorescence. *Laryngoscope*, **110**, 78–83.

25. H. Messmann, R. Knuchel, W. Baumler, A. Holstege, J. Scholmerich (1999). Endoscopic fluorescence detection of dysplasia in patients with Barrett's esophagus, ulcerative colitis, or adenomatous polyps after 5-aminolevulinic acid-induced protoporphyrin IX sensitization. *Gastrointest. Endosc.*, **49**, 97–101.

26. D.J. Anthony, A.E. Profio, O.J. Balchum (1989). Fluorescence spectra in lung with porphyrin injection. *Photochem. Photobiol.*, **49**, 583–586.

27. D.M. Harris, J. Werkhaven (1987). Endogenous porphyrin fluorescence in tumors. *Lasers Surg. Med.*, **7**, 467–472.

28. C. Fritsch, C. Abels, A.E. Goetz, W. Stahl, K. Bolsen, T. Ruzicka, G. Goerz, H. Sies (1997). Porphyrins preferentially accumulate in a melanoma following intravenous injection of 5-aminolevulinic acid. *Biol. Chem.*, **378**, 51–57.

29. G. Goerz, S. Bunselmeyer, K. Bolsen, N.Y. Schürer (1996). Ferrochelatase activity in patients with erythropoietic protoporphyria and their families. *Br. J. Dermatol.*, **134**, 880–885.

30. A. Martin, W.D. Tope, J.M. Grevelink, J.C. Starr, J.L. Fewkes, T.J. Flotte, T.J. Deutsch, R.R. Anderson (1995). Lack of selectivity of protoporphyrin IX fluorescence for basal cell carcinoma after topical application of 5-aminolevulinic acid: implications for photodynamic treatment. *Arch. Dermatol. Res.*, **287**, 665–674.

31. W.D. Tope, E.V. Ross, N. Kollias, A. Martin, R. Gillies, R.R. Anderson (1998). Protoporphyrin IX fluorescence induced in basal cell carcinoma by oral delta-aminolevulinic acid. *Photochem. Photobiol.*, **67**, 249–255.

32. J.F. Kelly (1975). Haematoporphyrins in the diagnosis and treatment of carcinoma of the bladder. *Proc. R. Soc. Med.*, **68**, 527–528.

33. W. Stummer, S. Stocker, S. Wagner, H. Stepp, C. Fritsch, C. Goetz, A.E. Goetz, R. Kiefmann, H.J. Reulen (1998). Intraoperative detection of malignant gliomas by 5-aminolevulinic acid-induced porphyrin fluorescence. *Neurosurgery*, **42**, 518–526.

34. E.G. King, G. Man, J. le Riche, R. Amy, A.E. Profio, D.R. Doiron (1982). Fluorescence bronchoscopy in the localization of bronchogenic carcinoma. *Cancer*, **15**, 777–782.

35. K. Svanberg, I. Wang, S. Colleen, I. Idvall, C. Ingvar, R. Rydell, D. Jocham, H. Diddens, S. Bown, G. Gregory, S. Montan, S. Andersson-Engels, S. Svanberg (1998). Clinical multi-colour fluorescence imaging of malignant tumors initial experience. *Acta Radiol.*, **39**, 2–9.

36. A.M. Wennberg, F. Gudmundson, B. Stenquist, A. Ternesten, L. Molne, A. Rosen, O. Larko (1999). In vivo detection of basal cell carcinoma using imaging spectroscopy. *Acta Derm. Venereol.*, **79**, 54–61.

37. A. Orenstein, G. Kostenich, Z. Malik (1997). The kinetics of protoporphyrin fluorescence during ALA-PDT in human malignant skin tumors. *Cancer Lett.*, **120**, 229–234.

38. P. Becker-Wegerich, C. Fritsch, W. Neuse, K.W. Schulte, T. Ruzicka, G. Goerz (1995). Effektive Kryochirurgie oberflächlicher Hauttumoren unter photodynamischer Diagnostik. *H+G*, **70**, 891–895.

39. P. Becker-Wegerich, C. Fritsch, K.W. Schulte, M. Megahed, W. Neuse, G. Goerz, W. Stahl, T. Ruzicka (1998). Carbon dioxide laser treatment of extramammary Paget's disease guided by photodynamic diagnosis. *Br. J. Dermatol.*, **138**, 169–172.

40. Q. Peng, J. Moan, T. Warloe, V. Iani, H.B. Steen, A. Bjorseth, J.M. Nesland (1996). Build-up of esterified aminolevulinic-acid-derivative-induced porphyrin fluorescence in normal mouse skin. *J. Photochem. Photobiol.*, *B Biol.*, **34**, 95–96.

41. C. Fritsch, B. Homey, W. Stahl, P. Lehmann, T. Ruzicka, H. Sies (1998). Preferential relative porphyrin enrichment in solar keratoses upon topical application of δ-aminolevulinic acid methylester. *Photochem. Photobiol.*, **68**, 218–221.

42. C. Fritsch, J. Batz, K. Bolsen, K.W. Schulte, M. Zumdick, T. Ruzicka, G. Goerz (1997). Ex vivo application of δ-aminolevulinic acid induces high and specific porphyrin levels in human skin tumors: possible basis for selective photodynamic therapy. *Photochem. Photobiol.*, **66**, 114–118.

43. H. Lui, L. Hruza, D. McLean, M. Grossman, G. Hruza, K. Gelmon, N. Kollias, J. Wimberly, M. Gagel, R.R. Anderson (1995). Photodynamic therapy of malignant skin tumors with BPD verteporfin (benzoporphyrin derivative). *Laser Surg. Med.*, **7** (Suppl), 44.

44. L. Hruza, H. Lui, G. Hruza, D. McLean, M. Grossman, N. Kollias, J. Wimberly, M. Gagel, R.R. Anderson (1995). Response of psoriasis to photodynamic therapy using benzoporphyrin derivative monoacid ring A. *Laser Surg. Med.*, **7**, 43–44.

45. A.M. Richter, R. Chowdary, L. Ratkay, A.K. Jain, A.J. Canaan, H. Meadows, M. Obochi, D. Warefield, J.G. Levy (1993). Non-oncologic potentials for photodynamic therapy. *SPIE Proc.*, **2078**, 293–304.

46. M. Dellian, C. Richert, F. Gamarra, A.E. Goetz (1996). Photodynamic eradication of amelanotic melanoma of the hamster with fast acting photosensitizers. *Int. J. Cancer*, **65**, 246–248.

47. S. Karrer, C. Abels, R.M. Szeimies, W. Baumler, M. Dellian, U. Hohenleutner, A.E. Goetz, M. Landthaler (1997). Topical application of a first porphycene dye for photodynamic therapy – penetration studies in human perilesional skin and basal cell carcinoma. *Arch. Dermatol. Res.*, **289**, 132–137.

48. H.I. Pass (1993). Photodynamic therapy in oncology. Mechanism and clinical use. *J. Natl. Cancer Inst.*, **85**, 443–456.

49. S.W. Taber, V.H. Fingar, C.T. Coots, T.J. Wieman (1998). Photodynamic therapy using mono-L-aspartyl chlorin e6 (Npe6) for the treatment of cutaneous disease: a Phase I clinical study. *Clin. Cancer Res.*, **4**, 2741–2746.

50. G. Wagnieres, C. Hadjur, P. Grosjean, D. Braichotte, J.F. Savary, P. Monnier, H. van den Bergh (1998). Clinical evaluation of the cutaneous phototoxicity of 5,10,15,20-tetra (m-hydroxyphenyl)chlorin. *Photochem. Photobiol.*, **68**, 382–387.

51. R.M. Szeimies, R. Hein, W. Bäumler, A. Heine, M. Landthaler (1994). A possible new incoherent lamp for photodynamic treatment of superficial skin lesions. *Acta Dermatol Venereol (Stockh).*, **74**, 117–119.

52. R.M. Szeimies, C. Abels, C. Fritsch, S. Karrer, P. Steinbach, W. Bäumler, G. Goerz, A.E. Goetz, M. Landthaler (1995). Wavelength dependency of photodynamic effects after sensitization with 5-aminolevulinic acid in vitro and in vivo. *J. Invest. Dermatol.*, **105**, 672–677.

53. C. Fritsch, B. Verwohlt, K. Bolsen, T. Ruzicka, G. Goerz (1996). Influence of topical photodynamic therapy with 5-aminolevulinic acid on the porphyrin metabolism. *Arch. Dermatol. Res.*, **288**, 517–521.

54. J. Webber, D. Kessel, D. Fromm (1997). On-line fluorescence of human tissues after oral administration of 5-aminolevulinic acid. *J. Photochem. Photobiol. B*, **38**, 209–14.

55. A. Radu, M. Zellweger, P. Grosjean, P. Monnier (1999). Pulse oximeter as a cause of skin burn during photodynamic therapy. *Endoscopy*, **31**, 831–833.

56. P. Wolf, R. Fink-Puches, A. Reimann-Weber, H. Kerl (1997). Development of malignant melanoma after repeated topical photodynamic therapy with 5-aminolevulinic acid at the exposed site. *Dermatology*, **194**, 53–54.

57. R.P. Allen, D. Kessel, R.S. Tharratt, W. Volz (1992). Photodynamic therapy of superficial malignancies with Npe6 in man. *Photodynamic Therapy and Biomedical Lasers* (pp. 441–445). Elsevier Science Publishers.

58. R.B. Buchanan, J.A.S. Carruth, A.L. McKenzie, S.R. Williams (1989). Photodynamic therapy in the treatment of malignant tumors of the skin and head and neck. *Eur. J. Surg. Oncol.*, **15**, 400–406.

59. T.J. Dougherty (1981). Photoradiation therapy for cutaneous and subcutaneous malignancies. *J. Invest. Dermatol.*, **77**, 122–124.
60. J.S. McCaughan, J.T. Guy, W. Hicks, L. Laufmann, T.A. Nims, J. Walker (1989). Photodynamic therapy for cutaneous malignant neoplasms. *Arch. Surg.*, **124**, 211–216.
61. D.G. Pennington, M. Waner, A. Knox (1988). Photodynamic therapy for multiple skin cancers. *Plast. Reconstr. Surg.*, **82**, 1067–1071.
62. P.J. Robinson, J.A.S. Carruth, G.M. Fairris (1988). Photodynamic therapy: a better treatment for widespread Bowen's disease. *Br. J. Dermatol.*, **119**, 59–61.
63. S.M. Waldow, V.L. Rocco, I.K. Kohler, S. Wallk, T.F. Fritts (1987). Photodynamic therapy for treatment of malignant cutaneous lesions. *Laser Sur. Med.*, **7**, 451–456.
64. J. Feyh, R. Gutmann, A. Leunig (1993). Die photodynamische Lasertherapie im Bereich der Hals-, Nasen-, Ohrenheilkunde. *Laryngo-Rhino-Otol*, **72**, 273–278.
65. (a) H. Lui, S. Salasche, N. Kollias, J. Wimberly, T. Flotte, D. McLean, R.R. Anderson (1995). Photodynamic therapy of nonmelanoma skin cancer with topical aminolevulinic acid: a clinical and histologic study. *Arch. Dermatol.*, **131**, 737–738. (b) T.S. Mang, R. Allison, G. Hewson, W. Snider, R. Moskowitz (1998). A phase II/III clinical study of tin ethyl etiopurpurin (Purlytin)-induced photodynamic therapy for the treatment of recurrent cutaneous metastatic breast cancer. *Cancer J. Sci. Am.*, **4**, 378–384.
66. C.A. Morton, R.M. MacKie, C. Whitehurst, J.V. Moore, J.H. McColl (1998). Photodynamic therapy for basal cell carcinoma: effect of tumor thickness and duration of photosensitizer application on response [letter]. *Arch. Dermatol.*, **134**, 248–249.
67. O. Santoro, G. Bandieramonte, E. Melloni, R. Marchesini, F. Zunino, P. Lepera, G. De Palo (1990). Photodynamic therapy by topical meso-tetraphenylporphyrinesulfonate tetrasodium salt administration in superficial basal cell carcinomas. *Cancer Res.*, **50**, 4501–4503.
68. S.D. Shanler, W. Wan, J.E. Whitaker, T.S. Mang, C. Jones, B.D. Wilson, H.L. Stoll, S. Pincus, A.R. Oseroff (1993). Topical δ-aminolevulinic acid for photodynamic therapy of cutaneous carcinomas and cutaneous T-cell lymphoma. *J. Invest. Dermatol.*, **101**, 406(A).
69. K. Svanberg, T. Anderson, D. Killander, I. Wang, U. Stenram, S. Andersson-Engels, R. Berg, J. Johansson, S. Svanberg (1994). Photodynamic therapy of non-melanoma malignant tumors of the skin using topical δ-aminolevulinic acid sensitization and laser irradiation. *Br. J. Dermatol.*, **130**, 743–751.
70. T. Warloe, Q. Peng, J. Moan, H.L. Qvist, K.E. Giercksky (1992). Photochemotherapy of multiple basal cell carcinoma with endogenous porphyrins induced by topical application of 5-aminolevulinic acid. In: P. Spinelli, M. Dal Fante, R. Marchesini (Eds) *Photodynamic Therapy and Biochemical Lasers* (pp. 449–453). Elsevier, Amsterdam.
71. A.M. Wennberg, L.E. Lindholm, M. Alpsten, O. Larko (1996). Treatment of superficial basal cell carcinomas using topically applied delta-aminolaevulinic acid and a filtered xenon lamp. *Arch. Dermatol. Res.*, **88**, 561–564.
72. B.D. Wilson, T. Mang, H. Stoll, C. Jones, M. Cooper, T.J. Dougherty (1992). Photodynamic therapy for the treatment of basal cell carcinoma. *Arch. Dermatol.*, **128**, 1597–1601.
73. A.C. Kubler, T. Haase, C. Staff, B. Kahle, M. Rheinwald, J. Muhling (1999) Photodynamic therapy of primary nonmelanomatous skin tumors of the head and neck. *Lasers Surg. Med.*, **25**, 60–68.
74. R. Rifkin, B. Reed, F. Hetzel, K. Chen (1997). Photodynamic therapy using SnET2 for basal cell nevus syndrome: a case report. *Clin. Ther.*, **19**, 639–641.
75. I. Wang, B. Bauer, S. Andersson-Engels, S. Svanberg, K. Svanberg (1999). Photodynamic therapy utilising topical delta-aminolevulinic acid in non-melanoma skin malignancies of the eyelid and the periocular skin. *Acta Ophthalmol. Scand.*, **77**, 182–188.

76. Y. Harth, B. Hirshowitz, B. Kaplan (1998). Modified topical photodynamic therapy of superficial skin tumors, utilizing aminolevulinic acid, penetration enhancers, red light, and hyperthermia. *Dermatol. Surg.*, **24**, 723–726.

77. A.F. Hurlimann, G. Hanggi, R.G. Panizzon (1998). Photodynamic therapy of superficial basal cell carcinomas using topical 5-aminolevulinic acid in a nanocolloid lotion. *Dermatology*, **197**, 248–254.

78. R. Fink-Puches, H.P. Soyer, A. Hofer, H. Kerl, P. Wolf (1998). Long-term follow-up and histological changes of superficial nonmelanoma skin cancers treated with topical delta-aminolevulinic acid photodynamic therapy. *Arch. Dermatol.*, **134**, 821–826.

79. A.M. Soler, T. Warloe, J. Tausjo, A. Berner (1999). Photodynamic therapy by topical aminolevulinic acid, dimethyl sulphoxide and curettage in nodular basal cell carcinoma: a one-year follow-up study. *Acta Derm. Venereol.*, **79**, 204–206.

80. C.M. Jones, T. Mang, M. Cooper, B.D. Wilson, H.L. Stoll (1992). Photodynamic therapy in the treatment of Bowen's disease. *J. Am. Acad. Dermatol.*, **27**, 979–982.

81. D.J. Robinson, P. Collins, M.R. Stringer, D.I. Vernon, G.I. Stables, S.B. Brown, R.A. Sheehan-Dare (1999). Improved response of plaque psoriasis after multiple treatments with topical 5-aminolaevulinic acid photodynamic therapy. *Acta Derm. Venereol.*, **79**, 451–455.

82. C.A. Morton, C. Whitehurst, H. Moseley, J.H. McColl, J.V. Moore, R.M. Mackie (1996). Comparison of photodynamic therapy with cryotherapy in the treatment of Bowen's disease. *Br. J. Dermatol.*, **135**, 766–771.

83. G.I. Stables, M.R. Stringer, D.J. Robinson, D.V. Ash (1997). Large patches of Bowen's disease treated by topical aminolevulinic acid photodynamic therapy. *Br. J. Dermatol.*, **136**, 957–60.

84. E.W. Jeffes, J.L. McCullough, G.D. Weinstein, P.E. Fergin, S. Nelson, T.F. Shull, K.R. Simpson, L.M. Bukaty, W.L. Hoffman, N.L. Fong (1997). Photodynamic therapy of actinic keratosis with topical 5-aminolevulinic acid. A pilot dose-ranging study. *Arch. Dermatol.*, **133**, 727–732.

85. R.M. Szeimies, S. Karrer, A. Sauerwald, M. Landthaler (1996). Photodynamic therapy with topical application of 5-aminolevulinic acid in the treatment of actinic keratoses: an initial clinical study. *Dermatology*, **192**, 246–251.

86. J.M. Nauta, H.L.L.M. Van Leengoed, W.M. Star, et al. (1996). Photodynamic therapy of oral cancer - a review of basic mechanisms and clinical applications. *Eur. J. Oral Sci.*, **104**, 69–81.

87. A. Kubler, T. Haase, M. Rheinwald, T. Barth, J. Muhling (1998). Treatment of oral leukoplakia by topical application of 5-aminolevulinic acid. *Int. J. Oral Maxillofac Surg.*, **27**, 466–469.

88. G.I. Stables, M.R. Stringer, D.J. Robinson, D.V. Ash (1999). Erythroplasia of Queyrat treated by topical aminolevulinic acid photodynamic therapy. *Br. J. Dermatol.*, **140**, 514–517.

89. I.M. Stender, H.C. Wulf (1996). Photodynamic therapy with 5-aminolevulinic acid in the treatment of actinic cheilitis. *Br. J. Dermatol.*, **135**, 454–456.

90. P. Wolf, H. Kerl (1991). Photodynamic therapy on a patient with xeroderma pigmentosum. *Lancet*, **337**, 1613–1614.

91. C.C. Dierickx, M. Goldenhersh, P. Dwyer, A. Stratigos, M. Mihm, R.R. Anderson (1999). Photodynamic therapy for nevus sebaceus with topical delta-aminolevulinic acid. *Arch Dermatol.*, **135**, 637–40.

92. B. Kukielczak, B. Romanowska, J. Bryk (1999). Gamma radiation and MC540 photosensitization of melanoma in the hamster's eye. *Melanoma Res.*, **9**, 115–124.

93. A. Busetti, M. Soncin, G. Jori, M.A. Rodgers (1999). High efficiency of benzoporphyrin derivative in the photodynamic therapy of pigmented malignant melanoma. *Br. J. Cancer*, **79**, 821–4.

94. (a) P. Wolf, R. Fink-Puches, L. Cerroni, H. Kerl (1994). Photodynamic therapy for mycosis fungoides after topical photosensitization with 5-aminolevulinic acid. *J. Am. Acad. Dermatol.*, **31**, 678–680; (b) R. Ammann, T. Hunziker (1995). Photodynamic therapy for mycosis fungoides after topical photosensitization with 5-aminolevulinic acid. *J. Am. Acad. Dermatol.*, **33**, 541.

95. V.G. Schweitzer, D. Visscher (1990). Photodynamic therapy for treatment of AIDS-related oral Kaposi's sarcoma. *Otolaryngol Head Neck Surg.*, **102**, 639–649.

96. Z.P. Bernstein, B.D. Wilson, A.R. Oseroff, C.M. Jones, S.E. Dozier, J.S. Brooks, R. Cheney, L. Foulke, T.S. Mang, D.A. Bellnier, T.J. Dougherty (1999). Photofrin photodynamic therapy for treatment of AIDS-related cutaneous Kaposi's sarcoma. *AIDS*, **13**, 1697–704.

97. D. Gilson, D. Ash, I. Driver, J.W. Feather, S. Brown (1988). Therapeutic ratio of photodynamic therapy in the treatment of superficial tumors of skin and subcutaneous tissue in man. *Br. J. Cancer*, **58**, 665–667.

98. S.A. Khan, T.J. Dougherty, T.S. Mang (1993). An evaluation of photodynamic therapy in the management of cutaneous metastases of breast cancer. *Eur. J. Cancer*, **29**, 1686–1690.

99. C.P. Lowdell, D.V. Ash, I. Driver, S.B. Brown (1993). Interstitial photodynamic therapy. Clinical experience with diffusing fibres in the treatment of cutaneous and subcutaneous tumors. *Br. J. Cancer*, **67**, 1398–1403.

100. M. Lapes, J. Petera, M. Jirsa (1996). Photodynamic therapy of cutaneous metastases of breast cancer after local application of meso-tetra-(para-sulphophenyl)-porphin (TPPS4). *J. Photochem. Photobiol. B*, **36**, 205–207.

101. M.J. Kaplan, R.G. Somers, R.H. Greenberg, J. Ackler (1998). Photodynamic therapy in the management of metastatic cutaneous adenocarcinomas: case reports from phase 1/2 studies using tin ethyl etiopurpurin (SnET2). *J. Surg. Oncol.*, **67**, 121–125.

102. P. Hillemanns, M. Untch, F. Prove, R. Baumgartner, M. Hillemanns, M. Korell (1999). Photodynamic therapy of vulvar lichen sclerosus with 5-aminolevulinic acid. *Obstet. Gynecol.*, **93**, 71–74.

103. B. Kirby, C. Whitehurst, J.V. Moore, V.M. Yates (1999). Treatment of lichen planus of the penis with photodynamic therapy. *Br. J. Dermatol.*, **141**, 765–766.

104. T. Ouyang, Y. Cheng, X. Xing (1998). [Clinical application of photodynamic therapy combined with non-coherent light (red light) for treatment of port-wine stains]. *Chung Hua Cheng Hsing Shao Shang Wai Ko Tsa Chih*, **14**, 163–165.

105. X.X. Lin, W. Wang, S.F. Wu, C. Yang, T.S. Chang (1997). Treatment of capillary vascular malformation (port-wine stains) with photochemotherapy. *Plast. Reconstr. Surg.*, **99**, 1826–1830.

106. L. Jiang, Y. Gu, X. Li, X. Zhao, J. Li, K. Wang, J. Liang, Y. Pan, Y. Zhang (1998). Changes of skin perfusion after photodynamic therapy for port wine stain. *Chin. Med. J. (Engl.)*, **111**, 136–138.

107. M.W Berns, M. Rettenmaier, J. McCullough (1984). Response of psoriasis to red laser light (630 nm) following systemic injection of hematoporphyrin derivative. *Laser Surg. Med.*, **4**, 73–77.

108. H. Pres, H. Meffert, N. Sonnichson (1989). Photodynamic therapy of psoriasis palmaris et plantaris using topically applied hematoporphyrin derivative and visible light. *Dermatol. Monatsschr.*, **175**, 745–750.

109. W.H. Boehncke, W. Sterry, R. Kaufmann (1994). Treatment of psoriasis by topical photodynamic therapy with polychromatic light. *Lancet*, **343**, 801.

110. P. Collins, D.J. Robinson, M.R. Stringer, G.I. Stables, R.A. Sheehan-Dare (1997). The variable response of plaque psoriasis after a single treatment with topical 5-aminolaevulinic acid photodynamic therapy. *Br. J. Dermatol.*, **137**, 743–749.

111. M.R. Stringer, P. Collins, D.J. Robinson, G.I. Stables, R.A. Sheehan-Dare (1996). The accumulation of protoporphyrin IX in plaque psoriasis after topical application of 5-aminolevulinic acid indicates a potential for photodynamic therapy. *J. Invest. Dermatol.*, **107**, 76–81.

112. M. Raghunath, S. Beissert, T. Schwarz (2000). Successful treatment of chronic psoriatic plaques refractary to conventional therapy. In press.

113. W.H. Boehncke, T. Elshorst-Schmidt, R. Kaufmann (2000). Systemic photodynamic therapy is a safe and effective treatment for psoriasis. *Arch. Dermatol.*, **136**, 271–272.

114. S.G. Parekh, K.B. Trauner, B. Zarins, T.E. Foster, R.R. Anderson (1999). Photodynamic modulation of wound healing with BPD-MA and CASP. *Laser Surg. Med.*, **24**, 375–381.

115. E.V. Ross, R. Romero, N. Kollias, C. Crum, R.R. Anderson (1997). Selectivity of proto-porphyrin IX fluorescence for condylomata after topical application of 5-aminolae-vulinic acid: implications for photodynamic treatment. *Br. J. Dermatol.*, **137**, 736–742.

116. S. Karrer, R.M. Szeimies, C. Abels, U. Wlotzke, W. Stolz, M. Landthaler (1999). Epidermodysplasia verruciformis treated using topical 5-aminolaevulinic acid photody-namic therapy. *Br. J. Dermatol.*, **140**, 935–938.

117. N.J. Petrelli, J.A. Cebollero, M. Rodriguez-Bigas, T. Mang (1992). Photodynamic therapy in the management of neoplasms of the perianal skin. *Arch. Surg.*, **127**, 1436–1438.

Chapter 10

Photodynamic applications in neurosurgery

Herwig Kostron

Table of contents

Abstract

Photodynamic applications, such as photodynamic therapy (PDT) and most recently photodynamic diagnosis (PDD) are currently undergoing intensive clinical investigations as an adjunctive treatment modality for malignant brain tumors.

In this review clinical data are critically analysed for over 490 patients treated by HPD mediated PDT for malignant brain tumors following tumor resection in open clinical Phase I/II trials. Variations in the treatment protocols, photosensitisers and light dose make the evaluation statistically difficult; however, there is a clear trend towards prolonging median survival after one single photodynamic treatment as compared to standard therapeutic regimens. For primary glioblastomas multiforme WHO IV and recurrences the median survival was 22 and 9 months, respectively. This has to be compared to survival of 12 months after conventional treatment and 3 months for recurrent tumors. Other entities such as metastasis, skull base tumors or pituitary tumors responded completely in the majority of cases. PDT was generally well tolerated and side effects consisted of increased intracranial pressure and prolonged skin sensitivity to direct sunlight. Recently PDD was reported for intraoperative tumor detection and fluorescence guided resection which bears great potential for a better and more radical tumor removal.

The development of new sensitisers, new light delivery sources and improved dosimetry will increase the potential of photodynamic application in neurosurgery.

10.1 Introduction

The incidence of malignant brain tumors varies from 4/100000 in the USA to 14/100000 in the Scandinavian countries with an increase up to 70/100000 in the elderly population above 65 years [1].

Within the European Community around 17000 glioblastoma multiforme will be diagnosed and 12000 people will die of glioblastoma annually. The natural life span for patients suffering from glioblastoma multiforme WHO grade IV is around 3 months after diagnosis. The 5 year survival is under 1%. Recurrent glioblastoma demonstrate a median survival of 3 months. Current treatment regimes, such as surgery, chemotherapy and radiotherapy prolong the life span to a median survival of 15 months [2](for histological gradings and different survival of gliomas see Table 1) (Figure 1a, 1b).

Surgery is the first and most important step in the treatment of malignant gliomas and remains the mainstay in therapy. However, radical resection is rarely possible due to infiltrating growth into normal brain parenchyma. Recurrences arise mostly locally

Table 1. WHO gradings, therapeutic strategies, outcome of gliomateous tumors

Histology WHO	Therapy	Recurrence time	Median survival
Glioblastoma multiforme IV ⎫ Anaplastic Astrocytoma ⎭	Surg, CHT, XRT	6 months	15 months
Astrocytoma III	Surg, CHT, XRT	18 months	3 years
Astrocytoma II	Surg, XRT	3 years	6 years
Astrocytoma I	Surg	8 years	10 years

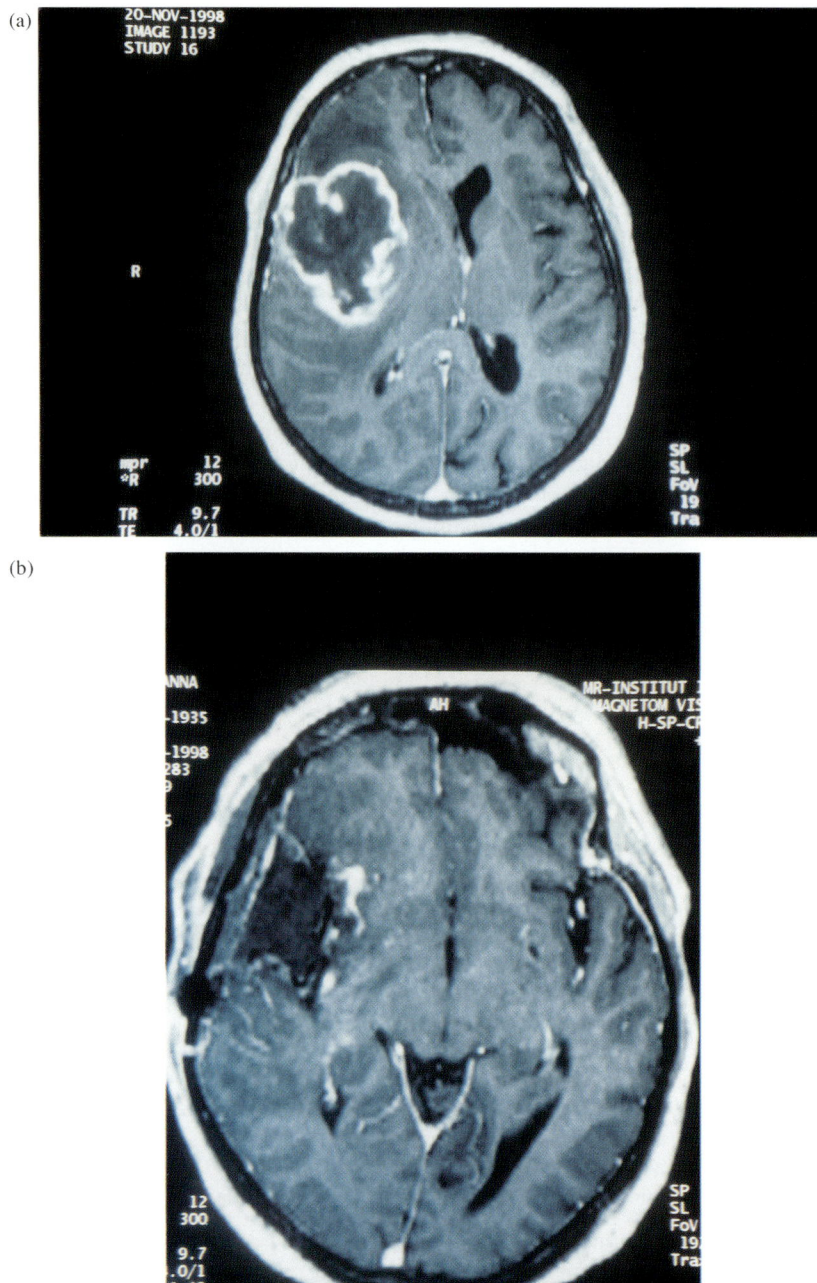

Figure 1. (a) Typical appearance of glioblastoma WHO IV right temporal lobe; note the shift over the midline and the peritumoral oedema. (b) The tumor after resection; note the existing oedema, most likely still containing residual tumor-guerrilla cells.

from tumor cells (guerrilla or satellite cells) embedded in the area of oedematous or normal brain adjacent to the tumor ("brain adjacent to tumor region" BAT) [3–5] (Figures 2a–2c).

New treatment modalities as well as intraoperative diagnosis have to be developed to enhance therapeutic selectivity and to reduce the amount of residual tumor. Intraoperative therapies currently include topical application of cytostatic agents or intraparenchymal application of gene modified cells such as Herpes Thymidine Kinase-suicide gene [6–8].

Photodynamic therapy (PDT) is currently undergoing intensive clinical investigations as an adjunctive treatment for malignant brain tumors. PDT is based on a higher accumulation of a photosensitiser in malignant over normal tissue with low systemic toxicity [9–11]. Subsequent light activation induces photooxidation followed by selective tumor destruction via vascular or direct cellular mechanisms [12–14].

Therefore PDT offers a more selective treatment of such malignancies as compared to other currently available treatment modalities and seems to be a logical concept for brain tumors infiltrating into normal brain [15,16]. Clinical studies demonstrated a benefit for the patients treated with PS mediated PDT in terms of prolongation of median survival time as well as quality of life [17–28].

In the past, intraoperative fluorescence guided delineation had been under investigation with fluorescine and tetracyclines or autofluorescence [29,30]; however, the sensitivity and specificity were too low to be of clinical significance, especially for neurosurgery. Modern intraoperative imaging techniques now include MRT and CT, neuronavigation and ultrasound, all very expensive tools.

The fluorescence properties of photosensitisers such as HPD, 5-Amino-Levulinic-Acid (5-ALA) and chlorin compounds such as meta-TetraHydroxy-PhenylChlorin (mTHPC) are exploited for Photodynamic Diagnosis (PDD) [31,32]. 5-ALA has been demonstrated to have a high selectivity and specificity for bladder tumors and skin lesions and is currently being introduced in clinical practice [33,34]. Intraoperative photosensitiser mediated photodynamic diagnosis (PDD) has been reported in first clinical use with promising results [35,36]. This intraoperative photodynamic diagnosis allows an optical delineation of normal and malignant tissue which facilitates intraoperative orientation and allows fluorescence guided resection which results in a more radical resection. This fact might affect survival by itself.

In cases where functional areas of the brain are infiltrated by tumor, tumor tissue has to be left behind in order not to impair neurological function and quality of life. Simultaneous intraoperative PDT offers a logical supplement to PDD according to the slogan "to see and to treat" [21,35].

The intention of this paper is to critically review the available data and draw conclusions the future potential of PDD and PDT for applications in neurosurgery.

10.2 Photosensitisers

The photosensitiser for brain tumors must have a different profile than for any other indication. The tumor burden is much larger and radical resection is seldom possible

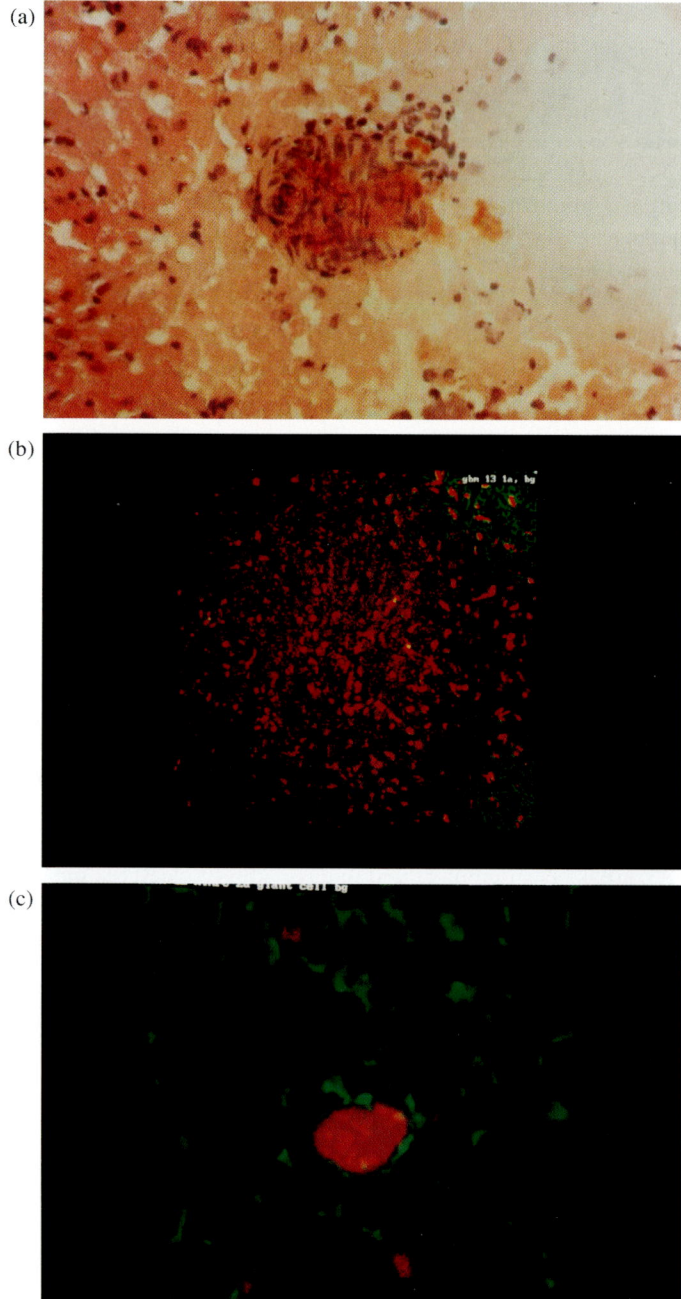

Figure 2. (a) HE histology demonstrates a tumor island within normal brain tissue. (b) Confocal laser scanning microscopy of a GBM after sensitisation with mTHPC fluorescence localised within the tumor cells, indicating selective uptake of the sensitiser. (c) Giant tumor cell imbedded in normal brain tissue, demonstrating selective uptake into tumor cells.

because of infiltration of tumor into normal brain structures of up to 3 cm and deeper [5].

The brain as an organ is unique and consists of three major sections which are the brain tissue, tumor tissue and the vasculature. The ideal neurosurgical sensitiser should: (1) have a high selectivity for tumor and tumor cells, sensitising only tumor tissue and tumor island (guerrilla cells) embedded in normal brain tissue; (2) have high phototoxicity; (3) be photoactivated in the near-infrared range, 700 nm and beyond; (4) have strong fluorescence properties; (5) not cross the blood–brain barrier (BBB); and (6) have no systemic toxicity.

The first generation photosensitiser Hematoporphyrin derivative (HPD) is a complex mixture of various porphyrins [9], which was used for most of the basic experimental work and almost exclusively in the clinical brain tumor studies. HPD has its optimum absorption between 628 and 632 nm and allows penetration depths up to 7 mm depending on the tissue [37,38]. The dose in clinical use is 2 mg kg^{-1} injected intravenously. Energies required range from 60 to 260 J cm^{-2}. The ratio of the concentration in tumor to normal brain ranges from 2.5–4:1 and in animal experiments it ranges up to 12:1 [39,40].

meta-Tetra-hydroxyphenyl-chlorin (mTHPC) has been used as a second generation sensitiser with a higher tumor to brain ratio of 10:1 at a dose of 0.15 mg kg^{-1} and light doses ranging up to 20 J cm^{-2} [41–45]. Experimental and first clinical results with mTHPC have demonstrated less skin toxicity and effective cytotoxicity at wavelengths of 652 nm and light doses of 20 J cm^{-2}.

Light sensitisation of the skin was observed for four weeks; however, skin toxicity could be more severe in the first two weeks. Next to mTHPC-mediated PDT trials there are also intensive investigations elsewhere for intraoperative diagnosis and fluorescence guided tumor resection. Bacteriochlorin A has been reported to be more effective at wavelengths of 735 nm; however, there are no clinical data [46].

5-ALA induced protoporhyrin IX produces excellent fluorescence for diagnostic purposes and is activated at 635 nm. Sufficient cell kill of superficial tumors is achieved at energies of up to 100 J cm^{-2}. The ratio of ALA concentration in tumor to normal brain is around 4:1; however, there are no clinical data for treatment [32,40,42,47]. Skin phototoxicity of ALA is only few hours after application.

A promising approach is the use of infrared sensitisers or sensitisers which are activated by ultrasound or by two-photon activation [48–50]. Chloraluminium-phthalocyanine (AlClPc), benzoporphyrin (BPD), tin ethyl purpurin (SnET2), texaphrin, methylene blue, bacteriochlorine and boronated porphyrins are further sensitisers with great potential for use in brain tumors [15,49,51–58].

The search for new sensitisers is intensive in order to increase tumor selectivity, fluorescence detection, phototoxicity, penetration depths and reduce skin toxicity.

10.3 PDT mechanisms

The mechanisms of PDT are based on photooxidative reactions and the primary target depends on the pathology, the absorbency and chemistry of the sensitiser and the incubation time [9,13,14].

Three different main mechanisms at the cellular level are involved in PDT mediated cytotoxicity. Parenteral injection, whether intravenously or (in animals) intraperitoneally, results in primary vascular damage [12], whereas the direct intratumoral injection results in a more pronounced direct cellular effect [59,60]. The third mechanism is mediated by cytokine modulation. Transforming growth factor, fibroblastic growth factor, interleukin-1 and interleukin-6 and also PDGF and TNF plays a role in mediating the photooxidative cytotoxic process [61,62]. Il-6, an auto-endocrine stimulator for glial tumors, is significantly reduced in cell cultures after PDT [63,64]. Oxidative stress activates early response genes [65] which further modulate apoptosis depending on the cell lines employed [55,66,68].

Since porphyrins are incorporated in lipoproteins, the low-density-lipoprotein (LDL) receptor pathway is one important factor for the selective accumulation of porphyrins by tumor cells and by subcellular structures such as lysosomes or mitochondria [69,73]. This mechanism is particularly important for brain tumors as malignant and reactive glia cells express significant amounts of LDL receptor related protein [72]. The effect of photoproducts must be eliminated further [31,43,74].

All these mechanisms depend strongly on the type of the tumor cells which are investigated. These mechanisms outlined above could well explain that the PDT effect exceeds by far the penetration depths of the activating light.

10.4 Effect of PDT on normal brain

Despite a high selectivity of PS towards tumor cells, injury to normal glia and neurones had been observed in the normal brain [53,59,75–77]. This damage depends on the type of sensitiser used, the concentration of sensitiser, time interval of sensitisation to light exposure and light density. An important role played by the blood–brain barrier is to protect the normal brain from toxic substances. Whereas the blood–brain barrier does not exist within the tumor, and the surrounding oedema, this barrier is intact in the brain embedding tumor island [5,40]. However, the evidence is mostly derived from experimental work. HPD is taken up by normal brain in a dose dependent fashion, varying between 0.2 μg g^{-1} to 1.2 μg g^{-1} wet weight brain tissue at a dose of 10 to 20 mg kg^{-1} bodyweight HPD [38,59]. Twenty mg kg^{-1} and 100 J cause death of the animals due to severe swelling [77]. Fluorescence diagnosis demonstrated the presence of the sensitiser mostly along the fibre bundles of the white matter. Upon light activation, first break down of blood–brain barrier, swelling of astrocytes and neurones are observed and after two days coagulation necrosis occurs in a dose dependent fashion at concentrations higher than 5 mg HPD [22]. Intratumoral or intraparenchymal application of the sensitiser does not exhibit selectivity [59,60]. Whereas in experimental studies high doses of HPD have been used, in clinically relevant doses of 2 to 5 mg kg^{-1} only a few reports have described morphological changes in the normal human brain [39].

Focal necrosis around the vessels 24 h after intraarterial injection was found in only one series [22], whereas oedema post PDT was described by almost all authors, starting with Perria in 1980 [27,78]. Muller reported a significant increase of intracranial pressure despite avoiding hyperthermic effects [23]. After stereotactic PDT, cerebral oedema was observed in most of the cases [18,26,76]. The amount of postoperative swelling correlates with the residual tumor volume, so the tumor resection has to performed to the utmost possible extent.

10.5 Interaction of PDT/chemotherapy/drug therapy/ hyperthermia

Since the introduction of modern chemotherapeutic agents to neuro-oncology, some tumors are responding to chemotherapy and it is the third mainstay of adjuvant therapy after surgery and ionising radiation [2].

There are experimental data which show a synergistic or a potentiating effect of chemotherapeutic agents on PDT [80]. The combination of ACNU and HPD-mediated PDT results in significantly increased cytotoxicity in the 9L gliosarcoma [81]. This effect could further be increased by adding hyperthermia as a third treatment modality. The RIF-1 tumor proved to be insensitive to PDT and to doxorubicin but sensitive to cisplatin with no increased cytotoxic effect when both modalities were combined. In the EMT-6 tumor model all three modalities showed a mild effect when used independently, doxorubicin enhanced significantly the effect of PDT, whereas cisplatin did not [82]. Mitomycin C potentiates the effect of PDT in adenocarcinoma [83]. There are no reported clinical data on the interaction of PDT and chemotherapy.

Steroids (dexamethasone) are widely used in neurosurgery for the treatment of tumor-associated oedema. Its mechanism in closing loose junctions in the endothelium and tightening the blood–brain barrier might interfere with the uptake of sensitisers into the tumor tissue. Patients on higher doses than 12 mg dexamethasone daily demonstrated a lower uptake of radiolabelled 111-In-Photofrin. Fluorescence was also decreased in animal models such as Lewis lung carcinoma and C6 glioma [73].

Furthermore an ameliorating effect of methylprednisolone on PDT was observed when given after light irradiation but not when given prior to irradiation [84].

A promising approach is the monoclonal antibody conjugate targeted PDT such as conjugates to anti-EGF or conjugates to anti-transferrin receptor [85,86] or the targeted therapy against the thymidin kinase system [6].

A synergistic effect of hyperthermia of 42 to 45°C was observed when delivered after or simultaneously to PDT [87–90] which could be of advantage in treating brain tumors. The PDT might be of advantage in the treatment of tumors not responding to chemotherapy because cells which express multi-drug resistance features are less likely to be cross-resistant to PDT [91].

10.6 Interaction of PDT and ionising radiation

Ionising radiation is a standard therapy in the postoperative course after surgery of malignant brain tumors [1,19,21,79].

We have investigated the interaction of HPD-mediated PDT and ionising radiation using the rat gliosarcoma model 9L which was subcutaneously implanted. In low doses of HPD after intraperitoneal injection an additive effect of both treatment modalities could be observed, whereas direct injection with high intratumoral concentration and high light doses of 120 J cm^{-2} and 4 Gy resulted in a significantly greater response, indicating a potentiating mechanism. The effect was more pronounced when PDT was followed by XRT within 30 min [21,22]. The underlying mechanism is thought to be the inhibition of the potential lethal damage induced by PDT.

In an early series at our own institution patients with glioblastomas had been treated with one single dose of 4 Gy of electrons within 30 min after PDT [21]. The results remained unchanged to that without immediate X-radiation treatment, therefore this treatment protocol has not been continued. Radiotherapy was commenced, in addition, in all patients with primary untreated glioblastomas within 10 days of surgery and PDT. We did not see any increased side effects from this radiotherapy. In another series conventional radiation treatment of 60 Gy was started up to four weeks after PDT with no side effects observed [19].

Gamma knife irradiation has not been used in combination with PDT.

10.7 Instrumentation

Studies had been initially performed with argon dye laser systems or xenon arc lamps with adequate filtering. Nd/YAG KTP frequency doubled dye lasers have come to market with low maintenance and high reliability, but high costs. A diode laser would be the best in terms of costs, small size and portability; however, it bears the disadvantage of only one wavelength. However, to also activate photo-products which are formed during light activation, conventional light sources with a wider spectrum could be beneficial.

Light delivery and dosimetry is a critical point in PDT and especially in neurosurgery where lesions which are not ideally geometrically shaped or superficial, but irregular and large in volume have to be treated [38].

An inflatable balloon is employed to facilitate dosimetry (Figure 3). In general the cavity or the balloon is filled with a 0.1% concentration of intralipid solution or lower to achieve a homogeneous light distribution [21,23]. The superficial light delivery is performed mostly by bare fibres. Interstitial light application is either performed by bare fibres or by cylindrical fibres, the latter allowing higher power density (350 mW cm^{-2}) without producing carbonisation at the fibre tip [19,20,23,26]. Intraoperative monitoring of PDT dosimetry has been reported [97].

10.8 Photodynamic diagnosis and fluorescence guided resection

In addition to the phototoxic properties, the fluorescence abilities of the photosensitiser can be used for optical discrimination of normal and malignant tissue, allowing intraoperative photodynamic diagnosis and fluorescence guided resection. This enables the surgeon to remove the tumor more radically, something that improves survival by itself.

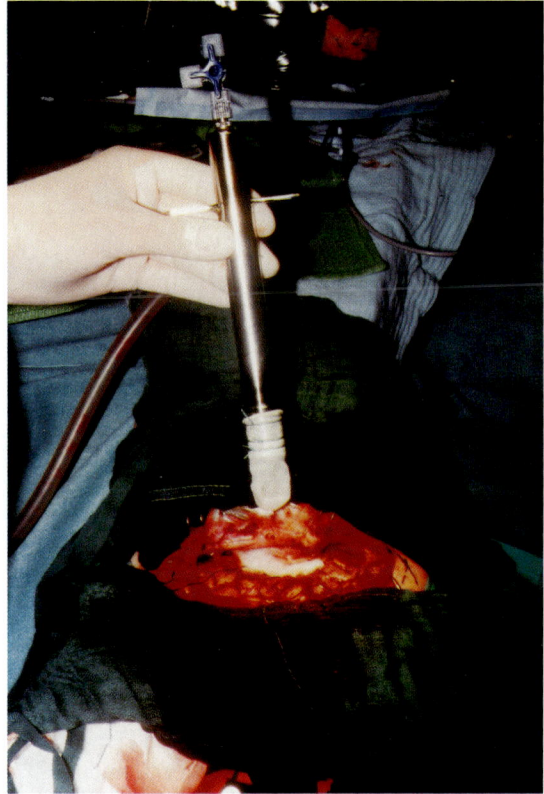

Figure 3. Balloon device for surface irradiation.

Intraoperative fluorescence is induced by a UV light source at 370–440 nm, which spectrally matches the main absorption peak of the sensitiser for fluorescence excitation and is delivered via a liquid light-guide to the surgical microscope (Figure 4).

The induced fluorescence light is collected by the same optical fibre, separated from the blue excitation light by passing through a dichromic beamsplitter and focused to the entrance slit of a modified CCD camera and a highly sensitive spectrograph in connection with an electric tuneable optical filter. In regions of faint fluorescence intensities (especially in the region of the tumor border), spatially resolved spectra can be taken with a very sensitive spectrometer to delineate tumor borders. The pictures are converted digitally by a standard frame grabber and processed in real time. A standard neurosurgical microscope can be modified accordingly for fluorescence detection, which is used for fluorescence guided resections. Normal white light illumination of the surgical microscope can be switched to blue light excitation with simultaneous observation by introducing a proper observation filter to block most of the light. The induced fluorescence can be seen directly through the observer light path [33,35].

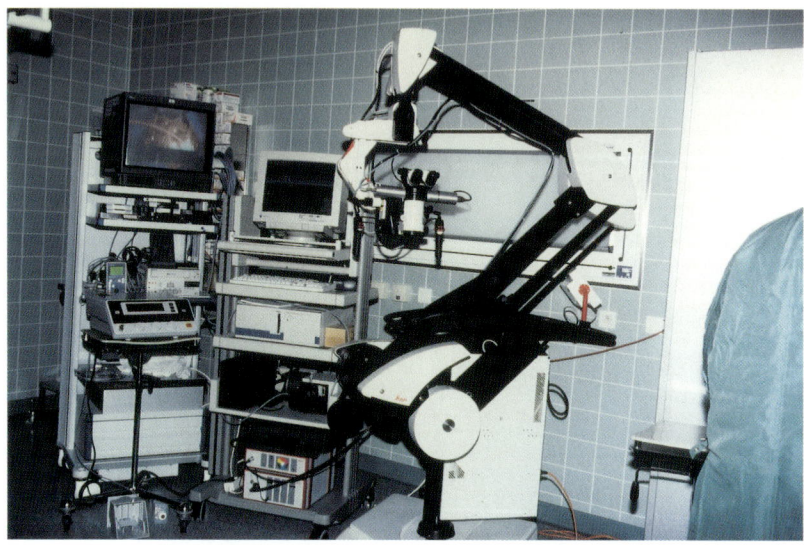

Figure 4. Technical set up for intraoperative fluorescence diagnosis PDD. Note the standard neurosurgical microscope, the fluorescence light emitting system, the spectrometer and the diode laser.

The confocal laser-scanning microscopy (Laser-scanning microscope 410 Zeiss) is performed on fresh frozen slices, the excitation wavelength is 488/514 nm of the internal argon laser, the fluorescence detection is performed at >590 nm within the red channel and phase contrast within the green channel (Figure 5a, 5b).

10.9 Methods and patients

Indications for PDT are recurrent malignant brain tumors. Slow growing tumors such as low grade astrocytomas or other benign lesions are not currently indications for PDT or PDD because of a lack of experimental and clinical data. Skull base tumors and pituitary tumors are a good indication for PDD/PDT because they are not protected by the BBB. Most recent case reports also include metastasis, malignant meningeomas and recurrent pituitary tumors.

Patients presenting with primary or recurrent brain tumors were enrolled in clinical Phase I/II trials which are detailed in Table 2. The patients received either various formulations of hematoporphyrin derivative (HPD, Photofrin I, DHE, Photofrin II, Photosan 3) or mTHPC 24 to 96 h prior to surgery.

At our institution the following protocol is used:

After fulfilling the criteria for informed consent the patients were sensitized with HPD (Photosan 3) 2.5 mg kg^{-1} or most recently mTHPC (Foscan®) 0.15 mg kg BW^{-1} 24–96 h prior to a standard craniotomy [20,21]. Steroids were withdrawn 2 to 3 days prior to sensitisation, so far tolerated. A standard craniotomy was performed and maximal tumor resection was performed. The anaesthesia consisted of a neuroleptanalgesia

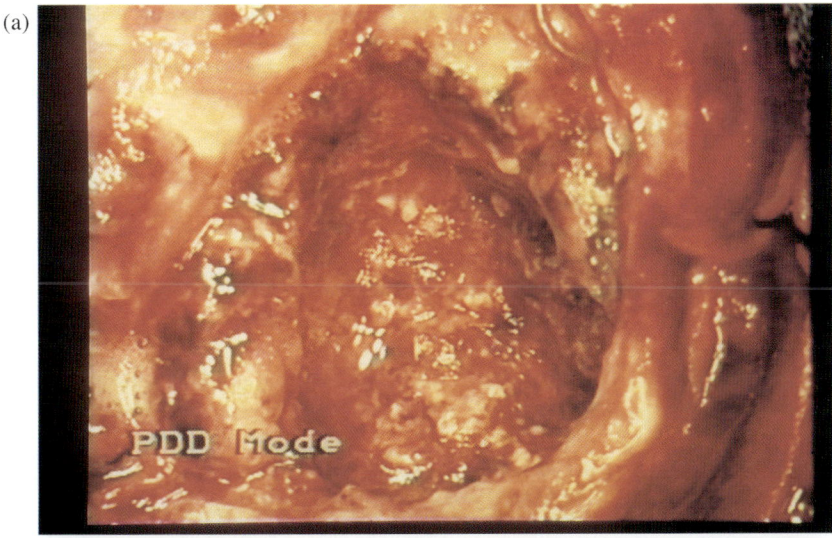

Figure 5. (a) Resection cavity under white light observation with "macroscopical radical resection". (b) Resection cavity under UV light, note the fluorescence at the depth of the resection cavity, indicating residual tumor.

avoiding barbiturates [93]. Patients sensitized by mTHPC underwent additional intra-operative photodynamic diagnosis and fluorescence guided resection.

PDT was performed by a KTP pumped dye laser (Laser Sonic) or a diode laser emitting at 652 nm (Diomed 1W). Light was delivered by bare fibres coupled into a modified balloon system (Lajat/Patrice), by a spherical distributor or by a 20 mm long cylinder for interstitial treatment (Medlight Switzerland). The power density varied from 2500 mW s^{-1} (KTP-dye) to 870 mW s^{-1} (diode laser). For interstitial

Table 2. HPD-mediated PDT; Clinical studies and results

Author	Histology	Drug	Light/method	Results
Perria (1980)				
recur $n = 9$	GBM	HP < 9 J cm^{-2}	HeNe laser	early recurrence
Laws (1981)				
recur $n = 44$ metastasis $n = 1$	HG glioma	HPD	Ar/dye	early recurrence
McCulloch (1984)				
$n = 10$ metastasis $n = 4$	HG glioma	HPD	Ar/dye/lamp	2/10 cr
Layat (1986)				
$n = 7$	GBM	HPD stereotactic	Arg-Dye	12 mo
Perria (1988)				
$n = 8$	HG glioma	HPD	Ar-dye/lamp	6 mo med. survival
Wharren (1991)				
$n = 6$	HG glioma	HPD 180 J cm^{-2}	lamp	10 mo
Powers (1991)				
$n = 6$	HG glioma	HPD 400 J cm^{-2}	stereotactic	4 mo
metastasis $n = 1$				1 (nr)
Originato (1993)				
recur $n = 15$	GBM	HPD		7 mo
HG glioma				
recur $n = 6$				7 mo
metastasis $n = 1$				1 (cr)
Kaye (1998)				
prim $n = 36$	GBM	HPD	goldvapor <260 J cm^{-2}	24 mo med. survival
recur $n = 39$				10 mo
Glioma III $n = 24$			N	
Muller (1996)				
recur $n = 64$	GBM	HPD > 1700 J	Xe-lamp	9 mo
prim $n = 45$		<175 J cm^{-2}		17 mo MS
metastasis $n = 5$				1 (nr), 4 (pr)
Kostron (1996)				
prim $n = 12$	GBM	HPD 240 J cm^{-2}	argon-dye	19 mo MS
recur $n = 39$	HPD			7 mo
metastasis $n = 5$				1 (cr), 1 (pr), 3 (nr)
meningeomas $n = 3$				3 (pr)
Kaneko (1999)				
$n = 28$	HG glioma 3 mg (kg BW)$^{-1}$	HPD	argon-dye	cr
metastasis $n = 5$				pr
Marks (1999)				
$n = 12$	pituitary	Photofrin	diode 632 nm	cr
Kostron (1998)				
prim $n = 2$	GBM	mTHPC	diode	9 mo
recur $n = 18$		0.15 mg kg^{-1}	652 nm	7 mo
metastasis $n = 4$				24 cr
skull-base $n = 4$				24 cr

GBM: glioblastoma multiforme; HG: high grade glioma; cr: complete response; recur: recurrent tumors.

treatment the power density was 350 mW s^{-1}. For treatments of HPD sensitized patients the light dose was increased from 20 J cm^{-2} for the first patients the majority receiving 240 J cm^{-2}. Treatment times ranged from 40 to 75 min and 9 to 20 min for the two types of sensitizer control of lasers, respectively.

During light irradiation 100% oxygen was used to ventilate patients and a bonus of 40 mg dexamethasone had been given after light treatment, followed by the generally used dose regimen for post operative steroids treatment (16 mg regressing doses for 10 days).

Postoperatively patients were kept in ambient room light. After mobilisation patients were slowly exposed to normal sunlight to allow bleaching of the PS and skin pigmenting. A light meter was provided for the patient's light exposure control. Patients were monitored by clinical exams and CT scans every 3 months.

10.10 Clinical trials and results (Table 2)

Since Perria started PDT for brain tumors in 1980, over 490 patients have been treated world-wide since then [15,18–21,24–28,47,94–96]. In general the patients were sensitised with HPD in various formulations and only one study used mTHPC as sensitiser [16]. The majority of the patients were sensitised parenterally and few patients received intratumoral or intra-arterial sensitisation [22].

Light irradiation was performed with various light sources such as photo-radiation lamps, filtered Xe lights, dye laser, gold vapour, KTP dye laser and diode lasers. Patients presenting with primary glioblastomas underwent 45 Gy of irradiation within four weeks. The light dose was initially low at 70 to 180 J cm^{-2}, which was finally increased to 240 J cm^{-2} for the majority of the patients. Muller conducted the only clinical trial engaging a light-dose escalation [23]. Most patients underwent standard craniotomies and open tumor resection. A stereotactic approach was chosen by several authors with somehow disappointing results [26,95]. Muller observed complete response with excellent survival in cases where a cystic geometric tumor cavity allowed a very homogeneous light distribution.

The results of the reported cases are difficult to evaluate because the histology is not always detailed. High grade gliomas consist of WHO grade IV and grade III gliomas (see Table 1).

Although over 490 patients have been treated world-wide, since Perria started PDT for brain tumors in 1980, this number is relatively low compared to the number of patients treated with lung or skin cancers. However, the incidence of brain tumors is lower and their treatment by PDT is more complicated since it requires highly sophisticated surgery and instrumentation.

10.10.1 HPD mediated PDT

10.10.1.1 Primary glioblastoma
In our personal series 12 primary glioblastomas (mean age 64 years) were treated with various sensitisers of the first generation (Photofrin I, HPD-Adelaide,

Photosan 3) 24 to 72 h prior to treatment [21,22,39,52–55,68,89,83,94,97]. The activation source was a PDT lamp (Swarovsky) in the first five patients, then followed by an argon-dye laser (Aurora-M) and a KTP dye laser. The light dose ranged from 15 to 260 J cm^{-2} delivered with a power density of up to 1600 mW s^{-1}. Immediately after light irradiation and wound-closure a single Co-60 irradiation of 400 rad was given to the tumor bed. These patients received in addition a conventional radiotherapy of 55–60 Gy as well as conventional chemotherapy (nitrosourea and cytosin arabinosid). The median time to recurrence was 13 months. Ten patients underwent retreatment with PDT without any other treatment because of recurrences. Their median survival was 10 months. There was no adverse effect from a second photosensitisation and photoradiation. The total survival of primary glioblastomas was, in our series, 19 months (range 0.5 to 27 months).

The largest series, encompassing more than 130 high grade gliomas, was reported by Kaye [19,20] . Primary gliomas received 45 Gy in addition with a median survival of 27 months. It is, however, not stated how many glioblastomas and anaplastic gliomas were involved. In a Canadian study 50 patients were treated with PDT and adjuvant radiotherapy of 45 Gy. The median survival was 9 months [23].

10.10.1.2 Recurrent glioblastomas

In our own series 39 patients were treated photodynamically for recurrent glioblastomas. Of those, 25 were first recurrences. The median time to first recurrence without any other treatment was seven months and the median survival time after PDT was 9 months (range 3 to 18 months). Fourteen patients were reoperated on after 12 months having presented with a first recurrence. No further treatment was commenced. These patients suffered another recurrence within 3 months. After a third surgical procedure and PDT the median time to recurrence was 6 months (3 to 8 months).

Kaye et al. reported a median survival of 18 months for patients with recurrent high grade gliomas and PDT. Muller (1998) reported on 64 recurrent glioblastomas which were subject to a dose escalation study. Patients receiving more than 1700 J had a median survival of 9.2 months, whereas those receiving less than that had a survival of 6.6 months [23].

Kaneko [18] had treated 22 patients mostly bearing glioblastomas with a median survival of 8 months. Fifteen recurrent glioblastomas were treated by Origitano [21] with multiple interstitial fibres on the basis of a computed 3-D image. The light dose was 100 J cm^{-2} with a median survival of 6 months; 2 patients lived longer than 13 and 16 months. In 1986 Lajat introduced the stereotactic approach for the first time in PDT and treated 7 recurrent glioblastomas photodynamically with a median survival of 12 months [89]. Stereotactically placed interstitial fibres with a light dose of up to 400 J cm^{-2} were also used by Powers [95] in 6 recurrent gliomas with a median survival of 2 months.

10.10.1.3 High grade gliomas WHO III

Thirty-three recurrent anaplastic astrocytomas were treated photodynamically with energies varying from 45 to 175 J cm^{-2} [19]. Kaye reported on 24 patients with good results not having reached median survival [19,28].

Malignant mixed oligo-astrocytomas and ependymomas had a 2 year survival of 37% and 75% respectively.

10.10.1.4 Brain tumors of other origin
Three recurrent malignant meningiomas were treated with 60, 120 and 260 J cm^{-2} and the median survival was 6, 15 and 23 months, respectively. The first patient died due to metastasis without evidence of local tumor growth.

10.10.1.5 Pituitary tumors
These are benign tumors; however, when they recur they produce anatomical malignancy.

Marks et al. treated 12 patients because of repeated recurrence and treatment failures. All cases responded favourably, all showing no progression or recurrence of tumors for $2^1/_2$ years [94,98].

10.10.1.6 Metastasis
Three patients suffered metastasis from melanotic melanoma which had a median survival after PDT of 12 months. One patient suffered from metastasis of a squamous carcinoma of the lung with two recurrences every 3 months despite radiotherapy after the first relapse. The tumor recurred again 6 months after reoperation and PDT. A second patient suffered from metastasis of an adenocarcinoma of unknown primary location with recurrence within 3 months after surgery and no adjuvant treatment. After reoperation, PDT and conventional radiotherapy of 45 Gy the patient has been recurrence free for over 7 years.

In addition to the above institutional series 15 metastasis were treated in various centres.

All except for melanoma metastasis showed complete or partial response.

10.10.1.7 HPD extraction from tissue
The extraction of hematoporphyrin from glioblastomas in 6 different patients resulted in significant variations ranging from 1.46 to 4.00 μg g^{-1} wet weight (mean concentration 2.78 $\pm$ 1.3 μg, $n = 6$). The normal tissue from BAT region contained 0.92 $\pm$ 0.2 μg g^{-1} (range 0.6 to 1.2 μg g^{-1}, $n = 3$) [34]. The concentration in another series contained higher HPD levels in the tumor (5.9 μg g^{-1} wet weight) and the concentrations in normal brain tissue amounted to 0.2 μg g^{-1} [15,23]. The same authors reported decreasing concentration in decreasing malignant grades, anaplastic glioma WHO grade III (2.4 μg g^{-1}) and in low grade glioma WHO grade II (1.6 μg g^{-1}) respectively.

The survival time for the patients in which the HPD concentrations were analysed at our institution [18] were as follows (the concentration of sensitiser in parentheses): 9 months (1.46 μg g^{-1}), 17 months (2.4 μg g^{-1}), 19 months (4.0 μg g^{-1}), 22 months (3.5 μg g^{-1}), 24 months (2.8 μg g^{-1}), and 27 months (3.2 μg g^{-1}).

Melanomas contained 2.10 μg g^{-1} and 0.90 μg g^{-1}. The corresponding survival time was 14 and 9 months, respectively. Two metastases of an adenocarcinoma (one from lung cancer, the other of unknown origin) contained 0.96 and 0.90 μg HPD per g of tumor, respectively. The first patient demonstrated a 6 months survival, the other patient has remained tumor free for 7 years.

10.10.2 mTHPC-mediated PDT

Twenty-two patients with brain malignancies were reported in this trial (see Table 3) [20].

Two patients with primary glioblastoma multiforme WHO grade 4 were treated in addition with 60 Gy of XRT and died 9 and 15 months afterwards. The patients with recurrent GBMs (n = 14) demonstrated a median time to progression of 4 months and a median survival time of 7 months, four patients are living with a follow up of up to 24 months. The patients with metastasis demonstrated complete tumor control in two cases for up to 28 months, 2 were progressing 6 months after treatment and were lost to follow up. One skull base tumor demonstrated a complete response with a follow up of 24 months, the other relapsed within 7 months.

All patients tolerated the mTHPC-mediated treatment well, one patient suffering brain swelling of the treated area.

Two patients experienced severe toxic reaction to sunlight due to unintentional exposure to direct sunlight. The sunburn required conventional treatment and was resolved within 6 days.

10.10.3 Photodynamic diagnosis and fluorescence guided resection

Eighteen out of 22 patients underwent mTHPC-mediated intraoperative photodynamic diagnosis and fluorescence guided resection. Two were false negative (selectivity 89%) because of too low a sensitivity of the primary instrumentation and three were false positive due to radiation necrosis and inflammatory reactions (specificity 84%). Additional spectroscopy applied in the last cases increased significantly the sensitivity of our system. Postoperative MR studies within 48 h revealed complete tumor removal in 10 cases, unintentional residual tumor in three cases, and in five patients tumor had to be left because of invasion in functional structures [30]. In no cases was normal brain resected (Table 4).

In another study 10 cases of ALA-mediated PDD were reported which demonstrated a specificity of 100% and a selectivity of 85% [36].

Table 3. *m*-THPC-PDT; patients and results

Histology	*n*	Time interval after 1 treatm	Time interval after PDT	Results survival median
Prim GBM	n = 2	0	8 mo	12 mo (9, 15)
Recur GBM	n = 14	6 mo (8,6 mo)	4 mo (4 mo)	9 mo (4 alive 24,14,12,10 mo)
Metastasis	n = 4	9 mo (10 mo)	6 mo (5 mo)	27 mo 2 cr
Skull base	n = 2	multiple recur	7 mo	24 mo 1 cr

GBM: glioblastoma multiforme; HG: high grade glioma; cr: complete response; recur: recurrent tumors.

Table 4. mTHPC extraction from tissue (μg per g tumor wet weight)

Histology	mTHPC	Tumor edge	Normal	FL
GBM	0.6 ± 0.1	0.03 ± 0.01	n.a.	++
GBM	1.0 ± 0.3		n.a	++
GBM	0.3 ± 0.04		n.a.	0
GBM	0.11 ± 0.03		n.a.	0
GBM	0.87 ± 0.3	0.17 ± 0.01	n.a.	n.a.
Meta	0.19 ± 0.04		n.a	+
Ca Skull base	1.98 ± 0.28	0.14 ± 0.01	n.a.	+++

Tissue extraction was performed by L. Lilges [92].

10.11 In vitro sensitivity testing

Biopsies of human brain tumors were subjected to in vitro sensitivity testing in primary cell cultures [64]. After incubation with increasing concentrations of HPD (Photosan 3) and light at 60 J cm^{-2} the surviving cell population was determined. HPD concentrations of 25 and 10 μg ml^{-1} of medium and 60 J cm^{-2} were necessary to achieve a growth inhibition of 75% in different glioblastomas. One meningioma, a benign tumor, was the most sensitive, demonstrating a 75% growth inhibition at only 2 μg ml^{-1}. Similar results for meningiomas were found by Marks [99]. Specimens of brain tumors of patients undergoing PDT were grown in culture pre and post radiation which demonstrated a significant decrease of growth after light irradiation. The interleukin-6 formation in one glioblastoma culture was 2400 ng ml^{-1} prior to radiation and significantly decreased to 24 ng ml^{-1} after irradiation of 240 J cm^{-2} [59]. The survival of this particular patient was 22 months.

10.12 Summary

The purpose of this review is to evaluate the current and future position of photodynamic applications in neurosurgery.

All centres reported an improved survival as compared to their historical control ranging from 12 to 27 months. All of these patients were treated with surgery and one single photodynamic treatment and in cases of primary tumors additionally with radiotherapy. Taking into consideration the median survival of 12 months after surgery and radiotherapy for patients with primary glioblastomas [1], the 3 to 18 months gained in additional lifespan after one single additional photodynamic treatment demonstrated that PDT bears the potential to be effective against malignant gliomas. Currently there is an ongoing randomised controlled trial conducted by Canadian and American investigators; however, the results will only be available in around 3 years time.

Despite an immediate histological and biological response after PDT, the tumors always recur locally and there is no difference in the recurrence patterns to after conventional treatment modalities, which are applied over 6 weeks for radiotherapy and up to 12 months for chemotherapy.

The temporary response of gliomas to PDT is owing to physical as well as biological factors. The latter will not be discussed here. We treat a large volume of tumor even after debulking, due to tumor infiltration into normal brain. This area – brain adjacent to tumor (BAT) – is always the site of tumor recurrence. In this critical area a low light dose and low sensitiser dose coincide. To overcome this problem and to circumvent the blood–brain barrier the sensitiser was injected directly into the tumor cavity, which resulted in significant higher intratumoral concentration with a ratio of 12:1 [22,60]; however, the clinical results did not improve, but, as outlined above, the survival of the patients did not correlate to the concentration of the sensitiser in the tumor. Therefore, most likely a low penetration of activating light is the most realistic reason for recurrences on that site. To the contrary, interstitial radiation as performed by Powers, Origitano and Kaneko improved depth penetration to about 2 cm and dosimetry at this critical area but did not improve the results. The best results were actually achieved by Kaye using a high dose of superficial irradiation. Muller irradiated patients within 12 h of sensitisation to prevent the rapid washout of sensitiser at the BAT region. However, only after increasing the light dose above 1700 J were the results improved. Furthermore the results were better if the shape of the resection cavity allowed a very homogeneous irradiation [23].

Since light dosimetry is critical for PDT response, underdosing might often be the cause of therapeutic failures; computer aided planning with 3D reconstruction of the tumor and an exact planning of dosimetry which can be archived by stereotactic interstitial fibres or by neuronavigation systems should be considered mandatory for treating brain lesions [38].

Ionising radiation was applied to patients suffering from primary tumors either almost simultaneously or consecutively within 4 weeks of PDT. There was no observed advantage or disadvantage for the patients despite the fact that an additive or synergistic effect was reported experimentally [21].

Hyperthemia as a sole modality has also been investigated as an adjunctive therapy for brain tumors, and also in combination with PDT. A synergistic response was observed when hyperthermia was applied simultaneously or within 30 min of PDT [84]. There are no clinical data, for the effect of chemotherapy and PDT; however, PDT could be beneficial in the treatment of tumors which are resistant to chemotherapy as could be demonstrated experimentally. Steroids should be avoided prior to sensitisation but might be advantageous after PDT [84].

Possible interactions of PDT with current standard treatment modalities such as ionising radiation and chemotherapy remain to be further investigated.

The PDT-specific side effects increased intracranial pressure and prolonged skin sensitivity was reported from all centres; however, this did not cause major problems for the patients. There are no data reported on the quality of life of the patients but our institutional experiences are favourable.

Future sensitisers for the brain must be tailored for cerebral lesions. The absorption must be, ideally, in the near-infrared range around 700 to 800 nm, which allows a penetration of up to 4 cm into the surrounding brain [21,33,40,81]. The use of carrier systems such as liposome-encapsulated porphyrins or phthalocyanines or combination of AB conjugates with second or third generation sensitisers increase selectivity and might enhance PDT-mediated effects significantly [26,50,57,63,87,91,100]. Furthermore two or more sensitisers might also be used in combination [95].

Intraoperative PDD and fluorescence guided resection have been reported that employ ALA for diagnostic reasons only [36] and mTHPC in combination with PDT [35]. Intraoperative visualisation is an important issue in brain tumor resection, allowing an utmost resection of the tumors which already translates into a longer survival. The combination of PDD and PDT, "*to see and to treat*", is currently the most promising approach to the treatment of malignant brain tumors, despite the fact that the first results are currently not superior to standard adjuvant therapy. This concept offers a one-time single treatment with an expected better quality of life. Further results are awaited.

The indications for PDT in neurosurgery are infiltrating high grade primary and recurrent gliomas. In the case of tumors not responding to radio- and chemotherapy, PDT might be a good second line treatment. Low grade gliomas demonstrate a significantly longer survival and may be not such good candidates for PDT. Less good indications are hard tumors such as meningiomas which can be more effectively vaporised by other surgical laser techniques. Also, tumors in delicate areas such as the brain stem must be excluded from PDT protocols, whereas in our own experience it can be used, for example, in the motorstrip without harming function. Besides the above tumors, pituitary tumors, cystic lesion at the skull base such as craniopharyngeomas, and metastatic lesions are good indications.

The development of PDT and especially PDD for clinical applications is just beginning. Currently only 4 photosensitisers are in clinical oncological use. Despite its still experimental status, the available data indicate that PDT bears the potential as a second line treatment for malignant brain tumors and might be the future first line treatment for recurrent brain tumors.

Acknowledgements

The experimental work conducted by the author was supported by a Grant from the Austrian National Bank Nr. 6394. We wish to thank Seehof Laboratories, Germany, for the donation of Photosan 3 and Scotia Pharmaceuticals, UK, for the donation of Foscan®. The CLSM work was performed in collaboration with A. Rueck, ILM, Ulm, Germany; mTHPC extraction was performed by Lothar Lilges, Canada. The fluorescence detection work was in part a thesis of A. Zimmermann, Innsbruck, with the support of a Grant of FWF P13458.

References

1. W.R. Shapiro (1989). Brain tumor cooperative goup trial 8001. *J. Neurosurg.*, **71**, 1–12.
2. B.A. Allison, P.H. Pritchard, J.G. Levy (1994). Evidence for low-density lipoprotein receptor-mediated uptake of benzoporphyrin derivative. *Br. J. Cancer*, **69**, 833–839.
3. K. Radhakrishnan, N.I Bohnen, L.T. Kurland (1994). Epidemiology of Brain Tumors. In: R.A. Morantz, J.W. Wals (Eds), *Brain Tumors* (pp. 1–18). Marcel Dekker Inc.

4. W. Baumler, C. Abels, S. Karrer, T. Weiss, H. Messmann, M. Landthaler (1999). Szeimies. Photo-oxidative killing of human colonic cancer cells using indocyanine green and infrared light. *Br. J. Cancer*, **80**, 360–363.

5. A. Giese, M. Westphal (1996). Glioma invasion in the central nervous system. *Neurosurg.*, **39**, 235–252.

6. E.H. Oldfield, Z. Ram, K.W. Culver, R.M. Blaese, H.L. DeVroom, W.F. Anderson (1993). Gene therapy for the treatment of brain tumors using intra-tumoral transduction with the thymidin kinase gene and intravenous gancoclovir. *Hum. Gene Ther.*, **4(1)**, 39–69.

7. A. Benzer, C. Putensen, H. Kostron (1989). Photodynamic therapy [letter] *Lancet*, Aug 12 2-8659, 382–383.

8. P. Riva, A. Arista, C. Sturiale, G. Franceschi, A. Spinelli, N. Riva, M. Casi, G. Moscatelli, M. Frattarelli (1994). Intralesional radioimmunotherapy of malignant gliomas. An effectice treatment in recurrent tumors. *Cancer*, **73**(3), 1076–1082.

9. T.J. Dougherty, C.J. Gomer, B.W. Henderson, G. Jori, D. Kessel, M. Korbelik, J. Moan, Qian-Peng (1998). Photodynamic therapy. *J. Natl. Care Institute*, **90**(12), 889–905.

10. M.R. Hamblin, E.L. Newman (1994). On the mechanism of the tumor-localising effect in photodynamic therapy. *J. Photochem. Photobiol. B*, **23**(1), 3–8.

11. R. Van Hillegersberg, W.J. Kort, J.H.P. Wilson (1994). Current status of photodynamic therapy in oncology. *Drugs*, **48**(4), 510–527.

12. B.W. Engbrecht, C. Menon, A.V. Kachur, S.M. Hahn, D.L. Fraker (1999). Photofrin-mediated PDT induces vascular occlusion and apoptosis in a human sarcoma xenograft model. *Cancer Res.*, **59**(17), 4334–4342.

13. M. Ochsner (1997). Photophysical and photobiological processes in the photodynamic therapy of tumors. *J. Photochem. Photobiol. B*, **39**(1), 1–18.

14. N.L. Oleinick, H.H. Evans (1998). The photobiology of photodynamic therapy: cellular targets and mechanisms. *Radiat. Res.*, **150**(5), S146–56.

15. J.S. Hill, S.B. Kahl, S.S. Stylli, Y. Nakamura, M.S. Koo, A.H. Kaye (1995). Selective tumor kill of cerebral glioma by PDT using a boronated porphyrin photosensizer. *Proc. Natl. Acad. Sci. U.S.A.*, **92**(26), 12126–12130.

16. Q. Chen, B.C. Wilson, M.O. Dereski, M.S. Patterson, M. Chopp, F.W. Hetzel (1992). Effects of light beam size on fluence distribution and depth of necrosis in superficially applied photodynamic therapy of normal rat brain. *Photochem. Photobiol.*, **56**, 379–384.

17. Y. Ji, D. Walstad, J.T. Brown, S.K. Powers (1992). Improved survival from intracavitary photodynamic therapy of rat glioma. *Photochem. Photobiol.*, **56**(3), 385–390.

18. S. Kaneko, H. Kobayashi, Y. Kohama (1999). Stereotactic intratumoral photodynamic therapy on malignant brain tumors. *Abstract, International Symposium on Photodynamic Therapy in Clinical Practice*, Innsbruck.

19. A.H. Kaye, J.S. Hill (1993). Photodynamic therapy of brain tumors. *Ann. Acad. Med. Singapore*, **22**(3 Suppl), 470–481.

20. H. Kostron, A. Obwegeser, R. Jakober, A. Zimmermann, A. Rueck (1998). Experimental and clinical results of mTHPC (Foscan®) mediated photodynamic therapy for malignant brain tumors. In: T.J. Dougherty (Ed.), *Optical Methods for Tumor Treatment and Detections: Mechanisms and Techniques in Photodynamic Therapy VII* (SPIE. Proc. Vol. 3247 pp. 40–45). International Society for Optical Engineering, Bellingham, WA.

21. H. Kostron, A. Obwegeser, M. Seiwald (1996). PDT in neurosurgery; a review. *J. Photochem. Photobiol. B*, **36**, 157–168.

22. H. Kostron, G. Weiser, E. Fritsch, V. Grunert (1987). Photodynamic therapy of malignant brain tumors: clinical and neuropathological results. *Photochem. Photobiol.*, **46**, 937–943.

23. P. Muller, B. Wilson (1998). Photodynamic therapy of supratentorial gliomas. In: Th. J. Dougherty (Ed.), *Optical Methods for Tumor Treatment and Detections: Mechanisms and Techniques in Photodynamic Therapy VII* (SPIE Proc. Vol. 3247, pp. 2–13).

24. D.P. Noske, J.G. Wolbers, H.J. Sterenborg (1991). Photodynamic therapy of malignant glioma. A review of literature. *Clin. Neurol. Neurosurg.*, **93**(4), 293–307.

25. F. Jiang, L. Lilge, M. Belcuig, G. Singh, J. Grenier, Y.L.I.M. Chopp (1998). Photodynamic therapy using Photofrin in combination with buthionine sulfoximine (BSO) to treat 9L gliosarcoma in rat brain. *Lasers Surg. Med.*, **23**(3), 161–166.

26. T.C. Origitano, M.J. Caron, O.H. Reichman (1994). Photodynamic therapy for intracranial neoplasms. Literature review and institutional experience. *Mol. Chem. Neuropathol.*, **21**(2–3), 337–352.

27. C. Perria, M. Carai, A. Falzoi, G. Orunesu, A. Rocca, G. Massarelli, N. Francaviglia, G. Jori (1988). Photodynamic therapy of malignant brain tumors: clinical results of, difficulties with, questions about, and future prospects for the neurosurgical applications. *Neurosurgery*, **23**(5), 557–563.

28. E.A. Popovic, A.H. Kaye, J.S. Hill (1996). Photodynamic therapy of brain tumors. *J. Clin. Laser Med. Surg.*, **14**(5), 251–261.

29. G. Bottiroli, A.C. Croce, D. Locatelli, R. Nano, E. Giombelli, A. Messina, E. Benericetti (1998). Brain tissue autofluorescence: an aid for intraoperative delineation of tumor resection margins. *Canc. Detect. Prev.*, 22-4, 330–339.

30. Y.G. Chung, J.A. Schwartz, C.M. Gardner, R. Sawya, S.L. Jaques (1994). Fluorescence of normal and cancerous brain tissue: the excitation/emission matrix. In: Robert R. Alfano (Ed.), *Advances in Laser and Light Spectroscopy to Diagnose Cancer and other Diseases* (SPIE Proc. Vol. 2135, pp. 66–75).

31. O. Kasselouri, Bourdon, D. Demore, C. Blais, P. Prognon, G. Bourg-Heckly, J. Blais (1999). Fluorescence and mass spectrometry studies of mTHPC photoproducts. *Photochem. Photobiol.*, **70**(3), 275–279.

32. J.C. Kennedy, S.L. Marcus, R.H. Potier (1996). Photodynamic therapy and photodiagnosis using endogenous photosensitisation induced by 5-aminolevulinic acid (ALA): mechanism and clinical results. *J. Clin. Laser Med. Surg.*, **15**(5), 289–304.

33. K.M. Hebeda, J.B. Wolbers, H.J.C.M. Sterenborg, W. Kamohorst, M.J.C. van Gemert, H.A.M. van Alphen (1995). Fluorescence localisation in tumor and normal brain after intratumoral injection of hematoporphyrin derivative into rat brain tumor. *J. Photochem. Photobiol. B*, **27**, 85–92.

34. Q. Peng, T. Warloe, K. Berg, J. Moan, M. Kongshaug, K.E. Giercksky, J.M. Nesland (1997). 5-Aminolevulinic acid-based photodynamic therapy. Clinical research and future challenges. *Cancer*, **79**(12), 2282–2308.

35. H. Kostron, A. Zimmermann, A. Obwegeser (1998). mTHPC-mediated photodynamic detection for fluorescence guided resection of brain tumors. Surgical-assist systems. In: S.M. Bogner, S.T. Charles, W.S. Grundfest, J.A. Harrington, A. Katzir, L.S. Lome, M.W. Vannier, R. von-Hanwehr (Eds), (SPIE Proc. Vol. 3262, pp. 259–264). International Biomedical Optics Society, Bellingham, WA.

36. W. Stummer, S. Stocker, S. Wagner, H. Stepp, C. Fritsch, C. Goetz, A. Goetz, R. Kiefmann, H.J. Reulen (1998). Intraoperative detection of malignant gliomas by 5-aminolevulinic acid-induced porphyrin fluorescence. *Neurosurgery*, **42**, 518–526.

37. M.O. Dereski, M. Chopp, J.H. Garcia, F.W. Hetzel (1991). Depth measurements and histopathological characterization of photodynamic therapy generated normal brain necrosis as a function of incident optical energy dose. *Photochem. Photobiol.*, **54**(1), 109–112.

38. S.H. Tudge, A.H. Kaye, J.S. Hill (1999). Modulation of light delivery in photodynamic therapy of brain tumors. *J. Clinical Neurosci.*, **6**(3), 227–232.

39. H. Kostron, B.W. Hochleitner, A. Obwegeser, M. Seiwald (1995). Clinical and experimental results of photodynamic therapy in neurosurgery. *SPIE*, **2371**, 126–128.

40. W. Stummer, A. Hassan, O. Kempski, C. Goetz (1996). Photodynamic therapy within edematous brain tissue: considerations on sensitiser dose and time point of laser irradiation. *J. Photochem. Photobiol. B*, **36**(2), 179–181.

41. L. Morlet, V. Vonarx-Coinsman, P. Lenz, M.T. Foultier, L.X de-Brito, C. Stewart, T. Patrice (1995). Correlation between meta(tetrahydroxyphenyl)chlorin (m-THPC) biodistribution and photodynamic effects in mice. *Photochem. Photobiol. B*, **28**(1), 25–32.

42. A. Obwegeser, R. Jakober, H. Kostron (1998). Uptake and kinetics of C-14 labelled m-THPC and 5-ALA in C-6 rat glioma model. *Br. J. Cancer*, **78**(6), 733–738.

43. Q. Peng, J. Moan, L.W. Ma, J.M. Nesland (1995). Uptake, localization, and photodynamic effect of meso-tetra(hydroxyphenyl)porphine and its corresponding chlorin in normal and tumor tissues of mice bearing mammary carcinoma. *Cancer Res.*, **55**(12), 2620–2626.

44. A.M. Ronn, J. Batti, C.J. Lee, D. Yoo, M.E. Siegel, M. Nouri, L.A. Lofgren, B.M. Steinberg (1997). Comparative biodistribution of meta-Tetra(Hydroxyphenyl)chlorin in multiple species: clinical implications for photodynamic therapy. *Lasers Surg. Med.*, **20**(4), 437–442.

45. R.B. Veenhuizen, M.C. Ruevekamp, T.J. Helmerhorst, P. Kenemans, F.A. Stewart (1997). Foscan mediated photodynamic therapy for a peritoneal cancer model: drug distribution and efficacy studies. *Int. J. Cancer*, **73**, 230–235.

46. J.P. Rovers, J.J. Schuitmaker, A.L. Vahrmeijer, J.H. Van-Dierendonck, O.T. Terpstra (1998). Interstitial photodynamic therapy with the second-generation photosensitiser bacteriochlorin in a rat model for liver metastases. *Br. J. Cancer*, **77**(12), 2098–2103.

47. J.C. Tsai, Y. Hsiao, L.J. Teng, C.T. Chen, M.C. Kao (1999). Comparative study on the ALA photodynamic effects of human glioma and meningioma cells. *Lasers Surg. Med.*, **24**(4), 296–305.

48. J.D. Bhawalkar, N.D. Kumar, C.F. Zhao, P.N. Prasad (1997). Two photon photodynamic therapy. *J. Clin. Laser Med. Surg.*, **15**(5), 201–204.

49. I. Fujishima, T. Sakai, T. Tanaka, H. Ryu, K. Uemura, Y. Fujishima, K. Horiuchi, N. Daikuzono, Y. Sekiguchi (1991). Photodynamic therapy using pheophorbide a and Nd:YAG laser. *Neurol. Med. Chir. Tokyo*, **31**(5), 257–263.

50. V. Shafirovich, A. Dourandin, N.P. Luneva, C. Singh, F. Kirigin, N.E. Geacintov (1999). Multiphoton near-infrared femtosecond laser pulse-induced DNA damage with and without the photosensitiser proflavine. *Photochem. Photobiol.*, **69**(3), 265–274.

51. G. Karagianis, J.S. Hill, S.S. Stylli, A.H. Kaye, N.J. Varadaxis, J.A. Reiss, D.R. Phillips (1996). Evaluation of porphyrin C analogues for photodynamic therapy of cerebral glioma. *Br. J. Cancer*, **73**(4), 514–521.

52. M.W. Leach, R.J. Higgins, J.E. Boggan, S.J. Lee, S. Autry, K.M. Smith (1992). Effectiveness of a lysyl chlorin p6/chlorin p6 mixture in photodynamic therapy of the subcutaneous 9L glioma in the rat. *Cancer Res.*, **52**(5), 1235–1239.

53. M.W. Leach, S. Khoshyomn, J. Bringus, S.A. Autry, J.E. Boggan (1993). Normal brain tissue response to photodynamic therapy using aluminum phthalocyanine tetrasulfonate in the rat. *Photochem. Photobiol.*, **57**(5), 842–845.

54. Y.S. Lee, R.D. Wurster (1995). Methylene blue induces cytotoxicity in human brain tumor cells. *Cancer Lett.*, **88**, 141–145.

55. L. Lilge, B.C. Wilson (1998). PDT of intracranial tissue: a preclinical study of four different photosensitizers. *J. Clin. Laser Med. Surg.*, **16**(2), 81.

56. M.H. Schmidt, K.W. Reichert II, K. Ozker, G.A. Meyer, D.L. Donohoe, D.M. Bajic, N.T. Whelan, H.T. Whelan (1999). Preclinical evaluation of benzoporphyrin derivative

combined with a light-emitting diode array for photodynamic therapy of brain tumors. *Pediatr. Neurosurg.*, **30**(5), 225–231.

57. K.W. Woodburn, Q. Fan, D. Kessel, Y. Luo, S.W. Young (1998). PDT of B16F10 murine melanoma with lutetium texaphrin. *J. Invest. Dermatol.*, **110**(5), 746–751.

58. S.W. Young, K.W. Woodburn, M. Wright, T.D. Mody, Q. Fan, J.L. Sessler, W.C. Dow, R.A. Miller (1996). Lutetium texaphyrin (PCI-0123): a near-infrared, water-soluble photosensitiser. *Photochem. Photobiol.*, **63**(6), 892–897.

59. T. Chanwitayanuchit, W. Bernwick, G. Weiser, E. Fritsch, M. Maier, H. Kostron (1999). Neuropathological changes in normal rat brain after intravenous or direct injection of haematoporphyrin derivative (HPD) followed by combination of light and fast electron therapy. *Asian J. Surg.*, **13**(1), 32–36.

60. K.M. Hebeda, W. Kamphorst, H.J.C.M. Sterenborg, J.G. Wobers (1998). Damage to tumor and brain by interstitial photodynamic therapy in the 9L rat tumor model comparing intravenous and intratumoral administration of the photosensitiser. *Acta Neurochirurg.*, **140**(5), 495–501.

61. D.A. Bellnier (1991). Potentiation of photodynamic therapy in mice with recombinant human tumor necrosis factor-alpha. *J. Photochem. Photobiol.*, **8**(2), 203–210.

62. D. Fanuel-Barret, T. Patrice, M.T. Foultier, V. Vonarx-Coinsmann, N. Robillard, Y. Lajat (1997). Influence of epidermal growth factor on photodynamic therapy of glioblastoma cells in vitro. *Res. Exp. Med. Berl.*, **197**(4), 219–233.

63. G. Kick, G. Messer, A. Goetz, G. Plewig, P. Kind (1995). Photodynamic therapy induces expression of interleukin 6 by activation of AP-1 but not NF-kappa B DNA binding. *Cancer Res.*, **55**(11), 2373–2379.

64. M. Plattner, W. Bernwick, H. Kostron (1993). HPD-PDT in vitro sensitivity testing for brain tumors. *SPIE*, **1616**, 182–185.

65. C.J. Gomer, S.W. Ryter, A. Ferrario, N. Rucker, S. Wong, A.M. Fisher (1996). PDT mediated oxidative stress can induce expression of heat shock proteins. *Cancer Res.*, **56**(10), 2355–2360.

66. X.Y. He, R.A. Sikes, S. Thomsen, L.W. Chung, S. Jacques (1994). Photodynamic therapy with Photofrin II induces programmed cell death in carcinoma cell lines. *Photochem. Photobiol.*, **59**(4), 468–473.

67. D. Kessel (1998). Mitochondrial photodamage and PDT induced apoptosis. *Photochem. Photobiol.*, **42**(2), 89–95.

68. L. Lilge, M. Portnoy, B.C. Wilson (1999). PDT-induced apoptosis in brain tissue in vivo. In: Th.J. Dougherty (Ed.), *Optical Methods for Tumor Treatment and Detection: Mechanisms and Techniques in Photodynamic Therapy VIII* (SPIE-Proc. Vol. 3592, pp. 28–36).

69. P.C. De-Smidt, A.J. Versluis, T.J. Van-Berkel (1993). Properties of incorporation, redistribution, and integrity of porphyrin-low-density lipoprotein complexes. *Biochemistry*, **32** (11), 2916–2922.

70. F. Jiang, L. Lilge, B. Logie, Y. Li, M. Chopp (1997). Photodynamic therapy of 9L gliosarcoma with liposome-delivered Photofrin. *Photochem. Photobiol.*, **65**(4), 701–706.

71. G. Jori, E. Reddi (1993). The role of lipoproteins in the delivery of tumor-targeting photosensitisers. *Int. J. Biochem.*, **25**(10), 1369–1375.

72. J. Leppala, M. Kallio, T. Nikula, P. Nikkinen, K. Liewendahl, J. Jaaskelainen, S. Savolainen, H. Gylling, J. Hiltunen, Callaway (1995). Accumulation of 99m-Tc low density lipoprotein in human malignant glioma. *Br. J. Cancer*, **71**(2), 383–387.

73. H.T. Whelan, L.H. Kras, K. Ozker, D. Bajic, M.H. Schmidt, Y. Liu, L.A. Trembath, F. Uzum, G.A. Meyer, A.D. Segura (1994). Selective incorporation of 111 In-labeled PHOTOFRIN by glioma tissue in vivo. *J. Neurooncol.*, **22**(1), 7–13.

74. H. Kostron, C. Plangger, E. Fritsch, H. Maier (1990). Photodynamic treatment of malignant brain tumors. *W. Klin. Woch.*, **102**(18), 531–535.

75. I. Yeung, L. Lilge, B.C. Wilson, T.Y. Lee, L. Stevens, A. Cenic (1997). Photodynamic therapy induced alterations of the blood-brain barrier transfer constant of a tracer molecule in normal brain. In: T.H.J. Dougherty (Ed.), *Optical Methods for Tumor Treatment and Detection: Mechanisms and Techniques in Photodynamic Therapy VI* (SPIE Proc. Vol. 2972, pp. 54–63). San Jose, CA.

76. Y. Ji, S.K. Powers, J.T. Brown, D. Walstad, L. Maliner (1994). Toxicity of photodynamic therapy with Photofrin in the normal rat brain. *Lasers Surg. Med.*, **14**(3), 219–228.

77. Y. Yoshida, M.O. Dereski, J.H. Garcia, F.W. Hetzel, M. Chopp (1992). Neuronal injury after photoactivation of photofrin II. *Am. J. Pathol.*, **141**(4), 989–997.

78. C. Perria, T. Capuzzo, G. Cavagnaro (1980). First attempts at the photodynamic treatment of human gliomas. *J. Neurosurg. Sci.*, 119–129.

79. N.M. Bleehen, J.M. Ford (1993). Radiotherapy, hyperthermia, and photodynamic therapy for central nervous system tumors. *Curr. Opin. Oncol.*, May; 5–3, 458–463.

80. G. Canti, A. Nicolin, R. Cubbedu, P. Taroni, G. Banddieramonte, G. Valentini. Antitumor efficacy of the combination of PDT and chemotherapy in murine tumors. *Cancer Lett.*, **125**(118), 39–44.

81. K.T. Chen, D.M. Hau, J.S. You, H.C. Pan, R.W. Wong (1995). Therapeutic effects of photosensitisers in combination with laser and ACNU on an in vivo or in vitro model of cerebral glioma. *Chin. Med. J. Engl.*, **108**, 98–104.

82. M.Y. Nahabedian, R.A. Cohen, M.F. Contino, T.M. Terem, W.H. Wright, M.W. Berns, A.G. Wile (1988). Combination cytotoxic chemotherapy with cisplatin or doxorubicin and photodynamic therapy in murine tumors. *J. Natl. Cancer Inst.*, **80**(10), 739.

83. L.W. Ma, J. Moan, H.B. Steen, V. Iani (1995). Anti-tumor activity of photodynamic therapy in combination with mitomycin C in nude mice with human colon adenocarcinoma. *Br. J. Cancer*, **71**(5), 950–956.

84. P.A. Cowled, L. Mackenzie, I.J. Forbes (1985). Potentiation of photodynamic therapy with haematoporphyrin derivatives by glucocorticoids. *Cancer Lett.*, **29**, 107–114.

85. H.W. Pogrebniak, W. Matthews, C. Black, A. Russo, J.B. Mitchell, P. Smith, J.A. Roth, H.I. Pass (1993). Targetted phototherapy with sensitiser-monoclonal antibody conjugate and light. *Surg. Oncol.*, **2**(1), 31–42.

86. K. Rittenhouse-Diakun, H. Van-Leengoed, J. Morgan, E. Hryhorenko, G. Paszkiewicz, J.E. Whitaker, A.R. Oseroff (1995). The role of transferrin receptor (CD71) in photodynamic therapy of activated and malignant lymphocytes using the heme precursor delta-aminolevulinic acid (ALA). *Photochem. Photobiol.*, **61**(5), 523–528.

87. M.O. Dereski, L. Madigan, M. Chopp (1995). The effect of hypothermia and hyperthermia on photodynamic therapy of normal brain. *Neurosurgery*, **36**(1), 141–145.

88. S. Kimel, L.O. Svaasand, M. Hammer-Wilson, V. Gottfried, S. Cheng, E. Svaasand, M.W. Berns (1992). Demonstration of synergistic effects of hyperthermia and photodynamic therapy using the chick chorioallantoic membrane model. *Lasers Surg. Med.*, **12**(4), 432–440.

89. T. Patrice (1991). The combination of PDT and hyperthermia is a promising method for tumor treatment. Challenges for photodynamic therapy in the treatment of gastrointestinal tumors. *J. Photochem. Photobiol. B*, **3**(4), 372–374.

90. C. Prinsze, L.C. Penning, T.M. Dubbelman, J. Van Steveninck (1992). Interaction of photodynamic treatment and either hyperthermia or ionizing radiation and of ionizing radiation and hyperthermia with respect to cell killing of L929 fibroblasts, Chinese hamster ovary cells, and T24 human bladder carcinoma cells. *Cancer Res.*, **52**(1), 117–120.

91. R. Hornung, H. Walt, N.E. Compton, K.A. Keefe, B. Jentsch, G. Perewusnyk, U. Haller, O.R. Kochli (1998). mTHPC mediated PDT does not induce resistance to chemotherapy, radiotherapy or PDT on human breast cancer cells in vitro. *Photochem. Photobiol.*, **68**(4), 569–574.

92. B.C. Wilson, A. Molchovsky, G.J. Czarnota, M.D. Sherar, M.C. Kolios, L. Lilge, R.S. Dattani, K.S. Osterman, K.D. Paulsen, P.J. Hoopes (1999). Monitoring tissue response to photodynamic therapy: the potential of minimally invasive electrical impedance spectroscopy and high-frequency ultrasound. In: Th.J. Dougherty (Ed.), *Optical Methods for Tumor Treatment and Detection: Mechanisms and Techniques in Photodynamic Therapy VIII* (SPIE Proc. Vol. 3592, pp. 73–82).

93. C.J. Gomer, N. Hayashi, A.L. Murphree (1987). The influence of sodium pentobarbital anesthesia on in vivo photodynamic therapy. *Photochem Photobiol.*, **46**(5), 843–846.

94. P.V. Marks (1995). Adjuvant therapy for pituitary adenomas: possible role of PDT. *Ann. Roy. Coll. Surg. Eng.*, **77**, 308–312.

95. S.K. Powers, S.S. Cush, D.L. Walstad, L. Kwock (1991). Stereotactic intratumoral photodynamic therapy for recurrent malignant brain tumors. *Neurosurgery*, **29**(5), 688–695.

96. A.J. Terzis, A. Dietze, R. Bjerkvig, H. Arnold (1997). Effects of photodynamic therapy on glioma spheroids. *Br. J. Neurosurg.*, **11**(3), 196–205.

97. L. Lilge, C. O'Carroll, B.C. Wilson (1997). A solubilization technique for photosensitiser quantification in ex vivo tissue samples. *J. Photochem. Photobiol. B*, **39**(3), 229–235.

98. R.W. Kirollos, P.V. Marks, U. Igbaseimokumo, A. Chakrabarty (1998). A preliminary experimental in vivo study of the effect of photodynamic therapy on human pituitary adenoma implanted in mice. *Br. J. Neurosurg.*, **12**(2), 140–145.

99. P.V. Marks, C. Furneaux, R. Shivvakumar (1992). An in vitro study of the effect of photodynamic therapy on human meningiomas. *Br. J. Neurosurg.*, **6**(4), 327–332.

100. J.S. Nelson, L.H. Liaw, R.A. Lahlum, P.L. Cooper, M.W. Berns (1990). Use of multiple photosensitisers and wavelengths during photodynamic therapy: a new approach to enhance tumor eradication. *J. Natl. Cancer Inst.*, **82**(10), 863–873.

101. H.T. Whelan, M.H. Schmidt, A.D. Segura, T.L. McAuliffe, D.M. Bajic, K.J. Murray, J.E. Moulder, D.R. Strother, J.P. Thomas, G.A. Meyer (1993). The role of photodynamic therapy in posterior fossa brain tumors. A preclinical study in a canine glioma model. *J. Neurosurg.*, **79**(4), 562–568.

Chapter 11

Clinical application of photomedical techniques in gynecology

P. Wyss and A. Degen

Table of contents

Photodynamic techniques exhibit a growing potential for gynecological applications. A wide variety of cutaneous and mucous lesions of the genital tract, both benign and malignant, or breast cancer recurrences on the chest wall may be successfully managed by this new treatment modality.

11.1 Breast cancer recurrences

Approximately 5–19% of breast cancer patients suffer from chest wall recurrences following classical primary treatment such as lumpectomy, mastectomy, radiation therapy, chemotherapy, or hormone manipulation therapy [1–5]. The 5-year survival rate of these patients is about 20% and only 3% will show no progression of the disease [6]. Breast cancer recurrences impact physical and psychological well-being substantially [7] by pain, odor and secretion, and remind patients constantly of the presence of a progressing disease. Often in this clinical palliative situation only a few conventional treatment options remain. Several authors have suggested PDT as a minimally invasive, selective treatment for breast cancer patients suffering chest wall recurrences that have not responded to other treatment options.

Since 1979, photodynamic treatments of more than 200 patients suffering from chest wall recurrences have been published. The first photoradiations for controlling local and regional recurrences of breast carcinoma were performed using haematoporphyrin derivative (HPD) [8]. Cutaneous and subcutaneous tumors were treated effectively without undue damage to overlying and adjacent skin. In 1987, dihaematoporphyrin ether (PF) was applied as photosensitizer in 18 patients with chest wall recurrences [9,10]. Complete response, reported on 105 patients photosensitized by PF, ranged between 7–91% [9–16].

One of the major side effects of these first generation photosensitizers is prolonged skin photosensitivity, requiring sun and light protection for 1–2 months. Another interesting approach utilises antibody-targeted photodynamic therapy. The use of zinc phthalocyanine antibody–MCA complex improved tumor selectivity, avoided skin phototoxicity and resulted in partial remission of the chest wall metastases [17].

Little success was achieved by local application of drugs such as 5-aminolevulinic acid (ALA) and meso-tetra-(hydroxyphenyl)porphyrins (TPPS4). This may be due to insufficient penetration and inhomogeneous distribution of the photosensitizers in the tumor. Among the group of newer compounds m-THPC exhibits promising properties for PDT. It is a pure compound that is approximately 100 times more effective than HPD at inducing tissue necrosis in animal models; it is more selective for malignant tissues and skin phototoxicity is reduced to 14–20 days. In a multi-centre study, tin ethyl etiopurpurin (SnET2) was applied to 45 patients suffering recurrences [17]. After a period of 2.4 and 8 weeks all patients showed remission and healing of the treated skin area. Despite the fact that SnET2 exhibits a terminal half-life of several days, no phototoxic skin reactions were reported.

Photodynamic therapy offers a minimaly invasive, low-cost treatment modality for recurrent breast cancer with few side effects, high patient satisfaction and with possible repetitive application since tissue does not evolve resistance to photo-dynamic therapy. An additional advantage of PDT is the ability to treat numerous, widespread lesions in one sitting.

11.2 Photodynamic therapy (PDT) for dysfunctional uterine bleeding

Approximately 4% of all women of reproductive age consult their physician on an annual basis because of menorrhagia and almost 20% of the 700000 hysterectomies performed annually in North America are for its treatment [18]. Endometrial ablation provides an alternative to hysterectomy for treating menorrhagia. Hysteroscopic procedures such as Nd:YAG laser coagulation [19], electrocautery with resectoscope [20] or rollerball [21] are used for endometrial-destruction [22–25]. These techniques may be associated with a variety of complications including uterine perforation, intravasation of uterine distending medium with cardiovascular overload, and may be followed by continued bleeding or sequestra-tion of endometrial tissue due to incomplete endometrial destruction [26–29]. A number of non-hysteroscopic techniques such as radiofrequency [30] and microwave devices [31], thermal [32] or multielectrode balloons [33], and hot water instillation [34] cause less complications. But these approaches are based on temper-ature elevations up to 90°C, requiring some form of anesthesia and have no intrinsic specificity for endometrium over myometrium and adjacent organs. In contrast to endometrial ablation by heat, PDT provides photooxidation-induced, selective endometrial tissue destruction [35,36].

11.2.1 PDT of the endometrium

Following pharmacokinetic studies of aminolevulinic acid-induced Protoporphyrin IX (PpIX) as photosensitizer in human endometrium [37], morphological and functional effects of photodynamic endometrial ablations in humans suffering dysfunctional uterine bleeding were evaluated at our clinic.

For studying photodynamically-induced, morphological changes in the endo-metrium, photodynamic endometrial ablation was performed in a premenopausal patient (age 38 years, P1) suffering from heavy menorrhagia and in a post-menopausal patient (age 70 years, P2) with long-lasting postmenopausal bleeding. Malignancy was ruled out by prior endometrial biopsy in the postmenopausal patient. Hysterectomy was scheduled 3 days (P1) and 5 months (P2), respectively, following photodynamic endometrial damage. Sections for histology were stained with hematoxylin-eosin.

In the functional study, a total of 10 sessions for photodynamic endometrial ablation were performed. One postmenopausal (P1) and five premenopausal patients (P2–P5) were treated. Malignancy was ruled out by prior diagnostic hysteroscopy and curettage. Daily bleeding patterns were recorded in a scale scoring between 1 (very weak bleeding) and 6 (extremely heavy bleeding) 1–3 months before PDT and during the follow-up. To evaluate the efficacy of photodynamical endometrial destruction the area under the curve (averaged per month) of the bleeding patterns was compared before and after PDT.

Neither general nor local anesthesia was needed during photodynamic treatment. Assessment of efficacy needs further follow up; however, all five patients have reported reduction of uterine bleeding up to the present [38]. Microscopic examinations of the photodynamically treated postmenopausal endometrium exhibited regions with normal endometrium as well as regions showing a thinned endometrial layer lacking glands. Complete destruction of the endometrium in the isthmic and corporal regions was found in the premenopausal patient showing partial destruction in the fundus. Evidently, reduction of uterine bleeding is based on morphological changes of the endometrium [39]. Patient satisfaction, considering the procedure and the results, was high. In conclusion, photodynamic endometrial ablation is feasible without general or local anesthesia and offers a minimaly invasive, non-hysteroscopic method for endometrial ablation indicated for dysfunctional uterine bleeding. In contrast to endometrial ablation by heat, PDT provides photooxidation-induced, selective endometrial tissue destruction. Further technical developments of the light-application system are being undertaken to achieve even more uniform and efficient endometrial destruction resulting in amenorrhea.

11.3 Intraepithelial neoplasia and human papillomavirus of the uterine cervix

The following section gives an overview of PDD and PDT studies in Cervical Intraepithelial Neoplasias (CIN) and Human Papilloma Virus (HPV) infections of the cervix. Since different light doses and photosensitizer concentrations were used in most studies, the results are not always comparable. The overview does not focus on the discussion of different photodynamic parameters, but aims to give a general impression of successes and drawbacks.

11.3.1 CIN I–II

Preliminary results of follow-ups up to 9 months after PDT using topically applied 5-aminolevulinic acid (ALA)12% showed a cytological improvement in the grading of the PAP smears in 19 out of 20 patients and the eradication of cervical HPV in 80% [40].

11.3.2 CIN high grade (CIN II–III)

Ten treatment cycles of PDT using 5-ALA 20% were performed in seven patients with high-grade CIN. After three months, a significant reduction in the size of the ectocervical CIN lesions was noted in only three patients. No significant improvement in CIN lesions was noted since cold knife conization revealed persistent CIN in all seven cases. ALA-PDT does not appear to be effective in treating high grade CIN [41].

In a Phase I study [42], 2 ml of a 1% solution of DHE (Photofrin) in a 4% Azone and isopropyl alcohol vehicle was applied to the cervix 24 h prior to PDT. In a follow up of 12 months, 8 out of 11 patients (73%) were treated successfully with PDT at an energy density of 100 to 140 J cm^{-2}. Lower light doses showed higher recurrence rates.

11.3.3 Cervical cancer

Carcinoma in situ (CIS): Photoradiation was performed in 56 patients (39 CIS and 17 dysplasia) 48 h after intravenous injection of 1.5–2 mg kg^{-1} of the photosensitizer Porfimer sodium (Photofrin). Complete response was reported in 54 (96.4%) of these patients [43].

A histologically verified small residual of a cervix carcinoma (primary FIGO III b) was photodynamically treated using m-THPC at a dose of 0.15 mg kg^{-1} body-weight and a total light dose of 20 J cm^{-2} (power density of 100 mW s^{-1}). Within 24 h necrosis occurred which was restricted to the tumor area. Wound healing was significantly delayed and survival times were disappointingly short [44].

11.3.4 Comment

Accessibility of the uterine cervix is easy and may favor the performance of PDT on the cervix without anesthesia.

Considering the low numbers of examined patients and the wide range of photodynamic parameters used such as photosensitizer concentrations, light doses and light application systems, future progress in photodynamic techniques requires unique consent study protocols and coordinated multi-centre studies.

11.4 Vulva diseases

Photodynamic therapy of the vulva region has been evaluated in the treatment of viral infection (Condylomata acuminatum and Herpes genitalis), of psoriasis, of precancerous tissue alteration (Lichen sclerosus) and in recurrences of gynecological tumors independent of previous treatment. The aim of the application of this technique was not only to make the efflorescence or the tumor masses disappear but also to relieve pain or aching as a palliative effect.

Systemic but also topical application of the drug has shown high complete response rates in a variety of skin diseases.

The fact that no anesthesia is required, altered lesions can be destroyed selectively, there is no cumulation of the treatment dose, and the cosmetic result compared to surgery or thermal methods is excellent, gives photodynamic therapy a status that surely needs to be further explored.

11.4.1 Herpes genitalis

The prevalence of genital herpes is increasing in several populations world-wide. Factors that may be contributing to this increase include greater numbers of sexual partners, the high frequency of asymptomatic infections, and poor use of safe sexual practices [45]. Transmission occurs via skin-to-skin or may also occur when the patient is unaware of the lesions or when lesions are not clinically apparent. Primary infections are frequently disabling with severe pain, dysuria, lymphadenopathy, and malaise. Recurrent infections being more localized and less painful, provide sources for transmission of the disease and may result in psychosexual dysfunction [46].

11.4.2 Photodynamic therapy of herpes genitalis

In the 1970s, clinical trials were performed using neutral red [47,48] or proflavine [49], a DNA-photosensitizing dye [50], for photodynamic inactivation treatment of genital herpes infection. The results in these studies were not promising and the routine use of this technique for genital herpes therapy has been discontinued. In vitro studies have shown that photodynamic action is a potent physicochemical process being extremely lethal for herpes virus [51]. Singlet oxygen produced during the photodynamic procedure has been recognized as a major virucidal factor [52]. However, HSV-2-induced dermatitis in mice was not responsive to hematoporphyrin photoinactivation treatment [53]. In another study, zinc phthalocyanine was the most potent dye per absorbed photon for inactivating HSV compared to Photofrin II, polyhematoporphyrin esters and proflavine sulfate [54]. But it has been suggested that zinc phthalocyanine increased the mutation frequency in the surviving virus [54]. Clinical application of photodynamic HSV inactivation using neutral red could not corroborate an increased risk of the development of cutaneous malignancies [55]. In guinea pigs infected with HSV, subsequent administration of ALA and exposure of the lesions to red light shortened the duration of vesicles' appearance from more than a week to a few days and reduced the HSV titer in the lesions [56]. Not long ago, numerous antiviral compounds, such as acyclovir, valacyclovir, and famciclovir, were approved by the FDA to treat primary and recurrent genital HSV infections. Since the application (oral) of these antiviral compounds is easy and their use can significantly alter the disease by decreasing the morbidity, mortality and transmission rates, photodynamic inactivation of genital herpes infection will hardly be established as a primary or routine therapy. Accordingly, patients who experience frequent or severe recurrences, those particularly troubled by their disease, and those who wish to reduce the frequency of asymptomatic infection generally prefer suppressive therapy [45].

11.5 Condylomata acuminata

Condylomata acuminata (genital warts) are sexually transmitted benign neoplasms caused by the human papilloma virus (HPV) that may involve the vulva, vagina, cervix, urethra, anal canal, and perineal skin [57]. The prevalence of HPV infection varies greatly, depending on the population studies and has increased in the past decade.

Condylomata acuminata are commonly associated with vaginitis, pregnancy, diabetes mellitus, oral contraceptive use, poor perineal hygiene, immunosupression, and sexual activity with multiple partners [58].

Molecular biological methods employing hybridization have identified more than a dozen types of viruses in topical genital condylomata acuminata [59,60].

Progression to VIN has been documented [61]. Malignant transformation into squamous cell carcinoma has also been observed [62]. The topical application of dilute podophyllin, or the judicious application of concentrated halogenated acetic acid (trichloracetic acid), are common approaches to the treatment of small vulvar condylomata. Electrodissection, surgical excision, cryosurgery, hot wire loop excision, and mainly CO_2-laser ablation have been used for large excisions [57].

11.5.1 Photodynamic treatment of Condylomata acuminata

Frank et al. [63] published in 1996 the results of the treatment of seven genital condyloma acuminatum lesions. There had been 20% ALA applied topically 14 h before light exposure of an argon dye laser with a dose of 100 J cm^{-2} at an intensity of 75 or 150 mW cm^{-2}. After three months four lesions showed a complete response.

Fehr et al. [64] in 1996 proved, in a group of 24 patients, the selective fluorescence of condyloma acuminatum after an unselective topical 5-aminolevulinic acid application. Furthermore the fluorescence 1h after application was higher with 2.5% ALA than with 20% ALA. The conclusion was that studies evaluating selective photodynamic destruction of condylomas are justified.

Similar results were published by Ross et al. [65]. He applied ALA in a concentration of 20% and measured fluorescence for 3 h in six patients and up to 6 h in nine patients. Also in this trial there was significantly greater fluorescence compared with adjacent normal skin.

Clinical application revealed promising results in which no scar formation and a high selectivity could be achieved [66].

11.6 Psoriasis of the vulva

Psoriasis is inherited as a simple autosomal dominant trait with incomplete penetrance. Psoriasis affects approximately 2% of the population of the United States. On the vulva, the disease typically involves the lateral aspects of the labia majora and genitocrural areas [67].

In addition to the established photochemotherapies PUVA (ultraviolet A range) therapy is used. But because of the potential of PUVA to induce malignancies (squamous cell carcinoma) [68,69] other approaches are also being explored.

11.6.1 Photodynamic treatment of psoriasis

In 1937, Silver reported the clinical use of hematoporphyrin and UV-light in the treatment of psoriasis [70,71]. After systemically injected hematoporphyrin derivative (HPD) in combination with red laser light (630 nm, 40 and 20 J cm^{-2}), psoriasis lesions in the mons pubis area responded vigorously in a patient treated for intraepithelial neoplasia of the vulva [72]. An improvement greater than 90% was achieved in 15 out of 19 patients after a low dose of HPD (1.0 mg kg^{-1}) and daily whole body irradiation with UVA light for 15 days [73]. The optimum irradiation time for psoriatic lesions is 6 h after topically applied ALA [74]. Several other authors treated psoriasis at different locations on the body and suggest that PDT with topical, applied photosensitizers is a well tolerated additional treatment [75].

11.7 Lichen sclerosus (Lichen sclerosus et athrophicus)

Lichen sclerosus is a dermatosis of unknown etiology characterized by progressive thinning of the epithelium, subepithelial edema with fibrin deposition, and an underlying zone of chronic inflammation within the dermis. It is a common cause of white epithelial changes on the vulva [76]. It is most frequently found on the genitalia. The exact prevalence is unknown. The disease is not racially confined, and although it is noted most commonly in postmenopausal Caucasian women, cases may first appear in children as young as 18 months of age [77–79]. In adult women, the clinical findings show typical thining and a whitened epithelium, which is usually symmetrical and involves the labia minora, clitoris, prepuce, frenulum and perineal body. In advanced cases, loss, agglutination and adhesions are found; stenosis of the introitus is common [80].

11.7.1 Photodynamic therapy of Lichen sclerosus

Hillemanns [81] evaluated the therapeutic effect of photodynamic therapy on vulvar lichen sclerosus. Twelve women were treated four to five hours after topical application of 10 ml of a 20% solution of 5-aminolevulinic acid. The irradiation was 80 J cm^{-2} and the irradiance 40–70 mW cm^{-2}. Light with a wavelength of 635 nm was delivered by an argon-ion pumped dye laser. Six to eight weeks after photodynamic therapy, pruritus significantly improved in ten of the twelve women. A prolonged effect of photodynamic therapy was reported, with a mean of 6.1 months.

11.8 Vulvar intraepithelial neoplasia (VIN)

The international Society for the Study of Vulvar Disease (ISSVD) introduced a new classification (1989) of vulvar intraepithelial neoplasia (VIN) terminology replacing previously used terms such as Bowen's disease, erythroplasia of Queyrat etc. VIN is a type of precancerous vulvar tissue abnormality. It is no longer considered as a problem of postmenopausal women, but can develop in any age.

The incidence of VIN is increasing in young woman [82]. VIN may progress to invasive cancer of the vulva in a few cases. Patients may be without symptoms or complain of pruritus (itching) or 'burning'. Raised brown, red, pink, or white lesions of various colors may be present. Treatment depends on the degree of the disease. VIN III can usually be treated successfully with surgical or laser removal. The prevailing evidence favors human papilloma virus (HPV) as a causative factor in genital tract carcinomas. Standard treatment in vulvar cancer is surgery. Because of the psychosexual consequences of a significant morbidity associated with standard radical vulvectomy, there is a definite trend towards vulvar conservation and individualized management of patients with early vulvar cancer.

Standard therapy includes laser ablation and excision producing good short-term results. Recurrences often require repeated treatments which may cause vulvar damage by scar formation.

11.8.1 Photodynamic treatment of VIN

Eighteen women with high-grade VIN were recruited for PDT using topically applied 5-aminolevulinic acid (ALA) 20% [83]. Success of treatment was evaluated histologically by directed, vulvar biopsy 3 months after PDT. Three out of eight women showed clearance of the disease while 16 out of 18 women reported relief from pruritus vulvae and vulvar discomfort.

Tokuda [84] described one patient with extensive Bowen's disease (VIN III) treated with PDT with no evidence of recurrence after 18 months, and another patient who had a negative biopsy 28 days after treatment. Complete response was reported by McCaughan [85] in one patient with the same disease treated by Photofrin-PDT.

Koren treated two patients suffering from recurrence or residual cancer of the vulva after conventional therapy [86]. Photodynamic therapy using Photofrin systemically resulted in partial response with recurrence 2–3 months following PDT and complete response (CR) lasting >12 months.

Hetzel reported on one partial remission and two complete responses following PDT of vulvar carcinoma (Stad. III) recurrences using systemically applied Photofrin [87]. In a study published by Lobraico, an 81% CR was achieved in 17 patients at 3 months post Photofrin-PDT for vulvar carcinoma in situ. CRs of 83% and 50% resulted after two and three PDT-sessions, respectively [88].

Since the number of treated patients is small and the photosensitizers used are different, concluding remarks may not be definitive. However, PDT provides a minimaly invasive, outpatient alternative to established therapies for VIN and vulvar carcinoma recurrences.

References

1. P. Wyss, J.C. Rageth, K. Kohler, C. Unger, E. Hochuli (1991). Prognosis of local-regional recurrence in breast carcinoma. *Schweiz. Rundsch. Med. Prax.*, **80**(20), 556–559.

2. M. Overgaard, P.S. Hansen, J. Overgaard, C. Rose, M. Andersson, F. Bach, M. Kjaer, C. Gadeberg, H.T. Mouridsen, M.B. Jensen, K. Zedeler (1997). Postoperative radiotherapy in high-risk premenopausal women with breast cancer who receive adjuvant chemotherapy. Danish Breast Cancer Cooperative Group 82b Trial. *New Engl. J. Med.*, **337**(14), 949–955.

3. P. Casolo, D. Mosca, C. Amorotti, A. Raspadori, B. Drei, P. Di Blasio, G. Colli, R. De Maria, G. De Luca, E. Ganz, D. Amuso (1997). Our experience in the surgical treatment of early breast cancer. Results of a prospective study of 204 cases. *Ann. Ital. Chir.*, **68**(2), 195–205.

4. P.H. Elkhuizen, M.J. van-de-Vijver, J. Hermans, H.M. Zonderland, C.J. van-de-Velde, J.W. Leer (1998). Local recurrence after breast-conserving therapy for invasive breast cancer: high incidence in young patients and association with poor survival. *Int. J. Radiat. Oncol. Biol. Phys.*, **40**(4), 859–867.

5. I. Gage, A. Recht, R. Gelman, A.J. Nixon, B. Silver, B.A. Bornstein, J.R. Harris (1995). Long-term outcome following breast-conserving surgery and radiation therapy. *Int. J. Radiat. Oncol. Biol. Phys.*, **33**(2), 245–251.

6. I. Fredriksson, G. Liljegren, L.G. Arnesson, S.O. Emdin, M. Palm-Sjovall, T. Fornander, J. Frisell, L. Holmberg (2001). Time trends in the results of breast conservation in 4694 women. *Eur. J. Cancer.*, **37**(12), 1537–1544.

7. J.M. Stevenson, P. Bochenek, K. Jamrozik, R.W. Parsons, M.J. Byrne (1997). Breast cancer in Western Australia in 1989. V: Outcome at 5 years after diagnosis. *Aust. N. Z. J. Surg.*, **67**(5), 250–255.

8. T. Dougherty, G. Lawrence, J.H. Kaufman, D. Boyle, K.R. Weishaupt, A. Goldfarb (1979). Photoradiation in the treatment of recurrent breast carcinoma. *J. Natl. Cancer. Inst.* **62**(2), 231–237.

9. M. Schuh, U.O. Nseyo, W.R. Potter, T.L. Dao, T.J. Dougherty (1987). Photodynamic therapy for palliation of locally recurrent breast carcinoma. *J. Clin. Oncol.*, **5**(11), 1766–1770.

10. S.M. Waldow, R.V. Lobraico, I.K. Kohler, S. Wallk, H.T. Fritts (1987). Photodynamic therapy for treatment of malignant cutaneous lesions. *Lasers. Surg. Med.*, **7**(6): 451–456.

11. J.S. McCaughan, Jr., J.T. Guy, W. Hicks, L. Laufman, T.A. Nims, J. Walker (1989). Photodynamic therapy for cutaneous and subcutaneous malignant neoplasms. *Arch. Surg.* **124**(2), 211–216.

12. R.B. Buchanan, J.A. Carruth, A.L. McKenzie, S.R. Williams (1989). Photodynamic therapy in the treatment of malignant tumors of the skin and head and neck. *Eur. J. Surg. Oncol.*, **15**(5), 400–406.

13. P.W. Sperduto, T.F. DeLaney, G. Thomas, P. Smith, L.J. Dachowski, A. Russo, R. Bonner, E. Glatstein (1991). Photodynamic therapy for chest wall recurrence in breast cancer. *Int. J. Radiat. Oncol. Biol. Phys.* **21**(2), 441–446.

14. S.A. Khan, T.J. Dougherty, T.S. Mang (1993). An evaluation of photodynamic therapy in the management of cutaneous metastases of breast cancer. *Eur. J. Cancer 2 A*, **29**(12), 1686–1690.

15. P. Baas, I.P. van Geel, H. Oppelaar, M. Meyer, J.H. Beynen, N. van Zandwijk, F. Stewart (1996). Enhancement of photodynamic therapy by mitomycin C: a preclinical and clinical study. *Br. J. Cancer*, **73**(8), 945–951.

16. S.W. Taber, V.H. Fingar, T.J. Wieman (1998). Photodynamic therapy for palliation of chest wall recurrence in patients with breast cancer. *J. Surg. Oncol.*, **68**(4): 209–214.

17. St. Schmidt (1996). Photodynamic laser therapy of advanced breast carcinomas. *Geburtshilfe Frauenheilkd.*, **56**(10), 153–156.

18. R. Pokras, V.G. Hufnagel (1988). Hysterectomy in the United States, 1965–84. *Am. J. Public Health*, **78**(7), 852–853.

19. M.H. Goldrath, T.A. Fuller, S. Segal (1981). Laser photovaporization of endometrium for the treatment of menorrhagia. *Am. J. Obstet. Gynecol.*, **140**(1), 14–19.
20. A. DeCherney, M.L. Polan (1983). Hysteroscopic management of intrauterine lesions and intractable uterine bleeding. *Obstetrics Gynecol.*, **61**(3), 392–397.
21. B. McLucas (1990). Endometrial ablation with the roller ball electrode. *J Reprod. Med.*, **35**(11), 1055–1058.
22. J.F. Daniell, B.R. Kurtz, R.W. Ke (1992). Hysteroscopic endometrial ablation using the rollerball electrode. *Obstetrics Gynecol.*, **80**(3:Pt 1), 329–332.
23. M. Nisolle, J. Donnez (1997). Alternative techniques of hysterectomy [letter]. *New Engl. J. Med.*, **336**(4), 291–292.
24. J. Donnez, R. Polet, P.E. Mathieu, E. Konwitz, M. Nisolle, F. Casanas-Roux (1996). Endometrial laser interstitial hyperthermy: a potential modality for endometrial ablation. *Obstetrics Gynecol.*, **87**(3), 459–464.
25. R.S. Neuwirth, A.A. Duran, A. Singer, R. MacDonald, L. Bolduc (1994). The endometrial ablator: a new instrument. *Obstetrics Gynecol.*, **83**(5:Pt 1), 792–796.
26. D. Molloy, P.T. Taylor (1994). Gynaecological surgery after endometrial ablation [see comments]. *Med. J. Aust.*, **161**(10), 604–606.
27. C.A. Witz, K.M. Silverberg, W.N. Burns, R.S. Schenken, D.L. Olive (1993). Complications associated with the absorption of hysteroscopic fluid media. Review [55 refs]. *Fertility Sterility*, **60**(5), 745–756.
28. P.G. Brooks (1992). Complications of operative hysteroscopy: how safe is it?. *Clin. Obstetrics Gynecol.*, **35**(2), 256–261.
29. R. Jedeikin, D. Olsfanger, I. Kessler (1990). Disseminated intravascular coagulopathy and adult respiratory distress syndrome: life-threatening complications of hysteroscopy. *Am. J. Obstetrics Gynecol.*, **162**(1), 44–45.
30. J.H. Phipps, B.V. Lewis, T. Roberts, M.V. Prior, J.W. Hand, M. Elder, S.B. Field (1990). Treatment of functional menorrhagia by radio-frequency-induced thermal ablation, *Lancet.*, **335**, 374–376.
31. N.C. Sharp, N. Cronin, I. Feldberg, M. Evans, D. Hodgson, S. Ellis (1995). Microwaves for menorrhagia: a new fast technique for endometrial ablation. *Lancet*, **346**, 1003–1004.
32. G.A. Vilos, E.C. Vilos, L. Pendley (1996). Endometrial ablation with a thermal balloon for the treatment of menorrhagia. *J. Am. Assoc. Gynecol. Laparosc.*, **3**(3), 383–387.
33. R.M. Soderstrom, P.G. Brooks, S.L. Corson, J. Dequesne, A. Gallinat, J.G. Garza-Leal, J.L. Iglesias-Benavides, P.D. Indman, J. Liu, H. van der Pas, R.A. Stern, C. Sutton, T.G. Vancaillie, K. Wamsteker (1996). Endometrial ablation using a distensible multi-electrode balloon. *J. Am. Assoc. Gynecol. Laparosc.*, **3**(3), 403–407.
34. M. Baggish, M. Paraiso, E.M. Breznock, S. Griffey (1995). A computer-controlled, continuously circulating, hot irrigating system for endometrial ablation. *Am. J. Obstet. Gynecol.*, **173**, 1842–1848.
35. M.J. Gannon, D.I. Vernon, J.A. Holroyd, M. Stringer, N. Johnson, S.B. Brown (1997). PDT of the endometrium using ALA. *SPIE*, **2972**, 2–13.
36. P. Wyss, M. Fehr, H. Van-den-Bergh, U. Haller (1998). Feasibility of photodynamic endometrial ablation without anesthesia. *Int. J. Gynaecol. Obstet.*, **60**(3), 287–288.
37. M. Fehr, P. Wyss, B.J. Tromberg, T. Krasieva, P.J. DiSaia, F. Lin, Y. Tadir (1996). Selective photosensitizer localisation in the human endometrium after intrauterine application of 5-aminolevulinic acid. *Am. J. Obstet. Gynecol.*, **175**, 1253–1259.
38. P. Wyss (2000). Photodynamic endometrial ablation: Morphological and functional results. In: P. Wyss, Y. Tadir, B. Tromberg, U. Haller (Eds), *Photomedicine in Gynecology and Reproduction*. Karger Verlag, 2000.

39. P. Wyss, R. Caduff, Y. Tadir, A. Degen, G. Wagnieres, V. Schwarz, U. Haller, M. Fehr (2003). Photodynamic endometrial ablation: Morphological study. *Lasers Surg. Med.*, **32**(4), 305–309.

40. F. Wierrani, A. Kubin, R. Jindra, M. Henry, K. Gharehbaghi, W. Grin, J. Soltz-Szotz, G. Alth, W. Grunberger (1999). 5-aminolevulinic acid-mediated photodynamic therapy of intraepithelial neoplasia and human papillomavirus of the uterine cervix-a new experimental approach. *Cancer Detect. Prev.*, **23**(4), 351–355.

41. P. Hillemanns, M. Korell, M. Schmitt-Sody, R. Baumgartner, W. Beyer, R. Kimmig, M. Untch, H. Hepp (1999). Photodynamic therapy in women with cervical intraepithelial neoplasia using topically applied 5-aminolevulinic acid. *Int. J. Cancer*, **81**(1), 34–38.

42. B.J. Monk, C. Brewer, K. Van Nostrand, M.W. Berns, J.L. McCullough, Y. Tadir, A. Manetta (1997). Photodynamic therapy using topically applied dihematoporphyrin ether in the treatment of cervical intraepithelial neoplasia. *Gynecol. Oncol.*, **64**(1), 70–75.

43. T. Muroya, Y. Suehiro, K. Umayahara, T. Akiya, H. Iwabuchi, H. Sakunaga, M. Sakamoto, T. Sugishita, Y. Tenjin (1996). Photodynamic therapy (PDT) for early cervical cancer. *Gan. To. Kagaku. Ryoho*, **23**(1), 47–56.

44. E. Krimbacher, A.G. Zeimet, C. Marth, H. Kostron (1999). Photodynamic therapy for recurrent gynecologic malignancy: a report on 4 cases. *Arch. Gynecol. Obstet.*, **262**(3–4), 193–197

45. L. Stanberry, A. Cunningham, G. Mertz, A. Mindel, B. Peters, M. Reitano, S. Sacks, A. Wald, S. Wassilew, P. Woolley (1999). New developments in the epidemiology, natural history and management of genital herpes. *Antiviral. Res.*, **42**(1), 1–14.

46. A. Mindel (1996). Psychological and psychosexual implications of herpes simplex virus infections. *Scand. J. Infect. Dis. Suppl.*, **100**, 27–32.

47. A.P. Roome, A.E. Tinkler, A.L. Hilton, D.G. Montefiore, D. Waller (1975). Neutral red with photoinactivation in the treatment of herpes genitalis. *Br. J. Vener. Dis.*, **51**(2), 130–133.

48. M.G. Myers, M.N. Oxman, J.E. Clark, K.A. Arndt (1976). Photodynamic inactivation in recurrent infections with herpes simplex virus. *J. Infect. Dis.*, **133**(Suppl), A145–150.

49. R.H. Kaufman, E. Adam, R.R. Mirkovic, J.L. Melnick, R.L. Young (1978). Treatment of genital herpes simplex virus infection with photodynamic inactivation. *Am. J. Obstet. Gynecol.*, **132**(8), 861–869.

50. C.D. Lytle, P.G. Carney, R.P. Felten, H.F. Bushar, R.C. Straight (1989). Inactivation and mutagenesis of herpes virus by photodynamic treatment with therapeutic dyes. *Photochem. Photobiol.*, **50**(3), 367–371.

51. M. Jarrat (1977). Photodynamic inactivation of herpes simplex virus. *Photochem. Photobiol*, **25**(4), 339–340.

52. K. Muller-Breitkreutz, H. Mohr, K. Briviba, H. Sies (1995). Inactivation of viruses by chemically and photochemically generates singlet molecular oxygen. *J. Photochem. Photobiol, B.* **30**(1), 63–70.

53. M. Perlin, J.C. Mao, E.R. Otis, N.L. Shipkowitz, R.G. Duff (1987). Photodynamic inactivation of influenza and herpes viruses by hematoporphyrin. *Antiviral Res.*, **7**(1), 43–51.

54. C.D. Lytle, P.G. Carney, R.P. Felten, H.F. Bushar, R.C. Straight (1989). Inactivation and mutagenesis of herpes virus by photodynamic treatment with therapeutic dyes. *Photochem. Photobiol.*, **50**(3), 367–371.

55. E.G. Friedrich, Jr., T. Masukawa (1975). Effect of povidone-iodine on Herpes genitalis. *Obstet. Gynecol.*, **45**(3), 337–339.

56. Z. Smetana, Z. Malik, A. Orenstein, E. Mendelson, Ben Hur-E. (1997). Treatment of viral infections with 5-aminolevulinic acid and light. *Lasers Surg. Med.*, **21**(4), 351–358.

57. P.J. Lynch (1985). Condylomata acuminata (anogenital warts) (Review). *Clin. Obstet. Gynecol*, **28**(1), 142–151.

58. G. von Krogh (1979). Warts: immunologic factors of prognostic significance (Review). *Int. J. Dermatol.*, **18**(3), 195–204.
59. L. Gissmann, H. zur Hausen (1980). Partial characterization of viral DNA from human genital warts (Condylomata acuminata). *Int. J. Cancer*, **25**(5), 605–609.
60. L. Gissmann, E.M. deVilliers, H. zur Hausen (1982). Analysis of human genital warts (condylomata acuminata) and other genital tumors for human papillomavirus type 6 DNA. *Int. J. Cancer*, **29**(2), 143–146.
61. J. Buscema, Z. Naghashfar, E. Sawada, R. Daniel, J.D. Woodruff, K. Shah (1988). The predominance of human papillomavirus type 16 in vulvar neoplasia. *Obstet. Gynecol.*, **71** (4), 601–606.
62. C.M. Ridley (1994). The aetiology of vulval neoplasia. *Br. J. Obstet. Gynaecol.*, **101**(8), 655–657.
63. R.G. Frank, J.D. Bos (1996). Photodynamic therapy for condylomata acuminata with local application of 5-aminolevulinic acid. *Genitourin. Med.*, **72**(1), 70–71.
64. M.K. Fehr, C.F. Chapman, T. Krasieva, B.J. Tromberg, J.L. McCullough, M.W. Berns, Y. Tadir (1996). Selective photosensitizer distribution in vulvar condyloma acuminatum after topical application of 5-aminolevulinic acid. *Am. J. Obstet. Gynecol.*, **174**(3), 951–957.
65. E.V. Ross, R. Romero, N. Kollias, C. Crum, R.R. Anderson (1997). Selectivity of proto-porphyrin IX fluorescence for condylomata after topical application of 5-aminolaevulinic acid: implications for photodynamic treatment. *Br. J. Dermatol.*, **137**(5), 736–742.
66. M.K. Fehr, R. Hornung, A. Degen, V.A. Schwarz, D. Fink, U. Haller, P. Wyss (2002). Photodynamic therapy of vulvar and vaginal condyloma and intraepithelial neoplasia using topically applied 5-aminolevulinic acid. *Lasers Surg. Med.* **30**(4), 273–279.
67. C.M. Ridley (1980). Skin disorders of the vulva. *Practitioner*, **224**(1343), 481–486.
68. R.S. Stern, E.J. Lunder (1998). Risk of squamous cell carcinoma and methoxsalen (psoralen) and UV-A radiation (PUVA). A meta-analysis. *Arch. Dermatol.*, **134**(12), 1582–1585.
69. R.S. Stern, E.J. Liebman, L. Vakeva (1998). Oral psoralen and ultraviolet-A light (PUVA) treatment of psoriasis and persistent risk of nonmelanoma skin cancer. PUVA Follow-up Study. *J. Natl. Cancer Inst.*, **90**(17), 1278–1284.
70. H. Silver (1937). Psoriasis vulgaris treated with Hematoporphyrin. *Arch. Dermatol. Syphilol.*, **36**, 1118–1119.
71. P.G. Calzavara-Pinton, R.M. Szeimies, B. Ortel, C. Zane (1996). Photodynamic therapy with systemic administration of photosensitizers in dermatology (Review). *Photochem. Photobiol. B*, **36**(2), 225–231.
72. M.W. Berns, M. Rettenmaier, J. McCullough, J. Coffey, A. Wile, M. Berman, P. DiSaia, G. Weinstein (1984). Response of psoriasis to red laser light (630 nm) following systemic injection of hematoporphyrin derivative. *Lasers Surg. Med.*, **4**(1), 73–77.
73. H. Berg, E. Bauer, F.A. Gollmick, W. Diezel, F. Böhm, H. Meefert, N. Sönnichsen (1985). Photodynamic hematoporphyrin therapy of psoriasi. In: G Jori, C. Perria (Eds), *Photodynamic Therapy in Tumors and Other Diseases* (pp. 337–343). Progetto Editore, Padova.
74. C. Fritsch, P. Lehmann, W. Stahl, K.W. Schulte, E. Blohm, K. Lang, H. Sies, T. Ruzicka (1999). Optimum porphyrin accumulation in epithelial skin tumors and psoriatic lesions after topical application of delta-aminolaevulinic acid (Review). *Br. J. Cancer*, **79**(9–10), 1603–1608.
75. W.H. Boehncke, R. Kaufmann (1996) [Photodynamic therapy at the threshold of clinical use in disseminated dermatoses]. (Review. German). *Hautarzt*, **47**(11), 825–831.
76. J. Hewitt (1986). Histologic criteria for lichen sclerosus of the vulva. *J. Reprod. Med.*, **31**(9), 781–787.

77. J. Berth-Jones, R.A. Graham-Brown, D.A. Burns (1991). Lichen sclerosus et atrophi-cusa–review of 15 cases in young girls. *Clin. Exp. Dermatol.*, **16**(1), 14–17.
78. E.G Friedrich, Jr. (1976). Lichen sclerosus (Review). *J. Reprod. Med. Sep.*, **17**(3), 147–154.
79. V. Loening-Baucke (1991). Lichen sclerosus et atrophicus in children. *Am. J. Dis. Child.*, **145**(9), 1058–1061.
80. H. Kowarz-Sokolowska (1975). [Pathology of connective tissue and nerve fibers in lichen sclerosus et atrophicans and in genital atrophies]. *Hautarzt.* **26**(11), 602–606 (German).
81. P. Hillemanns, M. Untch, F. Prove, R. Baumgartner, M. Hillemanns, M. Korell (1999). Photodynamic therapy of vulvar lichen sclerosus with 5-aminolevulinic acid. *Obstet. Gynecol.*, **93**(1), 71–74.
82. R.W. Jones, J. Baranyai, S. Stables (1997). Trends in squamous cell carcinoma of the vulva: the influence of vulvar intraepithelial neoplasia. *Obstet. Gynecol.*, **90**(3), 448–452.
83. P.L. Martin-Hirsch, C. Whitehurst, C.H. Buckley, J.V. Moore, H.C. Kitchener (1998). Photodynamic treatment for lower genital tract intraepithelial neoplasia. *Lancet*, **351**(9103), 645–646.
84. Y. Tokuda (1983). Primary skin cancers. In: Y. Hayata, T. J. Dougherty (Eds), *Lasers and Hematoporphyrin Derivative in Cancer* (pp. 88–96) Igaku-Shoin, Ltd., Tokyo.
85. J.S. McCaughan, Jr., H.F. Schellhas, J. Lomano, B.H. Bethel (1985). Photodynamic therapy of gynecologic neoplasm after presensitization with hematoporphyrin derivative. *Lasers Surg. Med.*, **5**(5), 491–498.
86. H. Koren, G. Alth (1996). Photodynamic therapy in gynecologic cancer. *J. Photochem. Photobiol B*, **36**(2), 189–191.
87. H. Hetzel, E. Muller-Holzner, C. Marth, H. Kostron (1993). [Photodynamic therapy in patients with recurrent gynecologic cancers]. *Geburtshilfe Frauenheilkd.*, **53**(5), 333–336 (German).
88. R.V. Lobraico, L.I. Grossweiner (1993). Clinical experiences with photodynamic therapy for recurrent malignancies of the lower female genital tract. *J. Gynecol. Surg.*, **9**(1), 29–34.

Chapter 12

Photodynamic therapy of the gastrointestinal tract

Kenneth K. Wang

Table of contents

12.1 Background

Photodynamic therapy has been applied to a wide variety of lesions within the gastrointestinal tract. Esophageal and gastric cancers are among the first tumors that were treated with photodynamic therapy. This is likely due to the ease of access of the hollow viscera to endoscopic equipment. Advanced endoscopic equipment can be used to place fiberoptic devices to photoradiate the esophagus, stomach, small intestine, and colon. In addition, the solid organs associated with the gastrointestinal tract such as the liver or pancreas are also accessible through small ducts that are attached to the small intestine. Photodynamic therapy for tumors was initially administered for palliation of patients with advanced disease but more recently has been used for curative treatment of superficial tumors. Dosages and methods of drug and light administration have traditionally been similar to those used in the lung. However, specialized instrumentation has been developed to permit treatment in the gastrointestinal tract. This includes balloons which permit flattening of the gastrointestinal mucosa and allow for greater uniformity of light distribution. The advantage of treatment in the gastrointestinal tract over pulmonary lesions has been the avoidance of 'clean-up' procedures to debride necrotic tumors after therapy. In the pulmonary system this is needed to decrease occlusion of the airways, while in the gastrointestinal tract the necrotic debris is actually excreted or digested by the remainder of the tract.

Current photodynamic therapy for the gastrointestinal tract is limited by the lack of more flexible diffusing fibers to allow easier passage through the endoscopes and photosensitizers that can cause significant toxicity in terms of producing nausea, pain, cutaneous photosensitivity, and strictures. Currently, only sodium porfimer (Photofrin II) is approved for use for esophageal cancers in Canada, Finland, France, Japan, the Netherlands, the United Kingdom, and the United States. It is also approved for use in gastric cancers in Japan.

12.2 Gastrointestinal photosensitizers

Current photosensitizers used in the gastrointestinal tract are members of a family of porphyrin and porphyrin-like molecules. They contain ring-like aromatic molecules, which are crucial in determining the ability to absorb visible light, converting absorbed light into other forms of chemical and physical energy, and to enhance thermodynamic and kinetic stability. They have a central space that is created by four pyrrole rings composed of four carbon atoms and one nitrogen atom. The nitrogen atoms face inward and coordinate metals to form nearly planar 1:1 ligand-to-metal complexes.

The most commonly used photosensitizers are hematoporphyrin derivative, porfimer sodium, and aminolevulinic acid (ALA). ALA is a prodrug that is converted in mitochondria into protoporphyrin IX. Each of these agents has a different absorption spectrum. It is a general principal in photodynamic therapy to choose the wavelength of light that has the greatest degree of absorption. Even small changes in wavelength can markedly decrease the ability of light to activate the drug.

Activating 5-aminolevulinic acid (5-ALA) at 635 nm appears to generate significantly more tissue necrosis than light of 632 nm [1].

The new texaphyrin derivatives are of special interest to gastroenterologists [2].
Relative to the naturally occurring porphyrin structure, texaphyrins possess a 20%
larger binding cavity with five (rather than four) nitrogen donor atoms. These
ligands create stable, non-labile n-5 complexes with large cations. Of particular
interest are metallotexaphyrin complexes formed from metals of the lanthanide
series as they absorb light of higher wavelength, which allows deeper tissue penetration. The aromatic 22 π-electron delocalization of texaphyrins induces a red-shift in
the Q-type absorption bands of the metallated complexes from the 600–650 nm
range typical for porphyrins to the optimal range of 700–760 nm. Metallotexa-
phyrins incorporating diamagnetic metals (metals with no unpaired electrons) such
as lutetium(III) are able to use the absorbed 700–760 nm photons to generate
relatively long-lived triplet state complexes that efficiently generate cytotoxic
singlet oxygen and thus cause more tissue damage.

12.3 Light sources for gastrointestinal applications

Light sources used for photodynamic therapy in gastroenterology have been laser
based because of the need to channel light into fiberoptic bundles in order to enhance
delivery to large surface areas. Since initial targets in the gastrointestinal tract were
tumors, it has also been necessary to select wavelengths that allowed for significant
tumor penetration. The depth of light penetration through human tissue is controlled
by the choice of the color (wavelength) of light and the optical properties of the
tissue [3]. It has been shown that longer wavelengths of visible light penetrate tissue
more effectively than shorter wavelengths. Between 600 and 680 nm, a large rise in
penetration occurs which eventually plateaus between 700 and 800 nm. This makes
red light the preferred choice for PDT. Green light, which can also activate most
photosensitizers that have been used in the gastrointestinal tract, has a wavelength
around 500 nm and produces significantly less tissue penetration.

A variety of laser sources have been applied to photodynamic therapy trials. These
have included dye lasers (Coherent Lambda Plus, Santa Clara, CA or Laserscope Model
630, San Jose, CA) which use one source of laser energy (argon or KTP-YAG) to drive
a dye (kiton red) to produce a red light. These laser systems usually require special
power outlets and water cooling to function properly. Recently, a solid state diode laser
has been produced (Diomed 630 PDT, Cambridge, UK) which can be operated from
standard power outlets and can be air-cooled. This diode laser can supply up to 2 W of
power at 630 nm, which is sufficient to activate porphyrin compounds.

Laser light sources may generate significant amounts of thermal damage in
addition to the photodynamic therapy, which may increase the amount of destruction
inflicted. To avoid the thermal effect, light should be delivered at fluence rates lower
than 150 mW cm^{-2} to avoid heat accumulation. Some investigators have preferred
to combine the photodynamic effect with the thermal effect since there is some
evidence that hyperthermia to 40–42° C may provide some beneficial synergistic
effects [4].

12.4 Light dosimetry

Prior to using PDT in the gastrointestinal tract, serious consideration must be given to tumor dimensions and optical properties. Analytical light dosimetry modeling has helped to approximate the ideal delivered light dose which ensures that necrosis occurs in all tumor regions.

The amount of light delivered to a tumor is called fluence and is measured in joules per square centimetre. The fluence used for gastrointestinal lesions is dependent upon the type and concentration of the PDT agent, the light wavelength used, and the actual intrinsic photosensitivity of the tumor. For Photofrin, clinical parameters have been established by employing a standard drug dose of 2.0 mg kg^{-1} administered intravenously. Photoradiation is performed by administration of light of 630 nm after a time delay of 48 h between injection and photoradiation. Photoradiation is commonly applied in the gastrointestinal tract using three different methods:

(1) Cylindrical surface (CS): a line source centered in a cylindrical lumen.
(2) Cylindrical insertion (CI): a line source embedded in the tumor tissue used for bulky tumors with protruding nodules [5].
(3) Front surface (FS) which is a uniform irradiance incident beam used most commonly for small discrete lesions.

For most applications in the gastrointestinal tract, we use a cylindrical surface light source. With the cylindrical light source, the depth of tissue necrosis (d_n) can be expressed as a function of the light dose, the lumen diameter, the optical penetration depth (g) and the diffuse reflectance coefficient (Rd). The diffuse reflectance coefficient and optical penetration depth can be measured using a white light reflectance system although tables have been created using typical human tissues and have been published. The depth of tissue necrosis can be related to these optical parameters using the formula $d_n = \delta \log_e (DG)$ where d_n is the depth of necrosis of tissue from the photoradiated tissue surface, δ is the optical penetration depth, D is the incident fluence scaled to the type and concentration of the PDT drug, the treatment wavelength and the intrinsic photosensitivity of the tumor, and G depends on the optical constants.

If a light dose of 250 J cm^{-1} fiber is used, as may be done in Barrett's esophagus with a superficial tumor, we can estimate the diameter of the esophagus to be 25 mm diameter, the Rd to be 0.25 and g to be 2.0 mm, and we are able to calculate a d_n of 2.5 mm [3]. Most clinical trials done with light of 630 nm have produced tumor necrosis to a depth of 5 mm and rarely as deep as 10 mm.

12.5 PDT in esophageal diseases

More than 10000 Americans will develop esophageal cancer this year with an overall 5-year survival rate of 4–7% and a 1 year survival of 20% [6,7]. Surgical intervention is able to provide 40% of patients with palliation provided resection can be performed but mortality rates associated with surgery may be as high as 7–29% [8]. PDT has been used extensively in the management of esophageal diseases.

Use of PDT in esophageal diseases was initially reported in the mid to late 1980s, in patients who failed conventional treatment for esophageal squamous or adenocarcinoma. McCaughan described 40 patients with esophageal carcinoma who underwent PDT using hematoporphyrin derivative [9]. Dysphagia to solids was relieved in 86% of the patients at one month after the treatment. Jin treated 45 patients with advanced esophageal cancer. Initial response was seen in 70.5% patients and complete response in 15% [10]. One quarter of the patients from the complete response group were alive more than 5 years later. A multi-center randomized trial with 236 patients compared PDT to thermal ablative laser therapy (Nd-YAG laser) in patients with advanced stage, partially obstructing esophageal cancer [11]. This study found that the two treatments offered a similar rate of relief of dysphagia but that PDT was able to produce a longer lasting tumor response, 32% at one month for PDT versus 20% for Nd-YAG. In this study fewer procedures were required for the PDT group.

In addition to advanced stage esophageal cancer, several studies have reported success in the treatment of superficial esophageal cancer with PDT. This is one of the most exciting areas of photodynamic therapy in the esophagus since 'cures' have been reported. Tien treated 13 patients with early squamous esophageal cancers. Using balloon cytology as their method of following patients, 12 of the 13 patients had negative cytology results for 21 to 32 months [12]. Fujimaki and Nakayama treated 15 lesions measuring between 1 and 7 cm in a total of 14 patients [13]. Of the 11 patients treated with PDT alone, complete response was obtained in 9 patients. Two patients with local recurrence were retreated; they remained disease free at 8 and 18 months. McCaughn has treated four stage 1 patients with esophageal cancer [9]. Three out of four patients had no recurrence. Tajiri treated six patients with superficial esophageal carcinoma with PDT [14]. Interestingly, in this report, one patient had no photoreaction but three out of the remaining five patients were disease free at 15 to 47 months. Patrice has reported treating four patients with in situ malignant lesions of the esophagus [15,16]. All patients had successful eradication of cancer for a follow-up of 15.2 months (mean). Okunaka used PDT for 20 esophageal cancer patients [17]. Six patients had superficial and 14 had advanced invasive cancer. Four of the superficial cancers and six of the advanced invasive cancers had complete to significant remission. An additional eight patients had partial response. Wagnieres treated 15 patients with early esophageal carcinoma and recurrence was noted in only three patients [18]. Monnier showed eradication of 12 early esophageal cancers out of 15 cancers at 60 months follow-up [19]. The recent data regarding role of PDT in early esophageal carcinoma is outlined in Table 1.

In the early 1990s, Dougherty found regression of Barrett's epithelium while treating patients for superficial esophageal cancer [26]. These encouraging results coupled with improvements in light control and delivery system led to several studies using PDT to treat Barrett's esophagus with an emphasis on decreasing the grade of dysplasia and extent of Barrett's epithelium. Therapy for Barrett's esophagus is usually conducted using a cylindrical diffusing fiber either within a balloon of 25 mm diameter which permits flattening of the esophageal folds or with a bare fiber [27]. The length of the segment of the esophagus treated is limited to 7 cm due to the pain and odynophagia that is associated with treatment of longer lengths [28]. Treatment parameters for photodynamic therapy for Barrett's esophagus with high

Table 1. Photodynamic therapy for superficial (early) esophageal cancers

Author	Patients & stage	Photosensitizer	Diffuser	Adjuvant therapy	Cancer eradication %	Follow up	Comments
Overholt [20]	13 T1/T2	Porfimer sodium	balloon	Nd:YAG	77	4–84 months	metastatic disease 2, stricture 34%
Wang [21]	8 Tis-T2	HpD	cylinder	none	75	12 months	
Savary [22]	24 Tis/T1a	Photofrin II, HpD, mTHPC	cylinder	none	84	3 months–8 yrs	SCC only, stenosis 8%, TOF 8%
Gossner [23]	22 Tis/uT1	5-ALA	cylinder balloon	none	77	1–30 months	nausea, elevated AST
Grosjean [24]	14[a] Tis/T1ab	Photofrin II	cylinder	radiation	75–83	6–49 months	514 & 630 nm light used PDT results with 514 nm = 630 nm
Spinelli [25]	22	HpD Photofrin II	balloon	Nd:YAG	60	5–75 months	SCC in 20 patients stricture 9%, phototoxicity 9%

ALA, Aminolevulinic acid; ALT, aspartate aminotransferase; HpD, hematoporphyrin derivative; mTHPC, 5,10,15,20-tetra(m-hydroxyphenyl)chlorin; SCC, Squamous cell carcinoma; TOF, tracheo esophageal fistula; T1, cancer involving the mucosa and/or submucosa; T2, cancer invading into muscularis propria but not through this layer. [a] Indicates number of tumors.

grade dysplasia using Photofrin are 200 J cm^{-1} fiber at a power of 400 mW cm^{-1}. This would mean that the total power output required from the laser for a 7 cm fiber would be 2.8 W (400 mW cm^{-1} × 7 cm). This is well within the capability of the Laserscope laser but exceeds the capability of the diode lasers and some Coherent tunable dye lasers. The current approved laser fibers for photodynamic therapy are only 1.0, 2.5, and 5.0 cm in length. After therapy, approximately 70–80% of the mucosa appears to become normal squamous epithelium with residual islands of non-dysplastic Barrett's in the remainder [29]. A summary of the experience with photodynamic therapy for Barrett's esophagus is shown in Table 2.

The complications that may be experienced after photodynamic therapy include stricture formation, chest pain, and cutaneous photosensitivity [20]. Serious dermal phototoxic side effects have been observed after administration of Photofrin with large scale erythema, edema, urticarial lesions and pruritus [20]. Adverse effects also include low-grade fever 33%, photosensitivity 19%, pleural effusion 28%, anemia 26%, constipation 33% and respiratory insufficiency 10% [30] for those who have also received radiation therapy. All patients undergoing PDT should be counseled to avoid sunlight exposure for at least one month due to the possibility of severe sunburns. These patients should also be advised that sunscreen is of little benefit as it only protects against ultraviolet light and not visible light. Sun protection should be in the form of light-proof clothing and gloves, large wide-brimmed hats, and sunglasses on sunny days.

PDT does have a number of advantages, including treatment of a large surface area in one session, a lack of smoke generation during therapy (which may block vision of personnel) and minimal thermal damage to the normal tissue surrounding the target lesion and the endoscope. To the inexperienced endoscopist, there may not appear to be any damage to the tissue noted during the procedure and hence there might be a tendency to give an increased amount of light exposure. This should be avoided as it results in increased damage to normal tissue and raises the likelihood of complications and serious adverse effects.

Although intuitively this approach appears to be very promising, there has not been evidence that the ablation of Barrett's esophageal mucosa actually reduces the risk of esophageal cancer [8]. In addition, it has been reported that it is common to find residual Barrett's mucosa underlying endoscopically normal appearing squamous mucosa after ablative therapy [31]. These findings emphasize the need to continue endoscopic surveillance of both residual Barrett's mucosa and regenerated squamous mucosa after photodynamic therapy.

12.6 Gastric cancers and PDT

Gastric cancer compromises about 10% of all cancers world-wide [33]. The majority of the malignancies of the stomach are adenocarcinomas. Lymphomas, carcinoid tumors and sarcomas make up the remaining gastric neoplasms. The most efficacious curative therapy to date has been surgical resection. In experienced hands, surgical resection for palliation has mortality as low as 8% and nearly 25% may survive two years after the operation [34]. However, patients with metastatic disease

Table 2. Photodynamic therapy for Barrett's esophagus (metaplasia and HGD)

Author	Patients	Photosensitizer	Diffuser	Adjuvant therapy	HGD response (%)	Barrett's response (%)	Comments
Overholt [20] 1999	73	Porfimer sodium	balloon	83% YAG	88	43	F/up 4–84 months stricture 34% phototoxicity 4% atrial fibrillation 3% thoracocentesis 2%
Wang [21] 1999	26	HpD	cylinder	none	88	35	F/up 24 months chest pain 81% stricture 27%
Gossner [23] 1998	10	ALA	cylinder 6 balloon 4	mTHPC	100	0	F/up 1–30 months nausea 47% AST elevation 65%
Barr [32] 1996	5	ALA	dilator	none	100	0	F/up 26–44 months Nondysplastic Barrett's island in depth 2 no complications

ALA, Aminolevulinic acid; ALT, aspartate aminotransferase; HGD, high grade dysplasia; HpD, hematoporphyrin derivative; mTHPC, 5,10,15,20-tetra(*m*-hydroxyphenyl)chlorin.

are not usually considered for resection because of the morbidity associated with gastrectomy. For these patients, the only alternatives have been chemotherapy with the possibility of radiation therapy.

For the unfortunate patients who are unable to have surgical resection, there is some data suggesting that PDT may be useful for palliation [25]. Utilizing photodynamic therapy in these patients has relieved obstruction due to malignancy as well as controlling hemorrhages. Responses of gastric cancers to photodynamic therapy have varied from 57% to 85% although there has been less than 20 patients in any of the series [14,35–37]. Recent studies examining the effectiveness of PDT in gastric carcinomas are listed in Table 3.

Photodynamic therapy has also been applied to the treatment of early gastric cancer. Photodynamic therapy appears to be superior to other modalities including thermal laser for this indication. This is particularly true for lesions with a depth of invasion estimated to be limited to the submucosa and where the margins of the lesion are unclear [38].

12.7 Role of PDT in pancreatic cancer

Pancreatic cancer has an incidence of approximately 28100 cases per year in the United States with deaths estimated at 25000 patients each year [40]. Although pancreatic cancers rate seventh in incidence of all malignancies, it is the fifth leading cause of cancer related deaths. This is because pancreatic cancer causes death in most of its victims. The incidence of the disease appears to have been increasing by 1% per year [41]. About 10% of these tumors are resectable with a less than 5% surviving more than a year. Thus the vast majority of patients receive palliation with conventional chemotherapy and radiation although there does not appear to be any increased survival with these modalities [42]. PDT was originally proposed in 1981 for use in pancreatic cancer by Holyoke [43]. The therapy is highly selective for tumor cells and is relatively benign to normal pancreatic cells due to their poor uptake of the photosensitizing agents [44]. The low normal tissue penetration of the photosensitizing agent may allow PDT to be useful for the palliative treatment of pancreatic carcinoma and as an adjuvant to radical tumor resection. This may make it possible to increase the benefit of tumor resections if there is a possibility of destroying residual tumor in the setting of surgically unresectable neoplastic tissue [45].

12.8 Photodynamic therapy in colorectal lesions

Colorectal cancer is one of the most common potentially lethal gastrointestinal diseases encountered in clinical practice and can be curable with proper management. In the United States in 1993, there were an estimated 152000 new cases of colorectal cancer and 57000 deaths from this disease [46]. Approximately 30–40% of patients with colorectal cancers are not candidates for aggressive surgical therapy because of distant metastases, extensive local tumor infiltration, poor general condition or refusal of the patient to undergo surgery. These are the patients who require

Table 3. Gastric cancer and photodynamic therapy

Author	Patients & stage	Photosensitizer	Diffuser/Laser	Adjuvant therapy	Cancer eradication %	Follow up	Comments
Ell [39] 1998	22	mTHPC	red light 652 nm	none	73	12–20 months	local pain 54% phototoxicity 31% Lauren's type poor response
Mimura [38] 1996	27 early 5 late stage	Photofrin II	excimer dye laser	none	100 for mucosal, 75% for submucosal	not separated for this group	superficial, depressed and nonulcerated – better response, only 22 patients accessible
Spinelli [25] 1995	7 Type II c and III	Photofrin I and II	Transparent tube and balloon	none	71	3–58 months	phototoxicity 28%
Jin M [10] 1994	39 advanced stage	HpD	30 mm cylindrical fiber	Nd:YAG	71.8%, complete response 23.1%	more than 4 weeks	phototoxicity

palliative therapy to ameliorate the symptoms of the disease. Palliative local surgery is associated with considerable morbidity and the figures for major procedures range between 30 and 45% [47]. PDT allows for symptomatic treatment with minimal complications.

A study has shown that ten patients who were treated with PDT for inoperable colorectal neoplasms had resolution of symptoms. Unfortunately, one of these patients suffered a hemodynamically significant hemorrhage. However, two other patients had tumor free periods of up to 28 months [48]. Another study involving 11 patients with unresectable disease also found significant improvement in symptoms in all patients with four patients having a prolonged disease-free interval [49]. These studies imply that PDT may even offer a potential cure for some inoperable tumors that are localized. Spinelli showed that PDT had a 84% eradication rate colorectal adenomas, with a morbidity and mortality of 5% and 0% [25]. However, sunburn due to photo-sensitization occurred in 7% of patients while another 6% had local complications.

It has been demonstrated that the collagen layer of the colonic wall remains intact even if full thickness necrosis is caused by the treatment. The colon affected by PDT usually heals predominantly by regeneration without substantial risk of scarring. These results are encouraging with 35% of local tumor cases in complete remission after therapy with PDT and partial remission achieved in 44% .

The results are suggestive of the beneficial role of PDT in colorectal tumors. The application is most suitable for the treatment of smaller tumors or for sterilization of areas containing microscopic residual tumor in resection margins. For bulkier tumors, the use of PDT would need to be in conjunction with surgery or thermal laser irradiation. There is evidence that PDT is suitable for palliative treatment of advanced rectal cancers; although there is only a limited experience with this and additional work is required.

12.9 Photodynamic therapy and cholangiocarcinoma

Nonresectable Bismuth type III and IV cholangiocarcinoma is an extremely difficult malignancy to deal with. Ortner has published data on nine patients who had non-resectable lesions who had not responded to palliative billiary stent placement (a bilirubin decrease of 50%) who underwent PDT [50]. After therapy, the bilirubin serum declined from 318 ± 72 to 103 ± 35 μmol L^{-1} ($p = 0.0039$) with no further increase in values over the follow-up period of two months. These patients had dramatic increases in the quality of life indices. The median survival time of 439 days was increased when compared to previously published data for these cancers. The study showed some evidence that PDT may be effective in restoring biliary drainage and improving quality of life in patients with nonresectable cholangiocarcinomas.

12.10 The future of PDT

Photodynamic therapy has previously been used in limited patient numbers at a large number of specialized centers. High equipment and photosensitizer costs as well as insufficient data regarding its efficacy have hindered its application in the past. As more experience is acquired in PDT, there should be a better understanding of the application of this treatment in mainstream gastroenterology. In addition, as the photosensitizers and light sources used become more advanced and costs decrease, the indications for PDT will expand. Currently there are several multi-center trials in development studying the role of PDT in patients with cholangiocarcinomas and hepatocellular carcinoma. These studies were preceded by animal cell line models where PDT did show some statistical significance in tumor necrosis.

Photodynamic therapy also may be beneficial in benign disorders. Our own group has recently presented evidence that photodynamic therapy may be of use in the treatment of gastrointestinal hemorrhage from radiation induced proctitis. Photodynamic therapy has the potential for use in a variety of different gastro-intestinal diseases. Cutaneous photosensitization should decrease with the develop-ment of new photosensitizers that have very rapid excretion. In addition, these new agents may well be more selective for specific tumor or mucosal types. The roles for photodynamic therapy have not yet been defined and applications may well extend to previously inaccessible regions of the gastrointestinal tract. Photodynamic therapy has a key advantage in that the photoradiation does not have to be targeted to a specific lesion. Mucosal ablation is certainly the key area of investigation at this time in the gastrointestinal tract and photodynamic therapy appears to be very well suited for this application.

References

1. Q. Peng, K. Berg , J. Moan, M. Kongshaug, J.M. Nesland (1997). 5-Aminolevulinic acid-based photodynamic therapy: principles and experimental research. *Photochem. Photobiol.*, **65**(2), 235–251.
2. J. Seeler (1994). Texaphyrin. *Chem. Res.*, **27**, 43–50.
3. L.I. Grossweiner (1997). PDT light dosimetry revisited. *J. Photochem. Photobiol. B - Biol.*, **38**(2–3), 258–268.
4. J.V. Moore, C.M. West, C. Whitehurst (1997). The biology of photodynamic therapy. *Phys. Med. Biol.*, **42**(5), 913–935.
5. W. Beyer (1999). Systems for light application and dosimetry in photodynamic therapy. *J. Photochem. Photobiol. B - Biol.*, **36**(2), 153–156.
6. R. Earlam, J.R. Cunha-Melo (1980). Oesophageal squamous cell carcinoma: I. A critical review of surgery. *Br. J. Surgery*, **67**(6), 381–390.
7. S. Narayan, M.V. Sivak, Jr. (1994). Palliation of esophageal carcinoma. Laser and photo-dynamic therapy. *Chest Surgery Clin. N. Am.*, **4**(2), 347–367.
8. R.E. Sampliner, R. Fass (1993) Partial regression of Barrett's esophagus–an inadequate endpoint. *Am. J. Gastroenterol.*, **88**(12), 2092–2094.
9. J.S. McCaughan, Jr., T.A. Nims, J.T. Guy, W.J. Hicks, T.E. Williams, Jr., L.R. Laufman (1989). Photodynamic therapy for esophageal tumors. *Arch. Surgery*, **124**(1), 74–80.
10. M. Jin, B. Yang, W. Zhang, Y. Wang (1994). Photodynamic therapy for upper gastro-intestinal tumors over the past 10 years. *Seminars Surgical Oncol.*, **10**(2), 111–113.
11. C.J. Lightdale, S.K. Heier, N.E. Marcon, et al. (1995). Photodynamic therapy with porfimer sodium versus thermal ablation therapy with Nd:YAG laser for palliation of esophageal cancer: a multicenter randomized trial. *Gastrointest. Endoscopy*, **42**(6), 507–512.

12. M.E. Tian, S.L. Qui, Q. Ji (1985). Preliminary results of hematoporphyrin derivative-laser treatment for 13 cases of early esophageal carcinoma. *Adv. Exp. Med. Bio.*, **193**, 21–25.

13. M. Fujimaki, K. Nakayama (1986). Endoscopic laser treatment of superficial esophageal cancer. *Seminars Surgical Oncol.*, **2**(4), 248–256.

14. H. Tajiri, H. Daikuzono, S.N. Joffe, Y. Oguro (1987). Photoradiation therapy in early gastrointestinal cancer. *Gastrointest. Endoscopy*, **33**(2), 88–90.

15. T. Patrice, M.T. Foultier, S. Yactayo, M.C. Douet, F. Maloisel, L. Le Bodic (1990). Endoscopic photodynamic therapy with haematoporphyrin derivative in gastro-enterology. *J. Photochem. Photobiol. B - Biol.*, **6**(1–2), 157–165.

16. T. Patrice, M.T. Foultier, S. Yactayo, et al. (1990). Endoscopic photodynamic therapy with hematoporphyrin derivative for primary treatment of gastrointestinal neoplasms in inoperable patients. *Digestive Dis. Sci.*, **35**(5), 545–552.

17. T. Okunaka, H. Kato, C. Conaka, H. Yamamoto, A. Bonaminio, M.L. Eckhauser (1990). Photodynamic therapy of esophageal carcinoma. *Surgical Endoscopy,* **4**(3), 150–153.

18. G. Wagnieres (1990). Photodynamic therapy of early cancer in upper aerodigestive tract and bronchi. *SPIE Int. Series*, IS **6**, 249–271.

19. P. Monnier (1990). Photodetection and photodynamic therapy of early squamous cell carcinoma of pharynx, esophagus and tracheo bronchial tree. *Lasers Med. Sci.*, **5**, 149–168.

20. B.F. Overholt, M. Panjehpour, J.M. Haydek (1999). Photodynamic therapy for Barrett's esophagus: follow-up in 100 patients. *Gastrointest. Endoscopy*, **49**(1), 1–7.

21. K. Wang (1999). Current status of photodynamic therapy of Barrett's esophagus. *Gastrointest. Endoscopy*, **49**(3 part 2), S20– S23.

22. J.R. Savary, P. Grosjean, P. Monnier, et al. (1998). Photodynamic therapy of early squamous cell carcinomas of the esophagus: a review of 31 cases. *Endoscopy*, **30**(3), 258–265.

23. L. Gossner, M. Stolte, R. Sroka, et al. (1998). Photodynamic ablation of high-grade dysplasia and early cancer in Barrett's esophagus by means of 5-aminolevulinic acid. *Gastroenterology*, **114**(3), 448–455.

24. P. Grosjean, G. Wagnieres, C. Fontolliet, H. van den Bergh, P. Monnier (1998). Clinical photodynamic therapy for superficial cancer in the oesophagus and the bronchi: 514 nm compared with 630 nm light irradiation after sensitization with Photofrin II. *Br. J. Cancer*, **77**(11), 1989–1895.

25. P. Spinelli, A. Mancini, M. Dal Fante (1995). Endoscopic treatment of gastrointestinal tumors: indications and results of laser photocoagulation and photodynamic therapy. *Seminars Surgical Oncol.*, **11**(4), 307–318.

26. T. Dougherty (1997). Photodynamic therapy for early esophageal cancer. Meeting of the International Photodynamic Therapy Association, April 1990, Buffalo, NY.

27. B.F. Overholt, M. Panjehpour (1990). Photodynamic therapy for Barrett's esophagus. *Gastrointest. Endoscopy Clinics N. Am.*, **7**(2), 207–220.

28. B.F. Overholt, M. Panjehpour (1996). Photodynamic therapy for Barrett's esophagus: clinical update. *Am. J. Gastroenterol.*, **91**(9), 1719–1723.

29. B.F. Overholt, M. Panjehpour (1996). Photodynamic therapy in Barrett's esophagus. *J. Clinical Laser Med. Surgery*, **14**(5), 245–249.

30. M. Dower (1994). QLT Photofrin recommended for US approval. *SCRIP, PJP publication*, 23.

31. L.R. Biddlestone, C.P. Barham, S.P. Wilkinson, H. Barr, N.A. Shepherd (1998). The histopathology of treated Barrett's esophagus: squamous reepithelialization after acid suppression and laser and photodynamic therapy. *Am. J. Surgical Pathol.*, **22**(2), 239–245.

32. H. Barr, N.A. Shepherd, A. Dix, D.J. Roberts, W.C. Tan, N. Krasner (1996). Eradication of high-grade dysplasia in columnar-lined (Barrett's) oesophagus by photodynamic therapy with endogenously generated protoporphyrin IX. *Lancet*, **348**(9027), 584–585.

33. D.M. Parkin, P. Pisani, J. Ferlay (1993). Estimates of the worldwide incidence of eighteen major cancers in 1985. *Int. J. Cancer*, **54**(4), 594–606.

34. J.R. Monson, J.H. Donohue, D.C. McIlrath, M.B. Farnell, D.M. Ilstrup (1991). Total gastrectomy for advanced cancer. A worthwhile palliative procedure. *Cancer*, **68**(9), 1863–1868.

35. Y. Hayata, H. Kato, H. Okitsu, M. Kawaguchi, C. Konaka (1985). Photodynamic therapy with hematoporphyrin derivative in cancer of the upper gastrointestinal tract. *Seminars Surgical Oncol.*, **1**(1), 1–11.

36. M.L. Jin (1986). Photo-chemistry therapy of gastrointestinal tumor – analysis of hematoporphyrin derivative (HpD) plus laser photodynamic therapy of 50 cases. *Chung-Hua Chung Liu Tsa Chih [Chin. J. Oncol.]*, **8**(1), 64–66.

37. Y. Ito, A. Kameya, T. Kano, S. Kobayashi, T. Kasugai, S. Hotta (1988). Indications and limitations of laser treatment for early gastric cancer and palliative treatments for malignant obstruction of the esophagus and stomach. *Gan to Kagaku Ryoho [Jpn. J. Cancer Chemother.]*, **15**(4 Pt 2–3), 1435–1439.

38. S. Mimura, Y. Ito, T. Nagayo, et al. (1996). Cooperative clinical trial of photodynamic therapy with photofrin II and excimer dye laser for early gastric cancer. *Lasers Surgery Med.*, **19**(2), 168–172.

39. C. Ell, L. Gossner, A. May, et al. (1998). Photodynamic ablation of early cancers of the stomach by means of mTHPC and laser irradiation: preliminary clinical experience. *Gut*, **43**(3), 345–349.

40. Society. AC. Cancer facts and figures 1990. *Am. Cancer Soc.*, 1990.

41. E. Silverberg, C.C. Boring, T.S. Squires (1990). Cancer statistics, 1990. *CA: Cancer J. Clinicians*, **40**(1), 9–26.

42. L.M. Morrell, A. Bach, S.P. Richman, P. Goodman, T.R. Fleming, J.S. MacDonald (1991). A phase II multi-institutional trial of low-dose N-(phosphonacetyl)-L-aspartate and high-dose 5-fluorouracil as a short-term infusion in the treatment of adenocarcinoma of the pancreas. A Southwest Oncology Group study. *Cancer*, **67**(2), 363–366.

43. E. Holyoke (1981). Surgical approaches to pancreatic cancer. *Cancer*, **47**, 1719–1723.

44. T. Mang (1991). Studies on the absence of photodynamic mechanism in normal pancreas. *SPIE*, **1426**, 188–199.

45. K.T. Moesta, P. Schlag, H.O. Douglass, Jr., T.S. Mang (1995). Evaluating the role of photodynamic therapy in the management of pancreatic cancer. *Lasers Surgery Med.*, **16**(1), 84–92.

46. C.C. Boring, T.S. Squires, T. Tong (1993). Cancer statistics, 1993. *CA: Cancer J. Clinicians*, **43**(1), 7–26.

47. M. Dohmoto, M. Hunerbein, P.M. Schlag (1996). Palliative endoscopic therapy of rectal carcinoma. *Eur. J. Cancer*, **32A**(1), 25–29.

48. H. Barr, N. Krasner, P.B. Boulos, P. Chatlani, S.G. Bown (1990), Photodynamic therapy for colorectal cancer: a quantitative pilot study. *Br. J. Surgery*, **77**(1), 93–96.

49. N. Krasner (1989). Laser therapy in the management of benign and malignant tumors in the colon and rectum. *Int. J. Colorectal Dis.*, **4**(1), 2–5.

50. M.A. Ortner, J. Liebetruth, S. Schreiber, et al. (1998). Photodynamic therapy of non-resectable cholangiocarcinoma. *Gastroenterology*, **114** (3), 536–542.

Chapter 13

Factors in the establishment and spread of photodynamic therapy

Thierry Patrice

Table of contents

13.1 Obstacles to the development and spread of cancer PDT

Although photodynamic therapy (PDT) is associated with many obvious advantages over other forms of treatment (radiotherapy, surgery), only a small percentage of patients to whom this treatment modality might be applied have yet been treated. This fact alone may indicate a degree of reticence on the part of clinicians to embrace this procedure, and it has even been argued by some detractors that the delay which has occurred between the publication of the earliest reports on PDT and the subsequent commercial development of this form of treatment is itself indicative of its limitations. In fact, PDT faces many difficulties which are certain to limit the speed and extent to which it is assimilated into modern therapeutics and the readiness with which it will be adopted in both developed and developing countries. It is appropriate to examine what these difficulties may be.

It is possible to identify three major categories of problem: (1) those arising out of the nature of PDT itself; (2) issues relating to acceptability; and (3) economic considerations.

13.1.1 Problems inherent in the nature of PDT

PDT is based on the administration of a chemical (photosensitizer), systemically or locally. This chemical accumulates in tissues for a certain amount of time before irradiation at a suitable wavelength. Herein lies one major problem with PDT. To obtain a clinically significant effect, irradiation must take place after an optimal delay following the administration of the photosensitizer. Since the duration of the irradiation is very short, relative to the incubation time (i.e., the injection–irradiation interval), it is necessary to be aware of the optimal delay in order to maximize the clinical therapeutic effect. However, this delay varies greatly from one experimental model to another, from one cell to another within a given model, and also probably between cells within a single tumor. An issue that is of little importance in cancer chemotherapy, where kinetics have little influence on efficacy, is of critical importance in PDT where efficacy depends crucially upon the sensitizer concentration and the mumber of cancer cells loaded with a lethal amount of sensitizer at the time of irradiation. Simple tools for the monitoring of tissue concentration still do not exist.

The main point of interest in PDT, as compared with other treatment approaches in oncology, lies in the relatively selective destruction of cancer tissues that can be achieved, leaving normal tissue as little affected as possible. Such selectivity is probably the result of several mechanisms, varying from one cell to another, at the same time obeying specific kinetic principles. When advocating PDT as a major new therapeutic modality it is usual to focus upon this specificity of action, but in the absence of definitive experimental data, or of a convincing explanation for its physiological/biochemical basis, it can be difficult to persuade sceptical clinicians to consider PDT as a viable therapeutic option. This may help to explain why PDT remains to some extent a controversial modality.

Clearly, the provision of a routinely available bioassay to quantify the presence of a sensitizer in normal and malignant tissues would help in the design of clinical trials. Is it really relevant (as has been done in most studies) to irradiate a whole group of patients at a fixed time following photosensitizer administration, and to analyse results in terms of survival, when not only it is unlikely that the irradiation has been performed after an appropriate delay, but also there would not be any repeats of the huge variation in the persistence of skin sensitivity between one patient and another? Because this problem is specific to PDT, and it has never previously been addressed, the experimental procedures needed to take these issues into account are yet to be perfected.

The PDT mechanism of cell death is also still discussed among PDT specialists: direct tumoral effect, vascular effect, apoptosis versus necrosis and so on. Such discussions also impair PDT credibility and prevent focus on the main PDT advantages and its efficacy.

13.1.2 Physician acceptance

It is characteristic of the human condition to regard all that is new as being potentially dangerous. This applies particularly to the feelings experienced by physicians facing PDT for the first time. Although this attitude varies in extent from one country or culture to another, it is true that PDT presents many features which raise certain doubts in addition to those linked to its intrinsic patterns.

Under some systems of payment for medical services, the financial reward received by a clinician is determined in part by the risk that the clinician takes during the treatment, and in part by the overall cost of the treatment itself. Compared with alternative forms of treatment, PDT is not only a relatively cheap technique, but is also associated with a high degree of safety. Whilst the majority of clinicians will, of course, wish to choose the procedure which maximizes benefit to the patients, irrespective of financial considerations, there may be some who would be reluctant to replace a technique (e.g., radical surgery), which is both familiar and, under some forms of health service provision, ensures an acceptable level of fees, with a new, but still experimental (and therefore to some degree mysterious) medical procedure. The physician has to learn new skills and, under certain systems of reimbursement, has to treat many patients to obtain a satisfactory financial return. This means that negotiations between healthcare systems and companies to determine the pricing of photosensitizer drugs, as well as the reimbursement to the patient, will be of critical importance. However, it is unlikely that any insurance company or healthcare system will accept a high level of reimbursement for a new and incompletely tested procedure. As a consequence of these considerations, the preferred indications for PDT, at least in the immediate future, will be those for which no satisfactory technique exists at the present time, or for which clinical results are particularly poor with the conventional treatments that are available.

If PDT is to be established as a widely used therapeutic modality, it will be important to emphasize its simplicity, and to institute practical courses leading to university diplomas (not company ones), to make physicians comfortable with PDT and with its medico-legal aspects. Here, too, problems arise. Teaching may be hindered by the fact that most of the literature relating to PDT is to be found in highly specialized photo-biology journals, being relatively under-represented in medical speciality journals. This is not surprising, since there has been a substantial delay between the initial description of PDT as a theoretical therapeutic concept and the appearance of the first marketed sensitizer drug. For many years, clinical trials designed to test PDT procedures were unavailable. Not only were the clinical results equivocal and difficult to interpret (groups of patients being small), but the credibility of PDT among clinicians was also impaired as it was argued, understandably, that, if commercial companies did not want to take the risk of supporting clinical trials for PDT, why should the academic/medical institutions do so? Such reluctance on the part of the companies was taken by many to mean that the technique was most probably inefficient when it is obviously not.

13.1.3 Economic considerations

13.1.3.1 The rate of development of PDT

We first have to ask whether, compared with the development of radiotherapy or simply with the delay involved in any new drug development in oncology, progress in the investigation and refinement of PDT has, in fact, been particularly slow. It is generally admitted that about 10 years are needed from the synthesis of a new compound to its commercial approval, and it would not be unrealistic that a further 10 years might be needed to demonstrate the clinical usefulness of drugs such as photosensitizers to which a new therapeutic concept (photodynamic therapy) has also to be applied. Photofrin was approved in France in 1996. During 1996–1998, 598 published papers were entered into the medline data bank. Between 1986 and 1988, 239 papers were published, as against only 11 during 1976–1978. It could thus be said that the development of PDT began during the early 1980s, i.e., 20 years ago, and that it has reached marketability within what we have defined above as a reasonable period of time. However, for those who are investing either time or money in the field such a rate of development is too slow.

13.1.3.2 Approval for laser technology

PDT is a physicochemical treatment requiring both a light source and a chemical. This means that not only does a pharmaceutical company have to obtain approval for a sensitizer as an anticancer drug, but also a laser manufacturer has to simultaneously get the same approval for the medical use of a laser light source. However, regulatory procedures differ widely from one country to another while research and development schedules depend upon the financial resources of companies and upon the priorities that they assign to their different commercial activities. Companies manufacturing laser devices are generally small, have limited amounts of money to spend and, before lasers were used in PDT, such companies had never been faced with the need to conduct clinical trials or with the huge expenses related to such trials. When PDT was first proposed in the 1990s as a significant clinical application for lasers, the manufacturers imagined that an alternative would be available to Nd-YAG lasers within one or two years. Many lost a substantial amount of money designing large machines that are totally obsolete today. Manufacturers were understandably disappointed, and PDT lost part of its commercial credibility. It was then up to pharmaceutical companies to demonstrate the feasibility of PDT, to push through the development of one or more sensitizers to the stage of approval, and to support laser manufacturers.

13.1.3.3 Identifying promising photosensitizers
Demonstrating the efficacy of PDT as a concept has always been a legitimate aim for researchers, but (at least in the early days) very few were conscious of the commercial consequences of publishing their results concerning sensitizers derived from physiologically active chemical structures. Finding efficient commercial protection for sensitizers has therefore been an additional preoccupation for capital risk companies. In the early era of PDT, it was difficult for investors to identify, among the impressive amount of published papers, what developments possessed good clinical (and hence commercial) potential. The criteria for deciding which putative photosensitizer to discard and which to keep in the company portfolio had not been clearly defined; nor was it easy to identify which apparently convincing and encouraging results merited additional financial efforts, and which ought realistically to be rejected. Until a sufficiently detailed corpus of background information had been assembled, investments could not be orientated with any degree of accuracy.

13.1.3.4 The emergence of new photosensitizers
For any pharmaceutical company to financially support the development of a specific sensitizer drug is to take a substantial risk: The commercialization of the sensitizer could well be jeopardized by the subsequent discovery of a new, and more effective, molecule. A company has to limit the risks by simultaneously developing several molecules, which is not possible for an emerging market such as PDT. In addition to committing financial resources to the investigation of just one putative agent, a company must also maintain awareness of parallel developments and be ready to change the focus of its activities if this seems appropriate. This degree of flexibility in research and development is necessarily financially demanding, but is important if commercial success is to be achieved. Big pharmaceutical companies should thus best be positioned for PDT development, but being big they are likely to develop, due to the pressure of shareholders, big potential markets and not emerging ones. In addition, those who invested in anti-cancer drugs will certainly not help the development of a competitive technology.

13.1.3.5 Clinical trials within a marketing framework
Although PDT originated in the work of investigators who were not involved in clinical practice, its subsequent development has been, and will continue to be, dependent upon its use and further exploration by clinicians. However, because PDT is an emerging technology with, as yet, only a small number of approved indications, there will be limited opportunities for clinicians to gain the necessary experience

with the technique and to become familiar with the scientific principles which are involved. Only as the range of indications for PDT broadens can this situation improve, but this will not occur unless support is provided for the institution of clinical trials. Thus, in addition to direct marketing, aimed at explaining PDT to potential users, indirect marketing is also needed in the form of support for clinical trials as well as for basic studies.

Physician acceptance of PDT will be based largely upon the results of clinical trials. When trials are conducted on chemotherapeutic agents for cancer treatment, toxicity and efficacy are usually tested in patients with large tumors. This is an inappropriate paradigm for testing PDT, because photosensitizers are generally toxic only when activated by light, and PDT is a local treatment aimed at treating cancers or tissues with limited depth invasion. The proper evaluation of the efficacy of any treatment in oncology must at least involve tumors for which the treatment is appropriate.

As has been noted earlier, PDT is most likely to be applied to patients for whom no established effective treatment procedure is available, or for whom such treatments have failed to result in clinical improvement. The rationale behind this is that, for patients with a significant lifespan, it is ethically more acceptable to offer a patient a treatment of known efficacy than to replace this with a treatment the efficacy of which is still uncertain. Only when no success has been achieved with the established treatments is it deemed appropriate to try new, experimental procedures; at this point, the argument is that doing anything for the patient is better than doing nothing. This approach, however, does not permit consideration of the cost-efficiency of the treatment that is being tested as survival will probably be short. In addition, hi-tech (hi-cost) treatment of patients that, having proved refractory to all available alternative therapies, represent the most severely ill cases, will increase the overall cancer care costs but with the likelihood that only relatively few patients will show marked clinical improvement. Hospital managers will be reluctant to pay for such palliative indications.

13.1.3.6 Cost-efficacy analysis

Clinical trials aimed at securing registration and approval for a new treatment are often biased toward assessing the safety of the treatment; we need, however, to bear in mind that risks are associated with the disease as well as with the treatment, and that a more clinically relevant index than treatment toxicity is the disease risk–treatment risk ratio.

Procedures for approval are so costly, relative to the expenditure needed in pre-approval development stages, that the market price of a new or different treatment procedure often bears little relationship to the actual production costs. Moreover, the approval requirements for treatments aimed at serious illnesses are generally more onerous than those directed towards conditions associated with less risk; they therefore take longer to complete (hence incurring greater delays before they can be brought to the market), and are more costly. This is clearly an unsatisfactory situation; it would surely be more appropriate to facilitate selectively the introduction of treatments for severe illnesses. This could be achieved by stressing the importance of the cost-efficacy of a treatment rather than efficacy alone, and of looking at the disease risk–treatment risk ratio. If this approach were adopted, it would provide the chance for PDT to be evaluated in direct comparison with radiotherapy and surgery, two procedures that have never been assessed in terms of cost-efficacy, and it would reduce the costs associated with bringing techniques such as PDT to approval.

13.2 Potential spread of PDT

Irrespective of the country in which it occurs, a diagnosis of cancer presents a dramatic situation for an individual and for that individual's family, as well as for society as a whole. The disease will inevitably be disruptive of the family and its economic equilibrium, leading to direct costs (treatments) as well as indirect ones (loss of productivity). Cancers are, however, generally more frequent amongst members of those social classes that are most socioeconomically deprived, as well as amongst those who live in the more polluted areas of the world, and whose general medical care and treatment is deficient, leaving patients either poorly treated, or even untreated, for reasons of cost. All these factors are particularly evident amongst the developing countries of the world (75% of the world's population, 52% of the cancer cases, countries that often also have to contend with the problems that arise from severe overpopulation. In such countries, and as a consequence of the overwhelming social and economic pressures that exist, cancer will become an increasingly serious challenge to the maintenance of such health services as are available, and its consequence will soon present insuperable difficulties unless a new, relatively inexpensive form of treatment can be developed which will reverse the otherwise inexorable trend [1].

People are not equal where diseases such as cancer are concerned. The reasons for the inequality lie not only in each individual's genetic makeup, but also in the legacy of medical conditions that each individual carries as a result of encountering various pathogenic agents in the course of their life. Inequalities in health are also due, in part, to the social inequalities present in our societies. Sanitary conditions are worse, mortality and morbidity rates (particularly amongst children) higher, and survival rates lower, in the most disadvantaged socioeconomic classes [2]. Similar, if not identical, differences can be seen between industrialized countries and the less developed ones. In 1981, the number of years of potential life lost as a result of cancer, per 1000 population in England, varied from 11 to 24, depending upon the professional qualifications of the patient [3]. Risk factors such as smoking, simply add a constant increase to the "social risk". Being a smoker and poor is more dangerous than being a rich smoker.

The most frequent cancers are between two and ten times more frequent in the developing countries than in countries that are members of the EEC. In men, the most frequent cancers are those which affect the stomach, oral cavity, liver, and oesophagus. In women, cancers of the cervix, breast, stomach, and oral cavity are the most common. Specific to developing countries—the following risk factors have been identified viz. betel quid chewing or bidi smoking for oral cancers and EBV for nasopharynx cancers (in southern China).

As a result of the proliferation of risk factors, and the progressive upward shift in the average age of the population, the absolute number of cases of cancer will continue to increase almost everywhere in the world (an increase of 27% is expected in Europe by 2010), but this increase will be substantially greater in the developing than in the developed countries; increases of 116% in Africa and 92% in Asia have been estimated [2]. However, in view of the fact that the average cost per patient of cancer treatment in Europe was estimated in 1991 as being US$3900 for radiotherapy, US$9100 for surgery, and US$15600 for chemotherapy (costs which will have risen considerably over the intervening years up to the present time), it is highly unlikely that developing countries (and possibly also some central and eastern European countries) will be able to afford the provision of adequate treatment for their cancer sufferers.

A policy of early cancer detection would then be effective, though only if a cost-effective treatment existed to treat the cancers detected in screening programmes and, of course, if an accurate and cheap system of detection could be developed. The discovery and development of a new, inexpensive therapy, not requiring long-term administration and/or hospitalization, would permit existing cancer sufferers to be treated, thereby reducing the social and economic burden which such patients would otherwise continue to represent.

All these aspects will be most important in the near future, reinforcing the potential interest in PDT which could be the first really cost-effective treatment. Diode laser sources are very cheap compared to X-ray machines, they are easy to use, and they do not need an engineer for their operation. In addition to laser diodes, cheap filtered lamps or LED arrays could also be used as light sources for some specific applications, though a laser is the simplest way of obtaining the highest possible power at a given wavelength. Light sources can be powered with 220 V or 110 V alternating current supplies or by mobile power generators, and can be constructed in such a way as to be quite insensitive to the climatic extremes often encountered in developing countries. As the equipment is compact, it can be transported easily between community clinics in the boot of any car. Its simplicity allows PDT to be performed anywhere that facilities are sufficient for establishing a firm diagnosis of cancer, and training a technician to operate the device may take little more than half a day. Lasers and sensitizers are already inexpensive compared to other forms of treatment, but it is likely that prices will fall further still in the future as development costs are progressively written off. In those developed countries in which the remuneration that a doctor receives is directly proportional to the cost of any treatment administered, the introduction of inexpensive cancer treatments, such as PDT, alongside or replacing the more expensive ones, may not be greeted with enthusiasm; economic pressures are, however, increasing and it is very likely that PDT will eventually be seen as a first rank treatment for cancers.

Finally, a major advantage of PDT is that the duration of treatment is, in most cases, very short, often permitting therapy to be performed on patients attending a clinic on an outpatient basis. The sensitizers that are used in PDT have very low toxicity in the absence of light, meaning that close patient follow-up need not be conducted, and an intensive care unit is not needed at all after drug administration.

Both the lasers and the sensitizers are nonpolluting. Unlike chemotherapy, the administration of PDT does not require the involvement of specially trained nurses.

As you can see, dear reader, I am fully convinced that PDT is a major medical technology of the future, despite the difficulties it faces!

References

1. T. Patrice (1999). Photodynamic therapy in developing countries. *Rev. Contemp. Pharmacother.*, **10**, 75–78.
2. M. Marmot, A. Feeney (1997). General explanations for social inequalities in health. In: M. Kogevinas, N. Pearce, M. Susser, P. Boffetta (Eds), *Social Inequalities and Cancer* (pp. 207–226). IART Scientific Publications, Lyon.
3. L. Tomatis (1997). Poverty and cancer. In: M. Kogevinas, N. Pearce, M. Susser, P. Boffetta, (Eds), *Social Inequalities and Cancer* (pp. 25–39). IARC Scientific Publications, Lyon.

Subject Index

ALA esters, 43
ALA, see Protoporphyrin IX
Anthraquinones, 71
Antiangiogenic, 121
Antibodies, 29, 30
Apoptosis, 12, 32
Auler, 6

Bacteriochlorophyll, 70
Barrett, 262
Bleaching, 13
Bowen, 200
BPD, 23, 30, 67
Brain tumours, 215
Breast cancers (see gynaecology), 243

Chlorin, 66
Condylomas, 247

Dermatology, see skin
Diffusers, 136
Dougherty, 7

Economics, 275
Esophagus, 262

Fluorescence, 5, 6, 161, 179, 223

Gastrointestinal, 259
Gliomas (see brain tumors), 215
Gynaecology, 243

Hematoporphyrin derivative, 6, 61
Hematoporphyrin, 6
History, 2

Immunological effects, 38
Interactions (with PDT), 221

Kessel, 8
Keratoses, 180

Laser, 132
Levulan, 92
Light, 21
Light sources, 129
Light scatter, 140
Light measurement, 144
Lipoproteins (LDL), 8, 28
Liposomes, 112
Lipson, 7
Lysosomes, 13

Necrosis, 32

Oesophagus, 260
Oxygen, 24

PCT, 3
Photofrin, 6, 36, 63
Photosensitizer dosimetry, 147
Phthalocyanines, 8, 68, 107
Polymers, 109
Protoporphyrin IX, 21,24, 39, 64, 83, 190, 192

Quinones, 71

Raab, 4

Schwartz, 7
Selectivity, 8, 27